Newnes
Circuit Calculations
Pocket Book

To my grandchildren Shelley Louise and Michael John Davies

Newnes
Circuit Calculations Pocket Book

with Computer Programs

Thomas J. Davies

Newnes
An imprint of Butterworth-Heinemann Ltd
Linacre House, Jordan Hill, Oxford OX2 8DP

 PART OF REED INTERNATIONAL BOOKS

OXFORD LONDON BOSTON
MUNICH NEW DELHI SINGAPORE SYDNEY
TOKYO TORONTO WELLINGTON

First published 1992

© Thomas J. Davies 1992

All rights reserved. No part of this publication
may be reproduced in any material form (including
photocopying or storing in any medium by electronic
means and whether or not transiently or incidentally
to some other use of this publication) without the
written permission of the copyright holder except in
accordance with the provisions of the Copyright,
Designs and Patents Act 1988 or under the terms of a
licence issued by the Copyright Licensing Agency Ltd,
90 Tottenham Court Road, London, England W1P 9HE.
Applications for the copyright holder's written permission
to reproduce any part of this publication should be addressed
to the publishers

British Library Cataloguing in Publication Data
A catalogue record for this book is
available from the British Library

ISBN 0 7506 0195 7

Printed and bound in Great Britain

Contents

Preface — viii
Note on the computer programs — ix

1 The d.c. voltage — 1
1.1 Introduction — 1
1.2 Units — 1

2 Resistors — 5
2.1 Introduction — 5
2.2 Units — 5
2.3 Resistance formulas — 5
2.4 Resistor values — 12
2.5 Resistor circuits — 16

3 D.c. circuits — 26
3.1 Current — 26
3.2 Units of current — 26
3.3 Ohm's law — 28
3.4 Power — 28
3.5 Types of circuits — 29
3.6 Measurements — 41
3.7 The Wheatstone bridge — 45

4 Network theorems — 50
4.1 Kirchhoff's laws — 50
4.2 Superposition theorem — 58
4.3 Thevenin's theorem — 62
4.4 Norton's theorem — 77
4.5 Maximum power transfer — 83

5 Time — 86
5.1 Introduction — 86
5.2 Units used — 86
5.3 Conversion of seconds to milliseconds, microseconds and nanoseconds — 87
5.4 Conversion of milliseconds, microseconds and nanoseconds to seconds — 87

6 The a.c. voltage — 89
6.1 Introduction — 89
6.2 Voltage generating — 89
6.3 Instantaneous value — 91
6.4 Peak, peak to peak and r.m.s. values — 92
6.5 Frequency — 94
6.6 Wavelength — 95
6.7 Average value for a sine wave — 97
6.8 Graphical determination of r.m.s. value of a sine wave — 98
6.9 Graphical determination of average and r.m.s. values of a non-sinusoidal waveform — 99

7 Capacitors — 102
7.1 Introduction — 102
7.2 Units — 102

	7.3	Capacitor formulas	103
	7.4	Capacitor circuits	110
	7.5	Reactance	122

8 Inductors — 125
- 8.1 Introduction — 125
- 8.2 Units — 125
- 8.3 Inductor formulas — 126
- 8.4 Inductor circuits — 132
- 8.5 Reactance (sinusoidal voltages only) — 135

9 D.c. transients — 139
- 9.1 Introduction — 139
- 9.2 C–R circuit — 139
- 9.3 L–R circuit — 150

10 Electromagnetism — 160
- 10.1 Force on a conductor, F — 160
- 10.2 E.m.f. generated, E — 161
- 10.3 Magnetomotive force, F — 163
- 10.4 Magnetic field strength, H — 163
- 10.5 Permeability — 164
- 10.6 Total flux — 164
- 10.7 Relative and absolute permeabilities — 167
- 10.8 Reluctance, R_m — 170

11 A.c. circuits — 173
- 11.1 Introduction — 173
- 11.2 Resistor, R — 173
- 11.3 Inductor, L — 173
- 11.4 Capacitor, C — 175
- 11.5 Series circuits — 177
- 11.6 Parallel circuits — 202

12 Phasors — 213
- 12.1 Introduction — 213
- 12.2 Addition — 217
- 12.3 Subtraction — 221
- 12.4 Multiplication — 224
- 12.5 Division — 225
- 12.6 Further addition — 226

13 Transformers — 229
- 13.1 Introduction — 229
- 13.2 Transformer action — 229
- 13.3 Maximum power transfer — 234

14 D.c. supplies — 242
- 14.1 Introduction — 242
- 14.2 Battery connections — 244
- 14.3 Battery charger — 249
- 14.4 Rectifier circuits — 253
- 14.5 Power supply unit — 255
- 14.6 Voltage regulators — 257

15 Transistor amplifiers — 262
- 15.1 Introduction — 262
- 15.2 Common emitter – static conditions — 263
- 15.3 Common emitter – dynamic conditions — 269

	15.4 Common base – static conditions	276
	15.5 Common base – dynamic conditions	277
	15.6 Common collector	281
16	**Operational amplifiers**	**286**
	16.1 Open loop amplifiers	286
	16.2 Closed loop amplifiers	286
	16.3 Summing amplifiers	289
	16.4 Subtracting amplifiers	292
17	**Oscillators**	**295**
	17.1 Introduction	295
	17.2 R–C phase shift oscillator	298
	17.3 Wien bridge oscillator	300
	17.4 Twin-T oscillator	301
	17.5 Non-sinusoidal oscillators	302
	17.6 Sawtooth waveform	305
	17.7 Triangular waveform	305
18	**Filters and attenuators**	**309**
	18.1 Introduction to filters	309
	18.2 The decibel	309
	18.3 Low pass filters	315
	18.4 High pass filters	319
	18.5 Introduction to attenuators	321
	18.6 Symmetrical T attenuator	322
	18.7 Asymmetrical T attenuator	324
	18.8 Symmetrical π attenuator	325
	18.9 Asymmetrical π attenuator	326
19	**Denary, binary and logic**	**329**
	19.1 The denary system	329
	19.2 The binary system	331
	19.3 Octal numbers	339
	19.4 Hexadecimal numbers	340
	19.5 Logic gates and truth tables	342
20	**Two- and three-phase systems**	**351**
	20.1 Introduction	351
	20.2 Two-phase supply	351
	20.3 Three-phase supply	354
	20.4 Three-phase star connected load	356
	20.5 Three-phase delta connected load	357
Appendix	*Symbols, abbreviations and definitions*	363
Answers to problems		365
Index		373

Preface

This book has been written for students embarking upon electronic and electrical engineering courses such as those offered by BTEC and the City and Guilds. The book will be a useful reference during the first three years in a technical college and in home study courses, and it contains most of the equations that will be encountered during that time.

The book is designed to be used in conjunction with one of the many theoretical books available. It contains over 400 worked examples, and 370 problems with answers. The worked examples have been carefully chosen to show how equations are transposed and used.

Over 300 computer programs are included which allow long and tedious problems to be solved in seconds. The inclusion of the programs will enable students to obtain experience, not only in using computers, but also in modifying programs and eventually writing their own.

I would like to thank the staff of the Bournemouth and Poole College of Further Education for their help and support, and in particular Paul Reaks for reading the whole work. I would also like to thank the publishers, and in particular Bridget Buckley, for helpful advice.

Finally I would like to thank my family for their support and especially my wife Enid, who typed the manuscript.

Note on the computer programs

The computer programs are written in GW-BASIC. Some minor modifications may be necessary to the programs for other BASIC dialects. If a syntax error is indicated, for example during loading, this may mean that special characters such as parentheses, commas or semicolons have been forgotten, or that too many characters have been included in a program line. Errors can easily be corrected by referring to the computer system manual.

Line 10 identifies the program number. Line 20 identifies the parameter or problem to be solved. The correct symbols have been used wherever possible. Some of the problems in this book, such as those in Chapter 4, do not readily lend themselves to computer programs, and some programs are beyond the scope of the book.

These programs give an opportunity to gain experience with computers, so that the reader can then modify them to write new programs for problems not covered in the text: for example, program 285 in Chapter 19 has been modified into program 293 by alteration and the inclusion of extra lines.

1 The d.c. voltage

1.1 Introduction

Electromotive force (E) and potential difference (p.d.) are both measured in volts. The symbol is V. Direct current (d.c.) sources include both primary and secondary batteries. A.c. voltages, which will be discussed in Chapter 6, can also be converted into d.c.

Voltages can be measured with an analogue instrument having a pointer which indicates the value of the voltage, or by a digital instrument which indicates the reading by a display of number. A d.c. voltage has a constant amplitude and its magnitude depends on the source used.

1.2 Units

It is important to examine the units used in the measurements of both large and small voltages, and to see how one unit can be converted into another.

(a) The main units
1. The volt, V
2. The kilovolt, kV, where $1 \text{ kV} = 10^3 \text{ V}$
3. The millivolt, mV, where $1 \text{ mV} = 1 \text{ V}/10^3 = 10^{-3} \text{ V}$
4. The microvolt, μV, where $1 \text{ } \mu\text{V} = 1 \text{ V}/10^6$ or 10^{-6} V

(2) is a multiple and is used to specify large voltages, whereas (3) and (4) are submultiples and are used to specify small voltages, typically less than 1 V.

$$\text{Since } 1 \text{ mV} = \frac{1 \text{ V}}{10^3} \text{ then } 10^3 \text{ mV} = 1 \text{ V}$$

$$\text{Since } 1 \text{ } \mu\text{V} = \frac{1 \text{ V}}{10^6} \text{ then } 10^6 \text{ } \mu\text{V} = 1 \text{ V}$$

It also follows that

$$10^3 \text{ mV} = 10^6 \text{ } \mu\text{V}$$

and

$$1 \text{ mV} = \frac{10^6}{10^3} \text{ } \mu\text{V} = 10^3 \text{ } \mu\text{V}$$

(b) Conversion of volts to millivolts and microvolts
Since

$$1 \text{ V} = 10^3 \text{ mV}$$
$$10 \text{ V} = 10 \times 10^3 \text{ mV} = 10^4 \text{ mV}$$
$$0.15 \text{ V} = 0.15 \times 10^3 \text{ mV} = 150 \text{ mV}$$
$$0.01 \text{ V} = 0.01 \times 10^3 \text{ mV} = 10 \text{ mV}$$

Since

$$1 \text{ V} = 10^6 \text{ } \mu\text{V}$$
$$5 \text{ V} = 5 \times 10^6 \text{ } \mu\text{V}$$
$$0.2 \text{ V} = 0.2 \times 10^6 \text{ } \mu\text{V} = 2 \times 10^5 \text{ } \mu\text{V}$$
$$0.00036 \text{ V} = 0.00036 \times 10^6 \text{ } \mu\text{V} = 360 \text{ } \mu\text{V}$$

Note that since

$$0.00036 \text{ V} \times 10^3 = 0.36 \text{ mV}$$

and since

$$1 \text{ mV} = 10^3 \text{ } \mu\text{V}$$

then

$$0.36 \text{ mV} = 0.36 \times 10^3 = 360 \text{ } \mu\text{V}$$

Table 1.1 shows five different voltages, each expressed in the three different units. Note the 10^3 relationship between all the columns and the 10^6 relationship between columns 1 and 3.

Table 1.1 Voltage units

V	mV	μV
0.001	1.0	1000
0.0025	2.5	2500
0.06	60	60×10^3
0.8	800	800×10^3
1.5	1.5×10^3	1.5×10^6

```
10 PRINT "PROG 1"
20 PRINT "THIS PROGRAM CONVERTS VOLTS TO MILLIVOLTS"
30 INPUT "ENTER VOLTAGE IN VOLTS" ; V
40 LET MV=V*10^3
50 PRINT "ANSWER = " MV "MILLIVOLTS"
```

```
10 PRINT "PROG 1"
20 PRINT "THIS PROGRAM CONVERTS VOLTS TO MICROVOLTS"
30 INPUT "ENTER VOLTAGE IN VOLTS" ; V
40 LET UV=V*10^6
50 PRINT "ANSWER = " UV "MICROVOLTS"
```

```
10 PRINT "PROG 1"
20 PRINT "THIS PROGRAM CONVERTS MILLIVOLTS TO MICROVOLTS"
30 INPUT "ENTER VOLTAGE IN MILLIVOLTS" ; MV
40 LET UV=MV*10^3
50 PRINT "ANSWER = " UV "MICROVOLTS"
```

(c) Conversion of millivolts and microvolts to volts

The examples in (b) showed that

$$10 \text{ mV} = 0.01 \text{ V} \times 10^3$$

giving

$$\frac{10 \text{ mV}}{10^3} = 0.01 \text{ V}$$

and

$$360 \text{ } \mu\text{V} = 0.00036 \text{ V} \times 10^6$$

giving

$$\frac{360 \ \mu V}{10^6} = 0.00036 \ V$$

Note also that

$$360 \ \mu V = 0.36 \ mV \times 10^3$$

and

$$\frac{360 \ \mu V}{10^3} = 0.36 \ mV$$

These examples show that

$$\frac{mV}{10^3} = V \ \frac{\mu V}{10^6} = V \ \frac{\mu V}{10^3} = mV$$

The following conversions can be carried out to column 3 of Table 1.1.

1 Microvolts to volts – divide by 10^6.

$$\frac{1000}{10^6} = 0.001 \ V$$

$$\frac{2500}{10^6} = 0.0025 \ V$$

$$60 \times \frac{10^3}{10^6} = 0.06 \ V$$

$$800 \times \frac{10^3}{10^6} = 0.8 \ V$$

$$1.5 \times \frac{10^6}{10^6} = 1.5 \ V$$

2 Microvolts to millivolts – divide by 10^3

$$\frac{1000}{10^3} = 1 \ mV$$

$$\frac{2500}{10^3} = 2.5 \ mV$$

$$60 \times \frac{10^3}{10^3} = 60 \ mV$$

$$800 \times \frac{10^3}{10^3} = 800 \ mV$$

$$1.5 \times \frac{10^6}{10^3} = 1.5 \times 10^3 \ mV$$

```
10 PRINT "PROG 2"
20 PRINT "THIS PROGRAM CONVERTS MILLIVOLTS TO VOLTS"
30 INPUT "ENTER VOLTAGE IN MILLIVOLTS" ; MV
40 LET V=MV/10^3
50 PRINT "ANSWER = " V "VOLTS"
```

```
10 PRINT "PROG 2"
20 PRINT "THIS PROGRAM CONVERTS MICROVOLTS TO VOLTS"
30 INPUT "ENTER VOLTAGE IN MICROVOLTS" ; UV
40 LET V=UV/10^6
50 PRINT "ANSWER = " V "VOLTS"
```

```
10 PRINT "PROG 2"
20 PRINT "THIS PROGRAM CONVERTS MICROVOLTS TO MILLIVOLTS"
30 INPUT "ENTER VOLTAGE IN MICROVOLTS" ; UV
40 LET MV=UV/10^3
50 PRINT "ANSWER = " MV "MILLIVOLTS"
```

(d) Units for large voltages

Using the unit kV means that large voltages can be expressed in a much simplified form, which the following examples show.

Since

$$10^3 \text{ V} = 1 \text{ kV}$$

then

$$133\,000 \text{ V} = 133 \times 10^3 = 133 \text{ kV}$$

i.e.

$$V = kV \times 10^3 \text{ and } \frac{V}{10^3} = kV$$

$$10\,000 \text{ V} = 10 \times 10^3 = 10 \text{ kV}$$

$$1500 \text{ V} = 1.5 \times 10^3 = 1.5 \text{ kV}$$

$$800 \text{ V} = 0.8 \times 10^3 = 0.8 \text{ kV}$$

```
10 PRINT "PROG 3",
20 PRINT "THIS PROGRAM CONVERTS VOLTS TO KILOVOLTS"
30 INPUT "ENTER VOLTAGE IN VOLTS" ; V
40 LET KV=V/10^3
50 PRINT "ANSWER = " KV "KILOVOLTS"
```

```
10 PRINT "PROG 3"
20 PRINT "THIS PROGRAM CONVERTS KILOVOLTS TO VOLTS"
30 INPUT "ENTER VOLTAGE IN KILOVOLTS" ; KV
40 LET V=KV*10^3
50 PRINT "ANSWER = " V "VOLTS"
```

Problems

1 Express the following in kilovolts: 750 V, 7500 V, 8250 V, 18 250 V.
2 How many volts are there in the following? 6.1 kV, 0.2 kV, 25.25 kV.
3 Express the following in both millivolts and microvolts: 0.6 V, 3.1 V, 0.0025 V.
4 Convert the following to volts and millivolts: 60 μV, 200 μV, 1600 μV.

2 Resistors

2.1 Introduction

Metals offer very little resistance to the flow of current. Copper is a good example and is referred to as a conductor. Other materials such as quartz, polythene and procelain offer very high resistance to the flow of current and are called insulators.

Intermediate between conductors and insulators are semiconductors, the most important being silicon and germanium used in the manufacture of transistors. Materials made into rods, films and wire form electronic components which are called resistors.

2.2 Units

The unit of resistance, R, is the ohm, which has the symbol Ω (Greek letter omega). The main units are:

1. The megohm MΩ where 1 MΩ = 10^6 Ω
2. The kilohm kΩ where 1 kΩ = 10^3 Ω
3. The microhm µΩ where 1 µΩ = 10^{-6} Ω

The conversion of one unit into another was dealt with in Chapter 1. It follows therefore that

$$\text{since } 1 \text{ µ}\Omega = \frac{1}{10^6} \Omega$$

$$10^6 \text{ µ}\Omega = 1 \Omega$$

and

$$\text{since } 10^3 \Omega = 1 \text{ k}\Omega$$

$$10^3 \text{ k}\Omega = 1 \text{ M}\Omega$$

2.3 Resistance formulas

The resistance of a conductor depends on dimensions, temperature and the material used. At a constant temperature

$$\text{Resistance } R = \frac{\rho l}{A}$$

where

l = The length of the conductor

A = The cross-sectional area

ρ = The resistivity, or specific resistance

ρ (Greek letter rho) is a factor which takes account of the conductor material.

(a) Resistivity
The unit of resistivity is the ohm metre (Ω m), and the value for copper as an example is approximately

$$1.72 \times 10^{-8} \text{ } \Omega \text{ m}$$

It is important in calculations to use the same units for ρ, l and A. The next three examples illustrate this point.

Example 1 A cube of copper shown in Figure 2.1 has sides 1 m in length, giving

$l = 1$ m and $A = 1$ m^2

$$R = \frac{1.72 \times 10^{-8} \times 1}{1} = 1.72 \, 10^{-8} \, \Omega$$

$1.72 \times 10^{-8} \, \Omega$ is the resistance of the cube, and this is constant at ambient temperature.

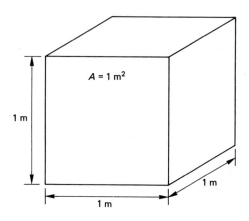

Figure 2.1

```
10 PRINT "PROG 4"
20 PRINT "RESISTANCE OF A CONDUCTOR"
30 INPUT "ENTER RESISTIVITY IN OHM METRE" ; RHO
40 INPUT "ENTER LENGTH IN METRES" ; L
50 INPUT "ENTER AREA IN SQUARE METRES" ; A
60 LET R=RHO*L/A
70 PRINT "R = " R "OHMS"
```

Example 2 If the problem is solved with centimetres (cm) as the units

$$R = 1.72 \times 10^{-8} = \frac{\rho \times 10^2}{10^2 \times 10^2}$$

(since 1 m = 10^2 cm).

this gives

$$1.72 \times 10^{-8} = \frac{\rho}{10^2}$$

then

$$\rho = 1.72 \times 10^{-8} \times 10^2$$
$$= 1.72 \times 10^{-6} \, \Omega \text{ cm}$$

Example 3 Solving the problem using millimetres (mm)

$$1.72 \times 10^{-8} = \frac{\rho \times 10^3}{10^3 \times 10^3}$$

(since $1 \text{ m} = 10^3 \text{ mm}$)

this gives

$$1.72 \times 10^{-8} = \frac{\rho}{10^3}$$

then

$$\rho = 1.72 \times 10^{-8} \times 10^3$$
$$= 1.72 \times 10^{-5} \, \Omega \text{ mm}$$

Example 4 Calculate the resistance of a copper wire 20 m long and 2 mm in diameter, given that the resistivity is

$1.72 \times 10^{-6} \, \Omega \text{ cm}$

This problem can be solved in three ways.

Using metres as units:

$$\rho = 1.72 \times 10^{-8} \, \Omega \text{ m}, \, l = 20 \text{ m}$$
$$A = \pi r^2 = \pi \times (10^{-3})^2 = \pi \times 10^{-6} \text{ m}^2$$
$$R = \frac{1.72 \times 10^{-8} \times 20}{\pi \times 10^{-6}}$$
$$= \frac{1.72 \times 10^{-2} \times 20}{\pi} = 0.109 \, \Omega$$

Using centimetres as units:

$$\rho = 1.72 \times 10^{-6} \, \Omega \text{ cm}$$
$$l = 20 \times 10^2 \text{ cm}$$
$$A = \pi r^2 = \pi \times (10^{-1})^2$$
$$= \pi \times 10^{-2} \text{ cm}^2$$
$$R = \frac{1.72 \times 10^{-6} \times 20 \times 10^2}{\pi \times 10^{-2}}$$
$$= \frac{1.72 \times 10^{-2} \times 20}{\pi} = 0.109 \, \Omega$$

Using millimetres as units:

$$\rho = 1.72 \times 10^{-5} \, \Omega \text{ mm}$$
$$l = 20 \times 10^3 \text{ mm}$$
$$A = \pi r^2 = \pi (1)^2 = \pi \text{ mm}^2$$
$$R = \frac{1.72 \times 10^{-5} \times 20 \times 10^3}{\pi}$$
$$= \frac{1.72 \times 10^{-2} \times 20}{\pi} = 0.109 \, \Omega$$

All three methods give the same answer, and the example shows the importance of expressing the given value of ρ in the correct unit.

Example 5 A wire 20 m long and 1.2 mm in diameter has a resistance of 6 ohms. Calculate the specific resistance.

$$\rho = \frac{R \times A}{l}$$

where

$R = 6$

$A = \pi r^2 = \pi \left(\frac{1.2 \times 10^{-1}}{2}\right)^2 \text{ cm}^2$

$l = 20 \times 10^2 \text{ cm}$

$\rho = \frac{6 \times \pi \times (0.6 \times 10^{-1})^2}{20 \times 10^2}$

$ = \frac{6 \times \pi \times 0.6^2 \times (10^{-1})^2}{20 \times 10^2} \; \Omega \text{ cm}$

$ = 3.39 \times 10^{-5} = 33.9 \times 10^{-6} \; \Omega \text{ cm}$

This problem has been solved using centimetres as the units. Using metres as the units would give an answer of $33.9 \times 10^{-8} \; \Omega$ m. Using millimetres as the units would give an answer of $33.9 \times 10^{-5} \; \Omega$ mm.

```
10 PRINT "PROG 5"
20 PRINT "SPECIFIC RESISTANCE"
30 INPUT "ENTER RESISTANCE IN OHMS" ; R
40 INPUT "ENTER AREA IN SQUARE CENTIMETRES" ; A
50 INPUT "ENTER LENGTH IN CENTIMETRES" ; L
60 LET RHO=R*A/L
70 PRINT "RHO = " RHO "OHM CMS"
```

Example 6 The resistance of a cable is 5 Ω. The diameter is 20 mm. Given that the specific resistance is $1.72 \times 10^{-8} \; \Omega$ m, calculate the length.

$$l = \frac{RA}{\rho}$$

where

$R = 5$

$A = \pi r^2 = \pi \times (10)^2 \times (10^{-3})^2 \text{ m}^2$

$\rho = 1.72 \times 10^{-8} \; \Omega \text{ m}$

$l = \frac{5 \times \pi \times 10^2 \times (10^{-3})^2}{1.72 \times 10^{-8}}$

$ = \frac{5 \times \pi \times 10^2 \times 10^{-6}}{1.72 \times 10^{-8}}$

$ = \frac{5 \times \pi \times 10^2}{1.72 \times 10^{-2}}$

$ = \frac{5 \times \pi \times 10^2 \times 10^2}{1.72} = 91\,325 \text{ m}$

```
10 PRINT "PROG 6"
20 PRINT "CONDUCTOR LENGTH"
30 INPUT "ENTER RESISTANCE IN OHMS" ; R
40 INPUT "ENTER AREA IN SQUARE METRES" ; A
50 INPUT "ENTER RESISTIVITY IN OHM METER" ; RHO
60 LET L=R*A/RHO
70 PRINT "LENGTH = " L "METRES"
```

Example 7 Find the diameter of a cable in millimetres whose length is 45 663 m, and whose resistance is 10 Ω. Take ρ as 1.72×10^{-8} Ω m.

$$A = \frac{\rho l}{R}$$

where

$$\rho = 1.72 \times 10^{-8} \ \Omega \text{ m}$$

$$l = 45\ 663 \text{ m}$$

$$R = 10 \ \Omega$$

$$A = \frac{1.72 \times 10^{-8} \times 45\ 663}{10}$$

$$= 78540 \times 10^{-9}$$

$$= 78.54 \times 10^{-6} \text{ m}^2$$

Since 10^6 mm^2 = 1 m^2

$$A = 78.54 \times 10^{-6} \times 10^6 = 78.54 \text{ mm}^2$$

Since

$$A = \pi r^2$$

$$r^2 = \frac{A}{\pi}$$

$$r = \sqrt{\left(\frac{A}{\pi}\right)} = \sqrt{\left(\frac{78.54}{\pi}\right)} = 5 \text{ mm}$$

Cable diameter = $2r = 2 \times 5 = 10$ mm

If the diameter had been asked for in centimetres, then

$$A = 78.54 \times 10^{-6} \times 10^4$$

$$= 0.7854 \text{ cm}^2 \text{ (since } 10^4 \text{ cm}^2 = 1 \text{ m}^2\text{)}$$

$$r = \sqrt{\left(\frac{A}{\pi}\right)} = \sqrt{\left(\frac{0.7854}{\pi}\right)} = 0.5 \text{ cm}$$

$$d = 1 \text{ cm (since } d = 2r\text{)}$$

If the diameter had been asked for in metres, then

$$A = 78.54 \times 10^{-6} \text{ m}^2$$

$$r = \sqrt{\left(\frac{A}{\pi}\right)} = \sqrt{\left(\frac{78.54}{\pi \times 10^6}\right)} = 5 \times 10^{-3}$$

$$= 0.005 \text{ m}$$

$$d = 2r = 0.01 \text{ m}$$

```
10 PRINT "PROG 7"
20 PRINT "CONDUCTOR RADIUS"
30 INPUT "ENTER RESISTIVITY IN OHM METRE" ; RHO
40 INPUT "ENTER LENGTH IN METRES" ; L
50 INPUT "ENTER RESISTANCE IN OHMS" ; R
60 LET A=RHO*L/R
65 PRINT "A = " A
70 LET RAD=SQR(A*10^6/3.142)
80 PRINT "RADIUS = " RAD "MILLIMETRES"
```

(b) Temperature coefficient of resistance

The resistance of a conductor also depends on its temperature. The resistance, for example, of pure metals increases with temperature, and the temperature coefficient α (Greek letter alpha) is the fractional increase of the resistance per degree increase of the temperature, above a defined temperature. Let T_1 be the defined temperature, and let R_1 and α_1 be the resistance and temperature coefficient respectively at T_1; then R_2 at T_2 is given by

$$R_2 = R_1[1 + \alpha_1(T_2 - T_1)]$$

For carbon, insulating materials and electrolytes, α will be negative, and hence R_2 will be less than R_1, showing that resistance decreases with increased temperature in these materials. The formula for R_2 can be transposed for the other four parameters.

1 Since

$$R_1[1 \times \alpha_1(T_2 - T_1)] = R_2$$

then

$$R_1 = \frac{R_2}{[1 + \alpha_1(T_2 - T_1)]}$$

2 Since

$$R_1[1 + \alpha_1(T_2 - T_1)] = R_2$$

then

$$l + \alpha_1(T_2 - T_1) = \frac{R_2}{R_1}$$

$$\alpha_1(T_2 - T_1) = \frac{R_2}{R_1} - 1 = \frac{R_2 - R_1}{R_1}$$

$$\alpha_1 = \frac{R_2 - R_1}{R_1(T_2 - T_1)}$$

3 Since

$$\alpha_1(T_2 - T_1) = \frac{R_2 - R_1}{R_1}$$

then

$$T_2 - T_1 = \frac{R_2 - R_1}{R_1 \alpha_1}$$

and

$$T_2 = \frac{R_2 - R_1}{R_1 \alpha_1} + T_1$$

4 From (3)

$$T_1 = T_2 - \frac{R_2 - R_1}{R_1 \alpha_1}$$

Example 8 A copper conductor has a resistance of 4 Ω at 20°C. What is the resistance at 40°C, if the temperature coefficient at 20°C is 0.0039/°C?

$$R_2 = R_1[1 + \alpha_1(T_2 - T_1)]$$
$$= 4[1 + 0.0039(40 - 20)] = 4.312 \ \Omega$$

```
10 PRINT "PROG 8"
20 PRINT "THIS PROGRAMME COMPUTES THE RESISTANCE"
30 PRINT "WHEN TEMPERATURE IS INCREASED"
40 INPUT "ENTER THE INITIAL  TEMPERATURE" ; T1
50 INPUT "ENTER RESISTANCE AT T1" ; R1
60 INPUT "ENTER TEMPERATURE COEFFICIENT AT T1" ; A
70 INPUT "ENTER THE FINAL TEMPERATURE" ; T2
80 LET R2=R1+R1*A*T2-R1*A*T1
90 PRINT "RESISTANCE = " R2 "OHMS"
```

Example 9 A current flow in a coil of copper wire raises the temperature from 20°C at switch on, to 50°C. If the resistance of the coil at 50°C is 100 Ω, calculate the resistance at 20°C. α = 0.0039/°C at 20°C.

$$R_1 = \frac{R_2}{[1 + \alpha(T_2 - T_1)]}$$
$$= \frac{100}{[1 + 0.0039(50 - 20)]}$$
$$= 89.5 \ \Omega$$

```
10 PRINT "PROG 9"
20 PRINT "THIS PROGRAM COMPUTES THE RESISTANCE"
30 PRINT "WHEN TEMPERATURE IS DECREASED"
40 INPUT "ENTER THE INITIAL TEMPERATURE " ; T2
50 INPUT "ENTER RESISTANCE AT T2" ; R2
60 INPUT "ENTER THE COEFFICIENT AT T1" ; A
70 INPUT "ENTER THE FINAL   TEMPERATURE" ; T1
80 LET R1=R2/((1+A*T2)-(A*T1))
90 PRINT "RESISTANCE = " R1 "OHMS"
```

Example 10 A conductor has a resistance of 10 Ω at 20°C, and a resistance of 10.6 Ω at 35°C. Calculate the value of the temperature coefficient of resistance at 20°C.

$$\alpha_1 = \frac{R_2 - R_1}{R_1(T_2 - T_1)}$$
$$= \frac{10.6 - 10}{10(35 - 20)}$$
$$= \frac{0.6}{150} = 0.004/°C$$

```
10 PRINT "PROG 10"
20 PRINT "THIS PROGRAM COMPUTES TEMPERATURE COEFFICIENT"
30 INPUT "ENTER INITIAL TEMPERATURE" ; T1
40 INPUT "ENTER RESISTANCE AT T1" ; R1
50 INPUT "ENTER FINAL TEMPERATURE" ; T2
60 INPUT "ENTER RESISTANCE AT T2" ; R2
70 LET A=(R2-R1)/(R1*T2-R1*T1)
80 PRINT "TEMPERATURE COEFFICIENT = " A "PER DEGREE C"
```

Example 11 The resistance of an aluminium conductor increases from 3 Ω to 3.2 Ω when heated. If the initial temperature was 20°C, calculate the final temperature given that the coefficient of resistance of aluminium is 0.004/°C at 20°C.

$$T_2 = \frac{R_2 - R_1}{R_1 \alpha_1} + T_1$$

$$= \frac{3.2 - 3}{3 \times 0.004} + 20$$

$$= 36.67°C$$

```
10 PRINT "PROG 11"
20 PRINT "THIS PROGRAM COMPUTES FINAL TEMPERATURE"
30 INPUT "ENTER INITIAL TEMPERATURE" ; T1
40 INPUT "ENTER RESISTANCE AT T1" ; R1
50 INPUT "ENTER TEMPERATURE COEFFICIENT" ; A
60 INPUT "ENTER RESISTANCE AT FINAL TEMPERATURE" ; R2
70 LET T2=(R2-R1)/(R1*A)+T1
80 PRINT "FINAL TEMPERATURE = " T2 "DEGREES C"
```

Example 12 A wire has a resistance of 6 Ω at 40°C. Calculate the temperature when the resistance was only 5 Ω assuming that α at that temperature was 0.005/°C.

$$T_1 = T_2 - \frac{R_2 - R_1}{R_1 \alpha_1} °C$$

$$= 40 - \frac{6 - 5}{5 \times 0.005} °C$$

$$= 0°C$$

```
10 PRINT "PROG 12"
20 PRINT "THIS PROGRAM COMPUTES INITIAL TEMPERATURE"
30 INPUT "ENTER RESISTANCE AT INITIAL TEMPERATURE" ; R1
40 INPUT "ENTER TEMPERATURE COEFFICIENT" ; A
50 INPUT "ENTER FINAL TEMPERATURE" ; T2
60 INPUT "ENTER RESISTANCE AT T2" ; R2
70 LET T1=T2-((R2-R1)/(R1*A))
80 PRINT "INITIAL TEMPERATURE = " T1 "DEGREES C"
```

2.4 Resistor values

The ohmic value of a resistor, and the tolerance on that value, are indicated by coloured bands on the body of the resistor towards one end as shown in Figure 2.2. Bands 1 and 2 indicate the first two significant figures of the rated value, band 3 is the multiplier, and band 4 is the tolerance.

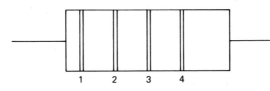

Figure 2.2

There are also five-band resistors where the first three bands indicate the first three significant figures, the fourth band the multiplier, and the fifth band the tolerance. Six-band resistors are the same as the five-band resistors, except that a sixth band is added to indicate the temperature coefficient.

(a) Colour code
Resistance values are often shown on circuit diagrams according to the BS 1852 resistance code. Since $1200\ \Omega = 1.2\ \text{k}\Omega$ this would be shown as 1K2 to avoid using the decimal point. Tolerances are indicated by a letter. $F = \pm 1\%$. $G = \pm 2\%$. $J = \pm 5\%$. $K = \pm 10\%$. $M = \pm 20\%$.

Table 2.1   Colour code for a four-band resistor

Colour	Band 1	Band 2	Band 3	Band 4
Black	0	0	$\times 10^0$	
Brown	1	1	$\times 10^1$	$\pm 1\%$
Red	2	2	$\times 10^2$	$\pm 2\%$
Orange	3	3	$\times 10^3$	
Yellow	4	4	$\times 10^4$	
Green	5	5	$\times 10^5$	
Blue	6	6	$\times 10^6$	
Violet	7	7	$\times 10^7$	
Grey	8	8	$\times 10^8$	
White	9	9	$\times 10^9$	
Gold			$\times 10^{-1}$	$\pm 5\%$
Silver			$\times 10^{-2}$	$\pm 10\%$
None				$\pm 20\%$

Table 2.2   Examples of resistor values

Band 1	Band 2	Band 3	Band 4	Value (Ω)
Orange	White	Red	Silver	$39 \times 10^2 = 3900 \pm 10\%$ (3.9 kΩ)
Yellow	Violet	Silver	Red	$47 \times 10^{-2} = 0.47 \pm 2\%$
Brown	Green	Black	Gold	$15 \times 10^0 = 15 \pm 5\%$
Brown	Grey	Brown	None	$18 \times 10^1 = 180 \pm 20\%$
Brown	Black	Red	Brown	$10 \times 10^2 = 1000 \pm 1\%$ (1 kΩ)
Blue	Grey	Orange	Silver	$68 \times 10^3 = 68\ 000 \pm 10\%$ (68 kΩ)
Brown	Red	Green	Red	$12 \times 10^5 = 1\ 200\ 000 \pm 2\%$ (1.2 MΩ)

Thus 1200 Ω ±5% would be shown as 1K2J. The examples looked at so far could be shown as follows:
3K9K = 3.9k ±10%. R47G = 0.47 Ω ±2%. 15RJ = 15 Ω ±5%. 180RM = 180 Ω ±20%. 1K0F = 1 kΩ ±1%. 68KK = 68 kΩ ±10%. 1M2G = 1.2 MΩ ±2%.

(b) Tolerances

Given a 470RK resistor the tolerance is

$$\pm \frac{10}{100} \times 470 = \pm 47 \ \Omega$$

This means that the value lies between

423 and 517R

A 470RJ resistor would have a tolerance of

$$\pm \frac{5}{100} \times 470 = \pm 23.5 \ \Omega$$

This means that the value lies between

446R5 and 493R5

Inspection of a component catalogue shows that resistors are produced in preferred values. Table 2.3 shows these standard values together with the lower and upper values for a 10% tolerance. The table shows that these standard values are chosen to avoid much overlap between tolerance spreads. Resistor values continue from the table by the addition of the appropriate number of zeros

100R, 120R, 150R, etc.

or by dividing by 10 or 100

1R, 1R2, 1R5, etc.

or

R1, R12, R15, etc.

Example 13 Given a 1M2G resistor, calculate the resistance spread.

$$\% \text{ tolerance} = \frac{2 \times 1.2 \times 10^6}{100} = \pm 24\,000$$

Table 2.3 Resistor preferred values

Standard value (R)	Lower limit (R)	Upper limit (R)
10	9	11
12	10.8	13.2
15	13.5	16.5
18	16.2	19.8
22	19.8	24.2
27	24.3	29.7
33	29.7	36.3
39	35.1	42.9
47	42.3	51.7
56	50.4	61.6
68	61.2	74.8
82	73.8	90.2

Resistors 15

The spread is 1 176 000 to 1 224 000, which equals 1M176 to 1M224.

```
10 PRINT "PROG 13"
20 PRINT "THIS PROGRAM COMPUTES RESISTANCE SPREAD"
30 INPUT "ENTER NOMINAL RESISTANCE VALUE" ; R
40 INPUT "ENTER PERCENTAGE TOLERANCE VALUE" ; T
50 LET A=R*T/100
60 PRINT "SPREAD IS " R-A "TO" R+A "OHMS"
```

Example 14 If a resistor with a tolerance of ±10% has a resistance spread of 30 Ω, what is the nominal value?

Let Y = nominal value

$$Y \times \frac{10}{100} = 15$$

$$Y = \frac{15 \times 100}{10} = 150R$$

```
10 PRINT "PROG 14"
20 PRINT "THIS PROGRAM COMPUTES NOMINAL VALUE"
30 INPUT "ENTER PERCENTAGE TOLERANCE VALUE" ; T
40 INPUT "ENTER RESISTANCE SPREAD" ; S
50 LET N=(S*100)/(2*T)
60 PRINT "NOMINAL VALUE IS " N "OHMS"
```

Example 15 The upper limit of a 10% resistor is 242R. Find the nominal value.

Let Y = nominal value, then

$$Y + \frac{(10Y)}{100} = 242R$$

$$Y + (0.1Y) = 242R$$

$$1.1Y = 242R, \text{ and } Y = \frac{242}{1.1} = 220R$$

```
10 PRINT "PROG 15"
20 PRINT "THIS PROGRAM COMPUTES NOMINAL VALUE"
30 INPUT "ENTER PERCENTAGE TOLERANCE VALUE" ; T
40 INPUT "ENTER UPPER LIMIT VALUE" ; U
50 LET N=(100*U)/(100+T)
60 PRINT "NOMINAL VALUE IS " N "OHMS"
```

Example 16 The lower limit of a 1% resistor is 990R. Find the nominal value.

Let Y = nominal value, then

$$Y - \frac{(1Y)}{100} = 990R$$

$$Y - (0.01Y) = 990R$$

$$0.99Y = 990R$$

and

$$Y = \frac{990}{0.99} = 1000 = 1K0$$

```
10 PRINT "PROG 16"
20 PRINT "THIS PROGRAM COMPUTES NOMINAL VALUE"
30 INPUT "ENTER PERCENTAGE TOLERANCE VALUE" ; T
40 INPUT "ENTER LOWER LIMIT VALUE" ; L
50 LET N=(100*L)/(100-T)
60 PRINT "NOMINAL VALUE IS" N "OHMS"
```

(c) Wattage

When current flows through a resistor heat is generated. The rate at which the heat is generated is measured in watts (W), and this parameter will be covered in Chapter 3. The physical size of the component determines the power rating, and this rating must not be exceeded otherwise it will overheat, and it could be damaged or even permanently destroyed. A typical 0.25 W carbon resistor would have a diameter of 2.5 mm and length 6.5 mm, whilst a 1 W resistor would have a diameter of 5.5 mm and length 14 mm.

2.5 Resistor circuits

(a) Series

1 Figure 2.3(a) shows two resistors connected in series. A resistance measurement between points A and B would give the combined resistance R, where

$$R = R_1 + R_2, \text{ i.e. } R = 47 + 470 = 517R.$$

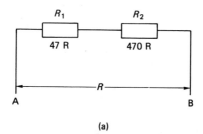

(a)

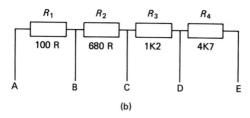

(b)

Figure 2.3

```
10 PRINT "PROG 17"
20 PRINT "THIS PROGRAM ADDS RESISTORS IN SERIES"
30 INPUT "ENTER THE VALUE OF R1" ; R1
40 INPUT "ENTER THE VALUE OF R2" ; R2
50 INPUT "ENTER THE VALUE OF R3" ; R3
60 LET R=R1+R2+R3
70 PRINT "TOTAL RESISTANCE = " R "OHMS"
```

2 Figure 2.3(b) shows four resistors connected in series. The resistances between the various points A to E are shown in Table 2.4.

Table 2.4 Resistance values

Points	Resistance Ω
AB	100R
AC	780R
AD	1K98
AE	6K68
BC	680R
BD	1K88
BE	6K58
CD	1K2
CE	5K9
DE	4K7

(b) Parallel

1 Figure 2.4 shows two resistors connected in parallel. The combined resistance R is given by

$$\frac{1}{R} = \frac{1}{R_1} + \frac{1}{R_2}$$

$$\frac{1}{R} = \frac{1}{100} + \frac{1}{150} = 0.01 + 0.0066$$

$$= 0.0166$$

Then

$$R = \frac{1}{0.0166} = 60R$$

```
10 PRINT "PROG 1·8"
20 PRINT "THIS PROGRAM COMPUTES THE COMBINED RESISTANCE"
30 PRINT "OF 2 RESISTORS IN PARALLEL"
40 INPUT "ENTER THE VALUE OF R1" ; R1
50 INPUT "ENTER THE VALUE OF R2" ; R2
60 LET A=(1/R1)+(1/R2)
70 LET R=1/A
80 PRINT "R = " R "OHMS"
```

2 Figure 2.5 shows three resistors in parallel.

$$\frac{1}{R} = \frac{1}{R_1} + \frac{1}{R_2} + \frac{1}{R_3}$$

$$= \frac{1}{120} + \frac{1}{220} + \frac{1}{330}$$

$$= 0.00833 + 0.00455 + 0.00303$$

$$= 0.01591$$

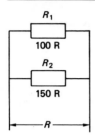

Figure 2.4

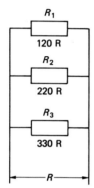

Figure 2.5

Then

$$R = \frac{1}{0.01591} = 62.85 = 62R85$$

```
10 PRINT "PROG 19."
20 PRINT "THIS PROGRAM COMPUTES R IN FIGURE 2.5"
30 INPUT "ENTER THE VALUE OF R1" ; R1
40 INPUT "ENTER THE VALUE OF R2" ; R2
50 INPUT "ENTER THE VALUE OF R3" ; R3
60 LET A=(1/R1)+(1/R2)+(1/R3)
70 LET R=1/A
80 PRINT "R = " R "OHMS"
```

For four resistors in parallel

$$\frac{1}{R} = \frac{1}{R_1} + \frac{1}{R_2} + \frac{1}{R_3} + \frac{1}{R_4}$$

For n resistors

$$\frac{1}{R} = \frac{1}{R_1} + \frac{1}{R_2} + \ldots + \frac{1}{R_n}$$

3 Returning to example (b)(1), since

$$\frac{1}{R} = \frac{1}{R_1} + \frac{1}{R_2}$$

then

$$\frac{1}{R} = \frac{R_1 + R_2}{R_1 R_2}$$

and

$$R = \frac{R_1 R_2}{R_1 + R_2}$$

giving

$$R = \frac{100 \times 150}{100 + 150} = 60\text{R}$$

```
10 PRINT "PROG 20"
20 PRINT "THIS PROGRAM IS AN ALTERNATIVE TO PROG 18"
30 INPUT "ENTER THE VALUE OF R1" ; R1
40 INPUT "ENTER THE VALUE OF R2" ; R2
50 LET R=(R1*R2)/(R1+R2)
60 PRINT "R = " R "OHMS"
```

This shows that R can be found by dividing the product by the sum. This alternative method can only be applied to two resistors in parallel.

Example 17 Two resistors R_1 and R_2 connected in parallel have a combined resistance of 277.913 Ω. If $R_1 = 470$ Ω, calculate the value of R_2.

$$R = \frac{R_1 R_2}{R_1 + R_2}$$

giving

$$R(R_1 + R_2) = R_1 R_2$$

Removing the bracket

$$RR_1 + RR_2 = R_1 R_2$$

and

$$RR_1 = R_1 R_2 - RR_2$$

then

$$RR_1 = R_2(R_1 - R)$$

and

$$R_2 = \frac{RR_1}{R_1 - R}$$
$$= \frac{277.913 \times 470}{470 - 277.913} = 680\text{R}$$

```
10 PRINT "PROG 21 "
20 PRINT "SOLUTION TO EXAMPLE 17"
30 INPUT "ENTER THE VALUE OF R1"; R1
40 INPUT "ENTER THE COMBINED RESISTANCE" ; R
50 LET R2=R*R1/(R1-R)
60 PRINT "R2 = " R2 "OHMS"
```

Example 18 Figure 2.6 shows four resistors connected in parallel. Calculate R.

$$\frac{1}{R} = \frac{1}{1000} + \frac{1}{1000} + \frac{1}{1000} + \frac{1}{1000}$$

$$\frac{1}{R} = 0.001 + 0.001 + 0.001 + 0.001$$

$$\frac{1}{R} = 0.004$$

giving

$$R = \frac{1}{0.004} = 250R$$

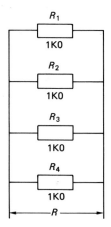

Figure 2.6

```
10 PRINT "PROG 22"
20 PRINT "THIS PROGRAM COMPUTES R IN FIGURE 2.6"
30 INPUT "ENTER THE VALUE OF R1" : R1
40 INPUT "ENTER THE VALUE OF R2" ; R2
50 INPUT "ENTER THE VALUE OF R3" ; R3
60 INPUT "ENTER THE VALUE OF R4" ; R4
70 LET A=(1/R1)+(1/R2)+(1/R3)+(1/R4)
80 LET R=1/A
90 PRINT "R = " R "OHMS"
```

This example shows that when all resistors in a parallel network have the same value, the equivalent resistance R can be found by dividing the resistance value by the number of resistors in the ladder, i.e.

$$\frac{1000}{4} = 250R$$

(c) Series–parallel

1 Figure 2.7 shows an example of a series–parallel circuit. To find the combined resistance R, first find the value of the parallel combination R_2 and R_3.

$$\text{Combined resistance} = \frac{R_2 R_3}{R_2 + R_3}$$

$$= \frac{220 \times 330}{220 + 330} = \frac{72\,600}{550} = 132R$$

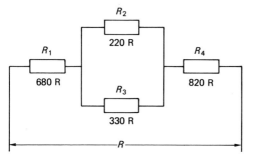

Figure 2.7

The circuit shown in Figure 2.8 is therefore equivalent to the circuit in Figure 2.7, and the total resistance is

$$R = 680 + 132 + 820$$
$$= 1632 = 1.632\text{k} = 1\text{K}632$$

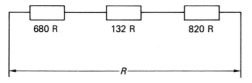

Figure 2.8

```
10 PRINT "PROG 23"
20 PRINT "THIS PROGRAM COMPUTES R IN FIGURE 2.7"
30 INPUT "ENTER THE VALUE OF R1" ; R1
40 INPUT "ENTER THE VALUE OF R2" ; R2
50 INPUT "ENTER THE VALUE OF R3" ; R3
60 INPUT "ENTER THE VALUE OF R4" ; R4
70 LET A=(R2*R3)/(R2+R3)
80 LET R=R1+A+R4
90 PRINT "R = " R "OHMS"
```

2 To find R in Figure 2.9, first find the equivalent resistances of R_1 and R_2, and R_3 and R_4.

$$\frac{22 \times 56}{22 + 56} = 15.79$$

$$\frac{27 \times 33}{27 + 33} = 14.85$$

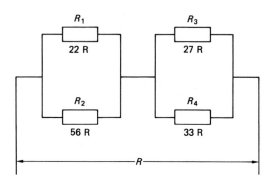

Figure 2.9

The circuit shown in Figure 2.10 is equivalent to the circuit shown in Figure 2.9, and the total resistance is

$R = 15.79 + 14.85$

$= 30.64$

$= 30R64$

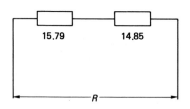

Figure 2.10

```
10 PRINT "PROG 24"
20 PRINT "THIS PROGRAM COMPUTES R IN FIGURE 2.9"
30 INPUT "ENTER THE VALUE OF R1" ; R1
40 INPUT "ENTER THE VALUE OF R2" ; R2
50 INPUT "ENTER THE VALUE OF R3" ; R3
60 INPUT "ENTER THE VALUE OF R4" ; R4
70 LET A=(R1*R2)/(R1+R2)
80 LET B=(R3*R4)/(R3+R4)
90 LET R=A+B
100 PRINT "R = " R "OHMS"
```

3 Figure 2.11 shows another type of a series–parallel arrangement. To solve for R, first combine the resistance of the series chain.

$$R_1 + R_2 + R_3 = 1500 + 1000 + 1500 = 4000$$

The circuit in Figure 2.11 is then equivalent to that of Figure 2.12.

$$R = \frac{4 \times 10^3 \times 10^6}{(4 \times 10^3) + 10^6} = 3984 = 3K984$$

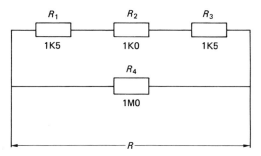

Figure 2.11

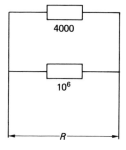

Figure 2.12

```
10 PRINT "PROG 25"
20 PRINT "THIS PROGRAM COMPUTES THE EQUIVALENT"
30 PRINT "RESISTANCE OF FIGURE 2.11"
40 INPUT "ENTER THE VALUE OF R1" ; R1
50 INPUT "ENTER THE VALUE OF R2" ; R2
60 INPUT "ENTER THE VALUE OF R3" ; R3
70 INPUT "ENTER THE VALUE OF R4" ; R4
80 LET A=R1+R2+R3
90 LET R=(A*R4)/(A+R4)
100 PRINT "R = " R "OHMS"
110 END
```

Problems

1 Calculate the resistance of a conductor 80 cm long and 1.5 mm in diameter, given that the specific resistance is 1.72×10^{-6} Ω cm.

2 If the diameter of the conductor in problem 2 was increased to 3.0 mm, what would be the reduction in the resistance?
3 A power cable connecting two properties 500 m apart has a diameter of 25 cm, and a resistance of 1 Ω. Calculate the specific resistance, and express the answer in Ω m, Ω cm, and Ω mm.
4 The resistance of a cable is 2 Ω. The diameter is 2 cm. Given that the specific resistance is 1.72×10^{-8} Ω m, calculate the length in metres.
5 Find the diameter of a cable in millimetres, given that the length is 50 km, the resistance 20 Ω, and $\rho = 1.72 \times 10^{-8}$ Ω m.
6 A conductor has a resistance of 2 Ω at 20°C. Calculate the resistance at 30°C, if the temperature coefficient at 20°C is 0.0039/°C.
7 The temperature of a coil of copper wire increases from 20°C to 40°C during a test. If the resistance of the coil at 40°C is 50 Ω, calculate the resistance at 20°C given that α at 20°C is 0.004/°C.
8 A conductor has a resistance of 9.5 Ω at 20°C and 10 Ω at 30°C. Calculate the value of α at 20°C.
9 The resistance of a cable increases from 2 Ω to 2.1 Ω when heated. If the initial temperature was 20°C, calculate the final temperature if α at 20°C is 0.004/°C.
10 A conductor has a resistance of 2 Ω at 30°C. What is the temperature when the resistance is only 1.5 Ω assuming that at that temperature α is 0.0038/°C?
11 A resistor has the following colour bands: red, violet, orange, silver. Calculate the resistance spread.
12 A resistor has a gold tolerance band, and a resistance spread of 1000 Ω. Calculate the nominal value.
13 The maximum value of a resistor with a red tolerance band is 693R6. Find the nominal value.
14 The lower limit of a 20% resistor is 800k. Find the nominal value.
15 A circuit has a 68R resistor connected in series with a 680R. What is the total resistance?
16 If two further resistors of 1K5 and 1M2 are connected in series with those in Problem 15, what is the total resistance?
17 A 47K resistor is connected in parallel with a 33K resistor. What is the combined resistance?
18 If a 22K resistor is added in parallel with the two in Problem 17, what is the equivalent resistance?

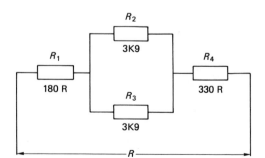

Figure 2.13

19 Two resistors connected in parallel have a combined resistance of 72 Ω. If one resistor has a value of 180R, calculate the value of the other.
20 Calculate the value of R in Figure 2.13, 2.14 and 2.15.

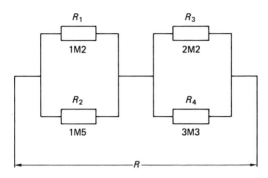

Figure 2.14

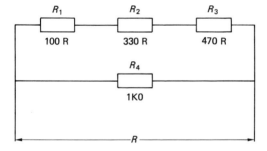

Figure 2.15

3 D.c. circuits

3.1 Current

Chapter 1 introduced d.c. voltages. The voltage between two points in a circuit is referred to as the potential difference and is the work done (W) in joules, to move 1 coulomb of charge (Q) between these points.

In general

$$V \text{ (volts)} = \frac{W \text{ (joules)}}{Q \text{ (coulombs)}}$$

Transposing the formula gives

$$Q = \frac{W}{V} \text{ and } W = VQ$$

A voltage produces current in a circuit. The unit of current is the ampere (A), and the quantity symbol is I. Current is the flow of electrons, and if a charge of 1 coulomb, which equals 6.241×10^{18} electrons, passes a given point in 1 second (t) then this constitutes a current of 1 ampere.

In general

$$I \text{ (amperes)} = \frac{Q \text{ (coulombs)}}{t \text{ (seconds)}}$$

Transposing the formula gives

$$Q = It \text{ and } t = \frac{Q}{I}$$

3.2 Units of current

The main units used are the following:

1. The ampere or amp (A)
2. The milliamp (mA), where

$$1 \text{ mA} = \frac{1}{10^3} \text{ A} = 10^{-3} \text{ A}$$

3. The microamp (µA), where

$$1 \text{ µA} = \frac{1}{10^6} \text{ A} = 10^{-6} \text{ A}$$

The relationship between milli and micro units was covered in Chapter 1, and therefore

$$1 \text{ mA} = \frac{1}{10^3} \text{ A and } 10^3 \text{ mA} = 1 \text{ A}$$

$$1 \text{ µA} = \frac{1}{10^6} \text{ A and } 10^6 \text{ µA} = 1 \text{ A}$$

$$10^3 \text{ mA} = 10^6 \text{ µA and } 1 \text{ mA} = 10^3 \text{ µA}$$

Conversions of units can be carried out, remembering that

$$\frac{mA}{10^3} = A, \quad \frac{\mu A}{10^6} = A, \quad \frac{\mu A}{10^3} = mA$$

Example 1 1.2 J are required to move 20 C of charge from point A to point B in a circuit. Find the potential difference between the two points.

$$V = \frac{W}{Q} = \frac{1.2}{20} = 0.06 \text{ V}$$

```
10 PRINT "PROG 26"
20 PRINT "THIS PROGRAM COMPUTES POTENTIAL DIFFERENCE"
30 INPUT "ENTER WORK DONE IN JOULES" ; W
40 INPUT "ENTER CHARGE IN COULOMBS" ; Q
50 LET V=W/Q
60 PRINT "POTENTIAL DIFFERENCE = " V "VOLTS"
```

Example 2 If a current of 12 A is measured at a circuit point over a period of 10 min, calculate the charge that has passed the point.

$$Q = I \text{ (A)} \times t \text{ (s)} = 12 \times 10 \times 60 = 7200 \text{ C}$$

Note. Since 1 ampere is 1 coulomb per second, the unit of charge could be called the ampere-second. A larger unit is the ampere hour (A h), which is equal to

$$60 \times 60 = 3600 \text{ C}$$

and

$$\frac{7200}{3600} \text{ C} = 2\text{A h}$$

```
10 PRINT "PROG 27"
20 PRINT "THIS PROGRAM COMPUTES CHARGE"
30 INPUT "ENTER CURRENT IN AMPS" ; I
40 INPUT "ENTER TIME IN SECONDS" ; T
50 LET Q=I*T
60 PRINT "CHARGE = " Q "COULOMBS"
```

Example 3 Calculate the time for 10 C to flow if the current is 20 mA.

$$t = \frac{Q}{I} = \frac{10}{20 \times 10^{-3}} = \frac{10 \times 10^3}{20} = 500 \text{ s}$$

or

$$\frac{500}{60} = 8.33 \text{ min}$$

```
10 PRINT "PROG 28"
20 PRINT "THIS PROGRAM COMPUTES TIME"
30 INPUT "ENTER CHARGE IN COULOMBS" ; Q
40 INPUT "ENTER CURRENT IN AMPS" ; I
50 LET T=Q/I
60 PRINT "TIME = " T "SECONDS"
70 LET T=Q/(I*60)
80 PRINT "TIME = " T "MINUTES"
```

Example 4 There is a flow of 20×10^3 C in a resistor over a period of 1 hour. Calculate the current.

$$I = \frac{Q}{t} = \frac{20 \times 10^3}{1 \times 60 \times 60} = 5.56 \text{ A}$$

```
10 PRINT "PROG 29"
20 PRINT "THIS PROGRAM COMPUTES CURRENT"
30 INPUT "ENTER CHARGE IN COULOMBS" ; Q
40 INPUT "ENTER TIME IN SECONDS" ; T
50 LET I=Q/T
60 PRINT "CURRENT = " I "AMPS"
```

3.3 Ohm's law

The current flowing in a conductor is proportional to the applied voltage. If the applied voltage is V (volts) and the resulting current is I (amperes), the relationship between V and I is

$$I = \frac{V}{R}$$

where R is the conductor resistance in ohms. This is known as Ohm's law. Since

$$I = \frac{V}{R}$$

transposing the formula gives

$$V = IR$$

and

$$R = \frac{V}{I}$$

The law can be applied to complete circuits and to individual components within the circuit.

3.4 Power

Power is the rate at which a conductor either absorbs or produces energy. The unit of power is the watt, with unit symbol W. The quantity symbol is P.

One watt is equal to 1 joule of energy released per second and therefore

$$P \text{ (watts)} = \text{joules per second}$$
$$= \text{volts} \times \text{coulombs per second}$$
$$= \text{volts} \times \text{amps}$$

i.e. $P = VI$, $V = \dfrac{P}{I}$ and $I = \dfrac{P}{V}$

The watt-second is a very small unit, and a more practical unit is the kilowatt-hour, used for example, for specifying the amount of electricity consumed in the home.

1 kilowatt-hour (kW h) = 1000 watts × 60 minutes
$$\times 60 \text{ seconds}$$
$$= 3\,600\,000 \text{ joules, or } 3.6 \times 10^6 \text{ J}$$

P which equals VI, can be expressed in two other ways.

Since

$$V = IR$$

then

$$P = IRI = I^2R$$

Since

$$I = \frac{V}{R}$$

then

$$P = V\frac{V}{R} = \frac{V^2}{R}$$

The main units used are

1. The watt (W)
2. The megawatt (MW) = 10^6 W
3. The kilowatt (kW) where 1 kW = 10^3 W
4. The milliwatt (mW) where 1 mW = $1/10^3$ W = 10^{-3} W
5. The microwatt (μW) where 1 μW = $1/10^6$ W = 10^{-6} W

It therefore follows that 1 MW = 10^3 kW. The relationships between the subunits are the same as those for voltages, covered in Chapter 1, and for current, covered in Section 3.2.

3.5 Types of circuits

(a) Series

Example 5 For the circuit shown in Figure 3.1 the following can be calculated:

$$I = \frac{V}{R} = \frac{30}{120} = 0.25 \text{ A}$$

$$P = VI = 30 \times 0.25 = 7.5 \text{ W}$$

The two other ways of calculating P are

$$P = I^2R = 0.25 \times 0.25 \times 120 = 7.5 \text{ W}$$

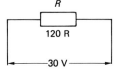

Figure 3.1

$$P = \frac{V^2}{R}$$

$$= \frac{30 \times 30}{120} = 7.5 \text{ W}$$

This is an example of the simplest possible circuit, which has four parameters, V, R, I and P. Given the values of two, the other two can be evaluated.

```
10 PRINT "PROG 30"
20 PRINT "THIS PROGRAM SOLVES EXAMPLE 5"
30 INPUT "ENTER VOLTAGE" ; V
40 INPUT "ENTER RESISTANCE" ; R
50 LET I=V/R
60 PRINT "CURRENT = " I "AMPS"
70 LET P=V*I
80 PRINT "POWER = " P "WATTS"
```

Example 6 In a circuit similar to Figure 3.1, $V = 120$ V, $I = 10$ A, calculate R and P.

$$R = \frac{V}{I} = \frac{120}{10} = 12 \text{ }\Omega$$

$$P = VI = 120 \times 10 = 1200 \text{ W}$$

```
10 PRINT "PROG 31"
20 PRINT "THIS PROGRAM SOLVES EXAMPLE 6"
30 INPUT "ENTER VOLTAGE" ; V
40 INPUT "ENTER CURRENT" ; I
50 LET R=V/I
60 PRINT "RESISTANCE = " R "OHMS"
70 LET P=V*I
80 PRINT "POWER = " P "WATTS"
```

Example 7 Given $V = 12$ V, $P = 24$ W, calculate I and R.

$$\frac{P}{V} = I = \frac{24}{12} = 2 \text{ A}$$

$$R = \frac{V}{I} = \frac{12}{2} = 6 \text{ }\Omega$$

```
10 PRINT "PROG 32"
20 PRINT "THIS PROGRAM SOLVES EXAMPLE 7"
30 INPUT "ENTER VOLTAGE" ; V
40 INPUT "ENTER POWER" ; P
50 LET I=P/V
60 PRINT "CURRENT = " I "AMPS"
70 LET R=V/I
80 PRINT "RESISTANCE = " R "OHMS"
```

Example 8 Given that $I = 20$ mA, $R = 150$ Ω, calculate V and P.

$$V = IR = 20 \times 10^{-3} \times 150 = 3 \text{ V}$$

$$P = VI = 3 \times 20 \times 10^{-3}$$

$$= 0.06 \text{ W} = 0.06 \times 10^3 \text{ mW} = 60 \text{ mW}$$

d.c. circuits 31

```
10 PRINT "PROG 33"
20 PRINT "THIS PROGRAM SOLVES EXAMPLE 8"
30 INPUT "ENTER CURRENT IN AMPS" ; I
40 INPUT "ENTER RESISTANCE" ; R
50 LET V=I*R
60 PRINT "VOLTAGE = " V "VOLTS"
70 LET P=V*I
80 PRINT "POWER = " P "WATTS"
```

Example 9 Given that $I = 10$ mA, $P = 100$ mW, calculate V and R.

$$V = \frac{P}{I} = \frac{100 \times 10^{-3}}{10 \times 10^{-3}} = \frac{100}{10} = 10 \text{ V}$$

$$R = \frac{V}{I} = \frac{10}{10 \times 10^{-3}} = 1000 \text{ } \Omega$$

```
10 PRINT "PROG 34"
20 PRINT "THIS PROGRAM SOLVES EXAMPLE 9"
30 INPUT "ENTER CURRENT IN MILLIAMPS" ; I
40 INPUT "ENTER POWER IN MILLIWATTS" ; P
50 LET V=P/I
60 PRINT "VOLTAGE = " V "VOLTS"
70 LET R=V/(I/1000)
80 PRINT "RESISTANCE = " R "OHMS"
```

Example 10 Given $R = 15$ kΩ, $P = 24$ μW, calculate V and I.

Since

$$P = \frac{V^2}{R}$$

then

$$V^2 = PR$$

and

$$V = \sqrt{(PR)} = \sqrt{(24 \times 10^{-6} \times 15 \times 10^{3})} = 0.6 \text{ V}$$

$$I = \frac{V}{R} = \frac{0.6}{15\,000} = 0.00004 = 4 \times 10^{-5} \text{ A} = 40 \text{ }\mu\text{A}$$

```
10 PRINT "PROG 35"
20 PRINT "THIS PROGRAM SOLVES EXAMPLE 10"
30 INPUT "ENTER RESISTANCE" ; R
40 INPUT "ENTER POWER IN MICROWATTS" ; P
50 LET V=SQR((P/1000000!)*R)
60 PRINT "VOLTAGE = " V "VOLTS"
70 LET I=V/R
80 PRINT "CURRENT = " I "AMPS"
```

Example 11 For the circuit shown in Figure 3.2, eight parameters can be evaluated.

$$I = \frac{V}{R}$$

where R is the equivalent resistance.

$$I = \frac{60}{120 + 150 + 330} = 0.1 \text{ A}$$

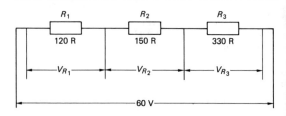

Figure 3.2

To find the voltage drop across each resistor apply Ohm's law.

$V_1 = IR_1 = 0.1 \times 120 = 12$ V

$V_2 = IR_2 = 0.1 \times 150 = 15$ V

$V_3 = IR_3 = 0.1 \times 330 = 33$ V

Note

$V_1 + V_2 + V_3 = V$

Circuit input power

$P = VI = 60 \times 0.1 = 6$ W

Let P_1, P_2 and P_3 equal the powers consumed by R_1, R_2 and R_3.

$P_1 = I^2 R_1 = 0.1 \times 0.1 \times 120 = 1.2$ W

$P_2 = I^2 R_2 = 0.1 \times 0.1 \times 150 = 1.5$ W

$P_3 = I^2 R_3 = 0.1 \times 0.1 \times 330 = 3.3$ W

Note

$P_1 + P_2 + P_3 =$ input power P

The power consumed by the resistors can also be calculated using the other two power formulas (see Example 5).

$P_1 = \dfrac{V_1^2}{R_1} = \dfrac{12^2}{120} = 1.2$ W

$P_1 = V_1 I = 12 \times 0.1 = 1.2$ W

$P_2 = \dfrac{V_2^2}{R_2} = \dfrac{15^2}{150} = 1.5$ W

$P_2 = V_2 I = 15 \times 0.1 = 1.5$ W

$P_3 = \dfrac{V_3^2}{R_3} = \dfrac{33^2}{330} = 3.3$ W

$P_3 = V_3 I = 33 \times 0.1 = 3.3$ W

```
10 PRINT "PROG 36"
20 PRINT "THIS PROGRAM SOLVES EXAMPLE 11"
30 INPUT "ENTER INPUT VOLTAGE"  ; V
40 INPUT "ENTER RESISTANCE R1" ; R1
50 INPUT "ENTER RESISTANCE R2" ; R2
60 INPUT "ENTER RESISTANCE R3" ; R3
70 LET I=V/(R1+R2+R3)
```

```
80  LET V1=I*R1
90  LET V2=I*R2
100 LET V3=I*R3
110 LET P=V*I
120 LET P1=V1*I
130 LET P2=V2*I
140 LET P3=V3*I
150 PRINT "INPUT CURRENT = " ; I "AMPS"
160 PRINT "V1 = " V1 "VOLTS"
170 PRINT "V2 = " V2 "VOLTS"
180 PRINT "V3 = " V3 "VOLTS"
190 PRINT "INPUT POWER = " P "WATTS"
200 PRINT "P1 = " P1 "WATTS"
210 PRINT "P2 = " P2 "WATTS"
220 PRINT "P3 = " P3 "WATTS"
```

Example 12 In the circuit shown in Figure 3.3 calculate the value of R_3, the total power consumed, and the power consumed by the individual resistors.

$$\text{Total resistance } R = \frac{V}{I} = \frac{35}{0.5} = 70 \ \Omega$$

$$R_3 = 70 - 15 - 22 = 33 \ \Omega$$

Input power $= VI = 35 \times 0.5 = 17.5$ W

Let P_1, P_2 and P_3 be the power consumed by R_1, R_2 and RR_3.

$$P_1 = I^2 R_1 = 0.5^2 \times 15 = 3.75 \text{ W}$$
$$P_2 = I^2 R_2 = 0.5^2 \times 22 = 5.5 \text{ W}$$
$$P_3 = I^2 R_3 = 0.5^2 \times 33 = 8.25 \text{ W}$$

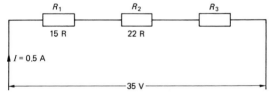

Figure 3.3

```
10  PRINT "PROG 37"
20  PRINT "THIS PROGRAM SOLVES EXAMPLE 12"
30  INPUT "ENTER INPUT VOLTAGE" ; V
40  INPUT "ENTER INPUT CURRENT" : I
50  INPUT "ENTER RESISTANCE R1" ; R1
60  INPUT "ENTER RESISTANCE R2" ; R2
70  LET R=V/I
80  LET R3=R-R1-R2
90  PRINT "R3 = " R3 "OHMS"
100 P=V*I
110 PRINT "INPUT POWER = " P "WATTS"
120 LET P1=I*I*R1
130 LET P2=I*I*R2
140 LET P3=I*I*R3
150 PRINT "P1 = " P1 "WATTS"
160 PRINT "P2 = " P2 "WATTS"
170 PRINT "P3 = " P3 "WATTS"
```

Other parameters can be calculated as shown in Example 11.

34 Circuit calculations pocket book

Example 13 Figure 3.4 shows three lamps which consume 40, 60 and 100 W when connected to a 250 V supply. Calculate the current, and the resistance of each lamp filament.

Total power consumed = 40 + 60 + 100 = 200 W

$$250I = 200$$

$$I = \frac{200}{250} = 0.8 \text{ A}$$

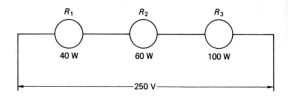

Figure 3.4

Since

$$I^2 R_1 = W_1$$

$$R_1 = \frac{W_1}{I^2} = \frac{40}{0.8^2} = 62.5 \text{ }\Omega$$

$$I^2 R_2 = W_2$$

$$R_2 = \frac{W_2}{I^2} = \frac{60}{0.8^2} = 93.75 \text{ }\Omega$$

$$I^2 R_3 = W_3$$

$$R_3 = \frac{W_3}{I^2} = \frac{100}{0.8^2} = 156.25 \text{ }\Omega$$

```
10 PRINT "PROG 38"
20 PRINT "THIS PROGRAM SOLVES EXAMPLE 13"
30 INPUT "ENTER INPUT VOLTAGE" ; V
40 INPUT "WATTAGE OF R1" ; W1
50 INPUT "WATTAGE OF R2" ; W2
60 INPUT "WATTAGE OF R3" ; W3
70 LET P=W1+W2+W3
80 LET I=P/V
90 PRINT "INPUT CURRENT = " I "AMPS"
100 LET R1=W1/(I*I)
110 LET R2=W2/(I*I)
120 LET R3=W3/(I*I)
130 PRINT "R1 = " R1 "OHMS"
140 PRINT "R2 = " R2 "OHMS"
150 PRINT "R3 = " R3 "OHMS"
```

(b) Parallel

Example 14 Figure 3.5 shows two resistors connected in parallel. The following parameters can be calculated:

d.c. circuits

$$I = \frac{V}{R}$$

where R is the equivalent resistance calculated as shown in Chapter 2.

$$R = \frac{1000 \times 1500}{1000 + 1500} = 600 \; \Omega$$

$$I = \frac{30}{600} = 0.05 \; \text{A}$$

$$I_1 = \frac{30}{1000} = 0.03 \; \text{A}, \; I_2 = \frac{30}{1500} = 0.02$$

Note that $I_1 + I_2 = I$.

Input power $P = VI = 30 \times 0.05 = 1.5 \; \text{W}$

Let P_1 and P_2 equal the powers dissipated in R_1 and R_2.

$$P_1 = I_1^2 R_1 = 0.03 \times 0.03 \times 1000 = 0.9 \; \text{W}$$

$$P_2 = I_2^2 R_2 = 0.02 \times 0.02 \times 1500 = 0.6 \; \text{W}$$

Note that $P_1 + P_2 = P$.

```
10 PRINT "PROG 3 9"
20 PRINT "THIS PROGRAM SOLVES EXAMPLE 14"
30 INPUT "ENTER INPUT VOLTAGE" ; V
40 INPUT "ENTER RESISTANCE R1" ; R1
50 INPUT "ENTER RESISTANCE R2" ; R2
60 LET R=(R1*R2)/(R1+R2)
70 LET I=V/R
80 PRINT "I = " I "AMPS"
90 LET I1=V/R1
100 LET I2=V/R2
110 PRINT "I1 = " I1 "AMPS"
120 PRINT "I2 = " I2 "AMPS"
130 LET P=V*I
140 PRINT "INPUT POWER = " P "WATTS"
150 LET P1=I1*I1*R1
160 LET P2=I2*I2*R2
170 PRINT "P1 = " P1 "WATTS"
180 PRINT "P2 = " P2 "WATTS"
```

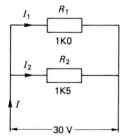

Figure 3.5

36 Circuit calculations pocket book

Example 15 For Figure 3.6 the following can be calculated:

$$\frac{1}{R} = \frac{1}{10} + \frac{1}{40} + \frac{1}{8}$$

$$= 0.1 + 0.025 + 0.125 = 0.25$$

Then

$$R = \frac{1}{0.25} = 4 \, \Omega$$

$$I = \frac{V}{R} = \frac{20}{4} = 5 \text{ A}$$

$$I_1 = \frac{V}{R_1} = \frac{20}{10} = 2 \text{ A}$$

$$I_2 = \frac{V}{R_2} = \frac{20}{40} = 0.5 \text{ A}$$

$$I_3 = \frac{V}{R_3} = \frac{20}{8} = 2.5 \text{ A}$$

Note that $I_1 + I_2 + I_3 = I$.

Input power $P = VI = 20 \times 5 = 100$ W

Let P_1, P_2 and P_3 equal the powers dissipated in R_1, R_2 and R_3.

$$P_1 = I_1^2 R_1 = 2^2 \times 10 = 40 \text{ W}$$

$$P_2 = I_2^2 R_2 = 0.5^2 \times 40 = 10 \text{ W}$$

$$P_3 = I_3^2 R_3 = 2.5^2 \times 8 = 50 \text{ W}$$

Note that $P_1 + P_2 + P_3 = P$.

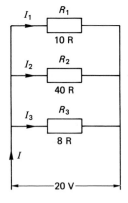

Figure 3.6

d.c. circuits 37

```
10 PRINT "PROG 40"
20 PRINT "THIS PROGRAM SOLVES EXAMPLE 15"
30 INPUT "ENTER INPUT VOLTAGE" ; V
40 INPUT "ENTER RESISTANCE R1" ; R1
50 INPUT "ENTER RESISTANCE R2" ; R2
60 INPUT "ENTER RESISTANCE R3" ; R3
70 LET A=(1/R1)+(1/R2)+(1/R3)
80 LET R=1/A
90 PRINT "COMBINED RESISTANCE = " R "OHMS"
100 LET I=V/R
110 PRINT "I= " I "AMPS"
120 LET I1=V/R1
130 LET I2=V/R2
140 LET I3=V/R3
150 PRINT "I1 = " I1 "AMPS"
160 PRINT "I2 = " I2 "AMPS"
170 PRINT "I3 = " I3 "AMPS"
180 LET P=V*I
190 PRINT "INPUT POWER = " P "WATTS"
200 LET P1=I1*I1*R1
210 LET P2=I2*I2*R2
220 LET P3=I3*I3*R3
230 PRINT "P1 = " P1 "WATTS"
240 PRINT "P2 = " P2 "WATTS"
250 PRINT "P3 = " P3 "WATTS"
```

Example 16 Three heaters connected in parallel are driven by a 40-V supply. The total power consumed is 200 W. If two of the heaters have resistance of 16 and 20 Ω, calculate the resistance of the third.

Since $P = \dfrac{V}{I}, I = \dfrac{P}{V} = \dfrac{200}{40} = 5$ A

Combined heater resistance $= \dfrac{V}{I} = \dfrac{40}{5} = 8$ Ω

$$\frac{1}{8} = \frac{1}{16} + \frac{1}{20} + \frac{1}{R_3}$$

$$\frac{1}{R_3} = \frac{1}{8} - \frac{1}{16} - \frac{1}{20} = 0.125 - 0.0625 - 0.05$$

$$= 0.0125, \text{ and } R_3 = \frac{1}{0.0125} = 80 \text{ Ω}$$

```
10 PRINT "PROG 41"
20 PRINT "THIS PROGRAM SOLVES EXAMPLE 16"
30 INPUT "ENTER SUPPLY VOLTAGE" ; V
40 INPUT "ENTER TOTAL POWER CONSUMED" ; P
50 INPUT "ENTER RESISTANCE R1" ; R1
60 INPUT "ENTER RESISTANCE R2" ; R2
70 LET I=P/V
80 LET R=V/I
90 LET A=(1/R)-(1/R1)-(1/R2)
100 LET R3=1/A
110 PRINT "RESISTANCE OF THIRD HEATER = " R3 "OHMS"
```

(c) Series–parallel

Example 17 To find the parameters of the circuit in Figure 3.7 it is necessary first to find the combined resistance R (see Chapter 2).

$$R = R_1 + \frac{R_2 R_3}{R_2 + R_3} + R_4$$

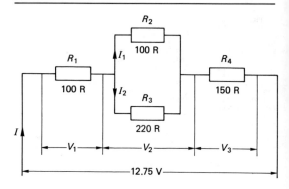

Figure 3.7

$$R = 100 + \frac{100 \times 220}{100 + 220} + 150$$

$$= 100 + 68.75 + 150 = 318.75 \ \Omega$$

$$I = \frac{V}{R} = \frac{12.75}{318.75} = 0.04 \ \text{A}$$

$$V_1 = IR_1 = 0.04 \times 100 = 4 \ \text{V}$$

$$V_2 = I\frac{R_2 R_3}{R_2 + R_3} = 0.04 \times 68.75 = 2.75 \ \text{V}$$

$$V_3 = IR_4 = 0.04 \times 150 = 6 \ \text{V}$$

Note that $V_1 + V_2 + V_3 = V$.

Since V_2 (2.75) is across both R_2 and R_3, then

$$I_1 = \frac{V_2}{R_2} = \frac{2.75}{100} = 0.0275 \ \text{A}$$

$$I_2 = \frac{V_2}{R_3} = \frac{2.75}{220} = 0.0125 \ \text{A}$$

Note that $I_1 + I_2 = I$.

Input power $P = V \times I = 12.75 \times 0.04 = 0.51$ W

Let P_1, P_2, P_3 and P_4 equal the powers dissipated in R_1, R_2, R_3 and R_4.

$$P_1 = I^2 R_1 = 0.04 \times 0.04 \times 100 = 0.16 \ \text{W}$$

$$P_2 = I_1^2 R_2 = 0.0275 \times 0.0275 \times 100 = 0.076 \ \text{W}$$

$$P_3 = I_2^2 R_3 = 0.0125 \times 0.0125 \times 220 = 0.034 \ \text{W}$$

$$P_4 = I^2 R_4 = 0.04 \times 0.04 \times 150 = 0.24 \ \text{W}$$

Note that $P_1 + P_2 + P_3 + P_4 = P$.

```
10 PRINT "PROG 42"
20 PRINT "THIS PROGRAM SOLVES EXAMPLE 17"
30 INPUT "ENTER INPUT VOLTAGE" ; V
40 INPUT "ENTER RESISTANCE R1" ; R1
50 INPUT "ENTER RESISTANCE R2" ; R2
60 INPUT "ENTER RESISTANCE R3" ; R3
70 INPUT "ENTER RESISTANCE R4" ; R4
80 LET R=R1+(R2*R3)/(R2+R3)+R4
90 PRINT "R = " R "OHMS"
100 I=V/R
110 PRINT "I = " I "AMPS"
120 LET V1=I*R1
130 LET V2=I*(R2*R3)/(R2+R3)
140 LET V3=I*R4
150 PRINT "V1 = " V1 "VOLTS"
160 PRINT "V2 = " V2 "VOLTS"
170 PRINT "V3 = " V3 "VOLTS"
180 LET I1=V2/R2
190 LET I2=V2/R3
200 PRINT "I1 = " I1 "AMPS"
210 PRINT "I2 = " I2 "AMPS"
220 LET P=V*I
230 LET P1=I*I*R1
240 LET P2=I1*I1*R2
250 LET P3=I2*I2*R3
260 LET P4=I*I*R4
270 PRINT "P = " P "WATTS"
280 PRINT "P1 = " P1 "WATTS"
290 PRINT "P2 = " P2 "WATTS"
300 PRINT "P3 = " P3 "WATTS"
310 PRINT "P4 = " P4 "WATTS"
```

Example 18 The combined resistance R for the circuit in Figure 3.8 is equal to R_1 in parallel with $R_2 + R_3$ (see Chapter 2).

$$R = \frac{24 \times 8}{24 + 8} = 6 \, \Omega$$

$$I = \frac{V}{R} = \frac{120}{6} = 20 \, \text{A}$$

$$I_1 = \frac{V}{R_1} = \frac{120}{24} = 5 \, \text{A}$$

$$I_2 = \frac{V}{R_2 + R_3} = \frac{120}{8} = 15 \, \text{A}$$

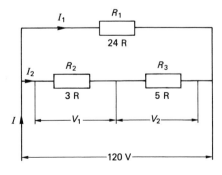

Figure 3.8

Note that $I_1 + I_2 = I$.

$$V_1 = I_2 R_2 = 15 \times 3 = 45 \text{ V}$$
$$V_2 = I_2 R_3 = 15 \times 5 = 75 \text{ V}$$

Note that $V_1 + V_2 = V$.

Input power $P = VI = 120 \times 20 = 2400$ W

Let P_1, P_2 and P_3 equal the powers dissipated in R_1, R_2 and R_3.

$$P_1 = VI_1 = 120 \times 5 = 600 \text{ W}$$
$$P_2 = V_1 I_2 = 45 \times 15 = 675 \text{ W}$$
$$P_3 = V_2 I_2 = 75 \times 15 = 1125 \text{ W}$$

Note that $P_1 + P_2 + P_3 = P$.

```
10 PRINT "PROG 43"
20 PRINT "THIS PROGRAM SOLVES EXAMPLE 18"
30 INPUT "ENTER INPUT VOLTAGE" ; V
40 INPUT "ENTER RESISTANCE R1" ; R1
50 INPUT "ENTER RESISTANCE R2" ; R2
60 INPUT "ENTER RESISTANCE R3" ; R3
70 LET R=R1*(R2+R3)/(R1+R2+R3)
80 PRINT "R = " R "OHMS"
90 LET I=V/R
100 LET I1=V/R1
110 LET I2=V/(R2+R3)
120 PRINT "I = " I "AMPS"
130 PRINT "I1 = " I1 "AMPS"
140 PRINT "I2 = " I2 "AMPS"
150 LET V1=I2*R2
160 LET V2=I2*R3
170 PRINT "V1 = " ; V1 "VOLTS"
180 PRINT "V2 = " ; V2 "VOLTS"
190 LET P=V*I
200 LET P1=V*I1
210 LET P2=V1*I2
220 LET P3=V2*I2
230 PRINT "P = " P "WATTS"
240 PRINT "P1 = " P1 "WATTS"
250 PRINT "P2 = " P2 "WATTS"
260 PRINT "P3 = " P3 "WATTS"
```

Example 19 In the circuit of Figure 3.9 find the value of R_4.

Equivalent resistance $R = \dfrac{V}{I}$

$$= \frac{60}{4} = 15 \text{ }\Omega$$

Combined resistance of R_1 and R_2

$$= \frac{12 \times 4}{12 + 4} = \frac{48}{16} = 3 \text{ }\Omega$$

Therefore

Combined resistance of R_3 and $R_4 = 15 - 3 = 12$ Ω

i.e. $12 = \dfrac{48 \times R_4}{48 + R_4}$

Evaluating for R_4 gives

$$12(48 + R_4) = 48 R_4$$
$$576 + 12 R_4 = 48 R_4$$

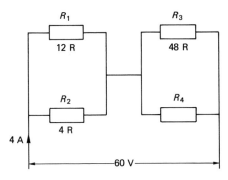

Figure 3.9

$$576 = 48R_4 - 12R_4$$
$$576 = 36R_4 \text{ and } R_4 = 16 \, \Omega$$

```
10 PRINT "PROG 44"
20 PRINT "THIS PROGRAM SOLVES EXAMPLE 19"
30 INPUT "ENTER INPUT VOLTAGE" ; V
40 INPUT "ENTER INPUT CURRENT" ; I
50 INPUT "ENTER RESISTANCE R1" ; R1
60 INPUT "ENTER RESISTANCE R2" ; R2
70 INPUT "ENTER RESISTANCE R3" ; R3
80 LET R=V/I
90 LET A=(R1*R2)/(R1+R2)
100 LET B=R-A
110 LET R4=(B*R3)/(R3-B)
120 PRINT "R4 = " R4 "OHMS"
```

3.6 Measurements

Various types of instruments exist for measuring both voltages and currents. Analogue instruments indicate the readings by means of a pointer, and electronic devices by a display of numbers.

Apart from the electronic and electrostatic types, all voltmeters are in effect milliammeters. The next examples will show how these meters can be adapted to measure both low and high voltages, the loading effect of a voltmeter on a circuit, and how they can be adapted to measure both low and high currents.

(a) Voltmeters

Example 20 A meter gives a full-scale deflection (f.s.d.) with a current of 1 mA and has a resistance R_m of 1 Ω. Calculate the series resistance R needed to enable it to read up to 1 V. The voltmeter arrangement is shown in Figure 3.10.

Applying Ohm's law

$$R + R_m = \frac{V}{I} = \frac{1}{1 \times 10^{-3}} = 10^3 \, \Omega$$

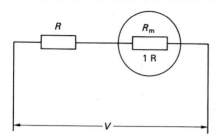

Figure 3.10

Hence

$$R = 10^3 - R_m = 10^3 - 1 = 999 \text{ }\Omega$$

Note

$$V_R = IR = 1 \times 10^{-3} \times 999 = 0.999 \text{ V}$$

and

$$V_{R_m} = IR_m = 1 \times 10^{-3} \times 1 = 0.001 \text{ V}$$
$$V_R + V_{R_m} = V$$

```
10 PRINT "PROG 45"
20 PRINT "THIS PROGRAM SOLVES EXAMPLES 20 AND 21"
30 INPUT "ENTER MAXIMUM VOLTAGE READING" ; V
40 INPUT "ENTER CURRENT IN MILLIAMPS" ; I
50 INPUT "ENTER METER RESISTANCE" ; RM
60 LET A=V/(I/1000)
70 LET R=A-RM
80 PRINT "SERIES RESISTANCE = " R "OHMS"
```

Example 21 Find the values of the series resistances required for the meter in Example 20 to read (i) up to 10 V and (ii) up to 100 V.

(i) $\quad R + R_m = \dfrac{V}{I} = \dfrac{10}{1 \times 10^{-3}} = 10 \times 10^3 = 10^4 \text{ }\Omega$

$\quad R = 10^4 - R_m = 10^4 - 1 = 9999 \text{ }\Omega$

(ii) $\quad R + R_m = \dfrac{V}{I} = \dfrac{100}{1 \times 10^{-3}} = 100 \times 10^3 = 10^5 \text{ }\Omega$

$\quad R = 10^5 - R_m = 99\,999 \text{ }\Omega$

(See program 45, above.)

Example 22 Calculate the voltages that would be indicated by the voltmeter considered in Examples 20 and 21, when measuring the voltage between points B and C in Figure 3.11 on (i) the 10-V range and (ii) the 100-V range.

(i) Since $R_1 = R_2$

$\quad V_{R_1} = V_{R_2} = 6 \text{ V}$

On the 10-V range the total meter resistance, 10 kΩ, is in parallel with R_2. The resistance between B and C becomes

$$\frac{10\,000}{2} = 5000\ \Omega$$

(see Chapter 2). The equivalent circuit is shown in Figure 3.12.

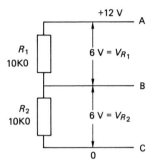

Figure 3.11

$$I = \frac{V}{R_1 + R_3} \text{ and } V_{R_3} = IR_3$$

$$V_{R_3} = \frac{V}{R_1 + R_3} R_3$$

$$= \frac{12}{10\,000 + 5000} \times 5000$$

$$= \frac{60\,000}{15\,000} = 4\ \text{V}$$

The difference between the actual voltage and the measured voltage is $6 - 4$, giving an error of -2 V, and this illustrates the loading effect of the voltmeter.

```
10 PRINT "PROG 46"
20 PRINT "THIS PROGRAM SOLVES EXAMPLE 22 (i) AND (ii)"
30 INPUT "ENTER INPUT VOLTAGE" ; V
40 INPUT "ENTER METER RESISTANCE" ; RM
50 INPUT "ENTER RESISTANCE R1" ; R1
60 INPUT "ENTER RESISTANCE R2" ; R2
70 LET VR1=(V*R1)/(R1+R2)
80 LET VR2=(V*R2)/(R1+R2)
90 PRINT "VR1 = " ; VR1 "VOLTS"
100 PRINT "VR2 = " ; VR2 "VOLTS"
110 LET R3=(RM*R2)/(RM+R2)
120 LET VR3=(V*R3)/(R1+R3)
130 PRINT "METER READING = " VR3 "VOLTS"
140 PRINT "VOLTAGE ERROR = " VR2-VR3 "VOLTS"
```

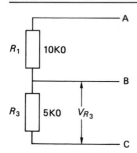

Figure 3.12

(ii) On the 100-V range the meter loading resistance is 100 kΩ in parallel with the 10 kΩ. The resistance between B and C becomes

$$\frac{10\,000 \times 100\,000}{10\,000 + 100\,000} = 9090.9 \ \Omega$$

$$V_{R_3} = \frac{V}{R_1 + R_3} R_3 = \frac{12 \times 9090.9}{10\,000 + 9090.9}$$

$$= 5.714 \ V$$

The error in this case is

$$5.714 - 6 = -0.286 \ V$$

which shows that errors in voltage measurements are reduced by using voltmeters with high input resistances.

The meter considered in the last three examples has an input resistance of 1000 Ω/V, which is low compared, for example, with the input resistance of electronic voltmeters.

(See program 46, above.)

(b) Ammeters

Example 23 An ammeter gives f.s.d. with a current I_c of 20 mA and has a coil resistance R_c of 5 Ω. Calculate the shunt resistance R_s required in parallel to enable the instrument to read up to 2 A. The circuit arrangement is shown in Figure 3.13.

$$V = I_c R_c = 20 \times 10^{-3} \times 5 = 0.1 \ V$$

Shunt current $I_{R_s} = I - I_c$

$$= 2 - (20 \times 10^{-3}) = 2 - 0.02 = 1.98 \ A$$

Therefore

$$R_s = \frac{V}{I_{R_s}} = \frac{0.1}{1.98} = 0.0505 \ \Omega$$

The effect of inserting an ammeter in a circuit is negligible provided the voltage loss across it, 0.1 V in this example, is low, compared with the voltage acting in the circuit.

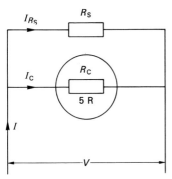

Figure 3.13

```
10 PRINT "PROG 47"
20 PRINT "THIS PROGRAM SOLVES EXAMPLE 23"
30 INPUT "ENTER F.S.D. CURRENT IN MILLIAMPS" ; IC
40 INPUT "ENTER COIL RESISTANCE" ; RC
50 INPUT "ENTER MAXIMUM CURRENT READING IN AMPS" ; I
60 LET V=IC*RC/1000
70 LET IRS=I-(IC/1000)
80 LET RS=V/IRS
90 PRINT "SHUNT RESISTANCE VALUE = " RS "OHMS"
```

(c) Wattmeter

Wattmeters are used to measure the true power consumed in a.c. and d.c. circuits. In a.c. circuits

True power = $VI \cos \phi$

where

V = the r.m.s. value of the applied voltage
I = the load current.
ϕ = the phase angle between the voltage and the current.

A.c. voltages are discussed in Chapter 6 and power in a.c. circuits in Chapter 11.

3.7 The Wheatstone bridge

The Wheatstone bridge, used to measure resistances over wide ranges, is shown in Figure 3.14. To find the value of the unknown resistance R_c, R_2 is adjusted for a zero current reading in the galvanometer G.

Example 24 Given that the resistor values for this condition are as shown, calculate the value of R_c. Since there is no current flowing in G there is no potential difference between points A and B. This means that $V_{R_1} = V_{R_2}$ and $V_{R_3} = V_{R_c}$.

$$V_{R_3} = I_1 R_3 \text{ and } I_1 = \frac{V}{R_1 + R_3} \text{ then } V_{R_3} = \frac{V R_3}{R_1 + R_3}$$

$$V_{R_c} = I_2 R_c \text{ and } I_2 = \frac{V}{R_2 + R_c} \text{ then } V_{R_c} = \frac{V R_c}{R_2 + R_c}$$

46 Circuit calculations pocket book

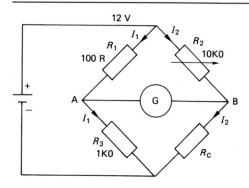

Figure 3.14

Equating these two results gives

$$\frac{VR_3}{R_1 + R_3} = \frac{VR_c}{R_2 + R_c}$$

Transposing and solving for R_c gives

$$VR_3(R_2 + R_c) = VR_c(R_1 + R_3)$$
$$VR_3R_2 + VR_3R_c = VR_cR_1 + VR_cR_3$$
$$VR_3R_2 = VR_cR_1 + VR_cR_3 - VR_3R_c$$
$$VR_3R_2 = VR_cR_1$$
$$\frac{VR_3R_2}{VR_1} = R_c = \frac{R_3R_2}{R_1}$$
$$R_c = \frac{10^3 \times 10^4}{10^2} = 10^5 = 100 \text{ k}\Omega$$

The result shows that when the bridge is balanced

$$\frac{R_1}{R_3} = \frac{R_2}{R_c}$$

giving

$$R_c = \frac{R_2R_3}{R_1}$$

Note that if the bridge is designed with $R_1 = R_3$ then $R_c = R_2$ and a reading of R_2 automatically gives the value of R_c.

```
10 PRINT "PROG 48"
20 PRINT "THE WHEATSTONE BRIDGE"
30 INPUT "ENTER RESISTANCE R1" ; R1
40 INPUT "ENTER RESISTANCE R2" ; R2
50 INPUT "ENTER RESISTANCE R3" ; R3
60 LET RC=R2*R3/R1
70 PRINT "VALUE OF UNKNOWN RESISTANCE = " RC "OHMS"
```

d.c. circuits 47

Problems

1. If 0.6 J move 25 C between two circuit points, what is the voltage between the two points?
2. A current of 0.5 A flows for 0.5 h in a conductor. Calculate the charge. Express the answer in coulombs and ampere-hours.
3. Calculate the time in minutes for 5 C of charge to flow in a resistor if the current is 10 mA.
4. One hundred and twenty coulombs of charge flow in a resistor during a period of 2 min. What is the current?
5. A voltage of 10 V is applied to a 150-Ω resistor. Calculate the current and the power consumed.
6. A resistor takes a current of 20 mA and consumes 20 mW. Find the value of the applied voltage and the resistor value.
7. Three resistors are connected in series across a 100-V supply. The resistor values are 330R, 470R and 1K0. Calculate: (a) the current, (b) the voltage drop across each resistor, (c) input power, (d) the power consumed by each resistor.
8. A current of 200 mA flows through three resistors connected in series across a 12-V supply. Two of the resistors have values of 10R and 15R. Calculate the value of the third, together with the power dissipated in each one.
9. Three lamps connected in series consume 6, 12 and 24 W when connected to a 24 V supply. Calculate the current flowing, and the resistance of each filament.
10. A 40R resistor is connected in parallel with a 120R resistor across a 6 V supply. Calculate the input current, the input power, the current taken by each resistor, and the power consumed by each resistor.
11. Three resistors with values of 100R, 300R and 600R are connected in parallel and are driven from a 100-V supply. Calculate (a) the input current, (b) the current flowing in each resistor, (c) the input power, (d) the power consumed by each resistor.
12. Three conductors connected in parallel consume a total of 500 W from a 25-V supply. Two conductors have resistances of 0.25 Ω and 0.5 Ω. Calculate the value of the third.
13. For the circuit shown in Figure 3.15 calculate: (a) I, (b) V_1, V_2, V_3, (c) I_1, I_2, (d) input power, (e) power consumed by each resistor.

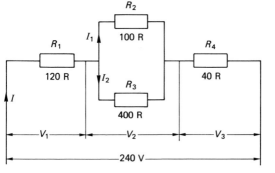

Figure 3.15

14 For the circuit shown in Figure 3.16 calculate: (a) I, (b) I_1, I_2, (c) V_1, V_2, (d) input power, (e) power dissipated in the individual resistors.

15 Find the value of R_4 in the circuit shown in Figure 3.17.

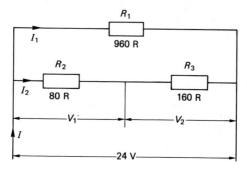

Figure 3.16

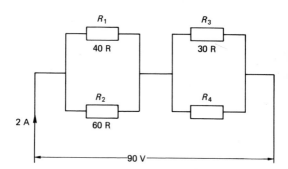

Figure 3.17

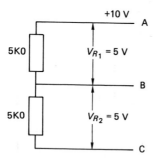

Figure 3.18

16 A meter gives a full-scale deflection with a current of 2 mA and has a resistance of 0.1 Ω. Calculate the series resistance needed to enable the meter to read up to 1 V.

17 Find the values of the series resistances required for the meter in Problem 16 to read (a) up to 5 V, (b) up to 10 V.

18 What would be the voltage indicated by a voltmeter on the 5-V range in the circuit shown in Figure 3.18, given that the loading is 1000 Ω/V? What would the reading be on the 10 V range?

19 An ammeter gives a full-scale deflection with a current of 10 mA. The coil resistance is 2 Ω. Calculate the shunt resistances needed to enable the meter to read currents of 1 A, 2 A and 10 A.

20 If the Wheatstone bridge shown in Figure 3.14 is balanced with $R_1 = 250$ Ω, $R_2 = 400$ Ω, and $R_3 = 450$ Ω, find the value of the unknown resistor R_c.

4 Network theorems

4.1 Kirchhoff's laws

In addition to Ohm's law, which was introduced in Chapter 3, these laws can be used to determine currents and voltage drops in d.c. networks.

(a) At any junction of resistances the sum of the currents flowing towards the junction is equal to the sum of currents flowing from it.

This means that the algebraic sum of currents meeting at any point in a network is zero.

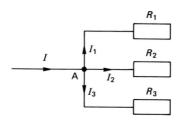

Figure 4.1

Consider Figure 4.1:

Current flowing towards A = I

Currents flowing away from A = $I_1 + I_2 + I_3$

Therefore

$$I = I_1 + I_2 + I_3$$

or

$$I - I_1 - I_2 - I_3 = 0$$

The law shows that there cannot be an accumulation of charge at any point in a circuit.

(b) In any closed circuit the sum of the e.m.f.'s, reckoned positive in the direction of the current, and negative in the opposite direction, is equal to the sum of the products of current and resistance, in every part of the closed circuit.

The law is borne out in Figure 4.2 where

The clockwise current is reckoned positive
The battery e.m.f. (12 V) is reckoned positive
The voltage drops (7 V and 5 V) are reckoned negative
Then 12 (e.m.f.) − 7(p.d.) − 5(p.d.) = 0

d.c. circuits **51**

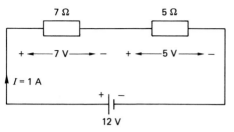

Figure 4.2

The law states that the algebraic sum of e.m.f.'s and p.d.'s in any circuit is zero.

Example 1 Use Kirchhoff's laws to find the values of currents I_1 and I_2 in Figure 4.3.

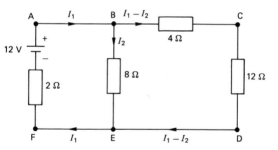

Figure 4.3

Having assigned the current directions I_1 and I_2, obviously the current through the 4-Ω and 12-Ω resistors is $I_1 - I_2$ (rule (a)). Since there are two unknowns two equations are needed.

In loop ABEF

$$12V = 2I_1 + 8I_2 \tag{4.1}$$

In loop ABCDEF

$$12V = 4(I_1 - I_2) + 12(I_1 - I_2) + 2I_1$$
$$= 4I_1 - 4I_2 + 12I_1 - 12I_2 + 2I_1$$
$$12V = 18I_1 - 16I_2 \tag{4.2}$$

Equation 4.1 × 2 $24 = 4I_1 + 16I_2$

Equation 4.2 $\underline{12 = 18I_1 - 16I_2}$

Adding $\overline{36 = 22I_1}$

$$I_1 = \frac{36}{22} = 1.636 \text{ A}$$

Substituting in equation (4.1)

$$12 = 2(1.636) + 8I_2$$

$$8I_2 = 12 - 2(1.636) = 8.728$$

$$I_2 = \frac{8.728}{8} = 1.091 \text{ A}$$

A very good method of checking whether the answers to the currents are correct is to calculate using Ohm's law the voltage drops across each resistor, and note carefully the polarity of each one. The results are shown in Figure 4.4. Note in particular the polarity of the voltage across the 2 Ω resistor. From the diagram it can be seen that

$$V_{AF} = 12 - 3.272 = 8.72 \text{ V}$$

$$V_{BE} = 8.72 \text{ V}$$

$$V_{BD} = 2.18 + 6.54 = 8.72 \text{ V}$$

These three voltages should of course have the same value since they are the voltages between the top and bottom rails of the network.

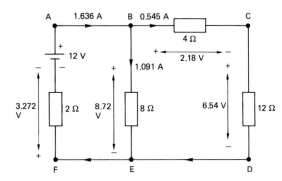

Figure 4.4

Example 2 Find the currents I_1, I_2 and I_3 in Figure 4.5.

Since there are three unknown quantities, three equations are needed. The current directions have been assigned. Note that there are clockwise currents which will be reckoned positive and anti-clockwise currents which will be reckoned negative.

In loop ACB which is closed

$$3I_2 - 6I_3 - 5(I_1 - I_2) = 0$$

Simplifying

$$3I_2 - 6I_3 - 5I_1 + 5I_2 = 0$$

$$8I_2 - 6I_3 - 5I_1 = 0$$

Network theorems 53

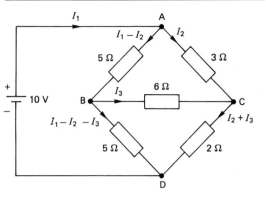

Figure 4.5

Multiplying by -1 and rearranging

$$5I_1 - 8I_2 + 6I_3 = 0 \tag{4.3}$$

in loop BCD which is closed

$$6I_3 + 2(I_2 + I_3) - 5(I_1 - I_2 - I_3) = 0$$
$$6I_3 + 2I_2 + 2I_3 - 5I_1 + 5I_2 + 5I_3 = 0$$
$$13I_3 + 7I_2 - 5I_1 = 0$$

or

$$5I_1 - 7I_2 - 13I_3 = 0 \tag{4.4}$$

In loop ACD and battery

$$3I_2 + 2(I_2 + I_3) = 10$$
$$3I_2 + 2I_2 + 2I_3 = 10$$
$$5I_2 + 2I_3 = 10 \tag{4.5}$$

Subtracting equation (4.3) from (4.4)

$$5I_1 - 7I_2 - 13I_3 = 0$$
$$\underline{5I_1 - 8I_2 + 6I_3 = 0}$$
$$I_2 - 19I_3 = 0$$
$$I_2 = 19I_3$$

Substituting for I_2 in equation (4.5)

$$5(19I_3) + 2I_3 = 10$$
$$97I_3 = 10$$
$$I_3 = \frac{10}{97} = 0.103 \text{ A}$$

Since $I_2 = 19I_3$

$$I_2 = 19(0.103) = 1.957 \text{ A}$$

Substituting in equation (4.3)

$$5I_1 - 8(1.957) + 6(0.103) = 0$$
$$5I_1 = 15.656 - 0.618$$
$$5I_1 = 15.038$$
$$I_1 = 3.0076 \text{ A}$$

Using the calculated currents and applying Ohm's law to the network, it is found that the voltage at point B (4.785 V) is higher than the voltage at point C (4.129 V) relative to point D. The assumed direction of I_3 was therefore correct, flowing from B to C. If I_3 had been negative, this would have indicated that the current was flowing in the opposite direction.

Example 3 Find the currents in the three branches of the network shown in Figure 4.6.

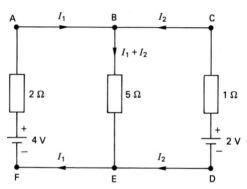

Figure 4.6

This network contains two batteries. The currents have been assigned assuming that both I_1 and I_2 are flowing from the positive plates of the batteries.

In loop ABEF

$$4 = 2I_1 + 5(I_1 + I_2)$$
$$4 = 2I_1 + 5I_1 + 5I_2$$
$$4 = 7I_1 + 5I_2 \tag{4.6}$$

In loop CBED where the voltage and current are acting anticlockwise

$$-2 = -1I_2 - 5(I_1 + I_2)$$
$$-2 = -1I_2 - 5I_1 - 5I_2$$
$$-2 = -6I_2 - 5I_1$$

or
$$2 = 5I_1 + 6I_2 \tag{4.7}$$

Equation 4.6 × 5 $20 = 35I_1 + 25I_2$

Equation 4.7 × 7 $14 = 35I_1 + 42I_2$

Subtracting $6 = - 17I_2$

$$I_2 = -\frac{6}{17} = -0.353 \text{ A}$$

Substituting in equation (4.6)

$$4 = 7I_1 + 5(-0.353)$$
$$7I_1 = 4 + 1.765 = 5.765$$
$$I_1 = \frac{5.765}{7} = 0.824 \text{ A}$$

The negative sign for I_2 indicates that the current is flowing in the opposite direction to the assumed direction. The circuit conditions are shown in Figure 4.7. Note the polarity of the voltages, and that the voltage between the top and bottom rails is 2.35 V, which is a confirmation that the calculated currents are correct.

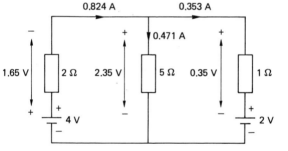

Figure 4.7

Since the current I_2 is flowing into the 2-V battery, the battery is being charged.

Example 4 Calculate the current flowing in the three branches in Figure 4.8.

Both current I_1 and I_2 are assumed to flow from the positive plates, and all currents therefore are assumed to be flowing clockwise.

In loop ABEF

$$20 - 10 = 15I_1 - 5I_2$$
$$10 = 15I_1 - 5I_2 \tag{4.8}$$

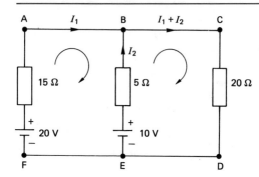

Figure 4.8

In loop ABCDEF

$$20 = 15I_1 + 20(I_1 + I_2)$$
$$20 = 15I_1 + 20I_1 + 20I_2$$
$$20 = 35I_1 + 20I_2 \qquad (4.9)$$

Equation 4.8 × 4 $\quad 40 = 60I_1 - 20I_2$

Equation 4.9 $\quad\underline{20 = 35I_1 + 20I_2}$

Adding $\quad\underline{60 = 95I_1}$

$$I_1 = \frac{60}{95} = 0.6316 \text{ A}$$

Substituting for I_1 in equation (4.8)

$$10 = 15(0.6316) - 5I_2$$
$$10 - 15(0.6316) = -5I_2$$
$$0.526 = -5I_2$$
$$I_2 = -\frac{0.526}{5} = -0.1052 \text{ A}$$

The minus sign indicates that the current is flowing in the opposite direction. The circuit conditions are shown in Figure 4.9. Note the voltage polarities, and that the voltage across each branch is 10.52 V.

Note also that when moving clockwise around the loop ABEF the 20-V battery is reckoned positive and the 10-V battery reckoned negative. The overall voltage is therefore $+20 - 10 = +10$ V.

Example 5 Find the value of I_1 and I_2 in Figure 4.10. Clockwise currents are assumed.

In loop ABEF

$$4 + 4 = 2I_1 + 4I_2$$
$$8 = 2I_1 + 4I_2 \qquad (4.10)$$

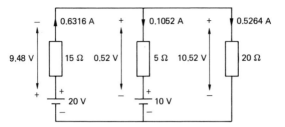

Figure 4.9

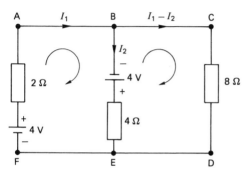

Figure 4.10

In loop BCDE

$$-4 = 8(I_1 - I_2) - 4I_2$$
$$-4 = 8I_1 - 8I_2 - 4I_2$$
$$-4 = 8I_1 - 12I_2 \quad (4.11)$$

Equation 4.10 × 4 $32 = 8I_1 + 16I_2$
Equation 4.11 $\underline{-4 = 8I_1 - 12I_2}$
Subtracting $\underline{36 = 28I_2}$

$$I_2 = \frac{36}{28} = 1.286 \text{ A}$$

Substituting in equation (4.10)

$$8 = 2I_1 + 4(1.286)$$
$$8 - 4(1.286) = 2I_1$$
$$I_1 = 1.428 \text{ A}$$

The assumed current directions were correct, and the circuit parameters are shown in Figure 4.11.

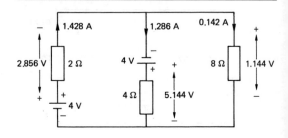

Figure 4.11

4.2 Superposition theorem

If a network of linear resistors contains more than one source of e.m.f., the current flowing at any point is the algebraic sum of the currents that would flow at that point, if each source was considered separately, with all other sources replaced at that time by resistances equal to their internal resistances.

Example 6 Find the currents in each branch of the circuit in Figure 4.12 using the superposition theorem. This circuit is identical to Figure 4.6.

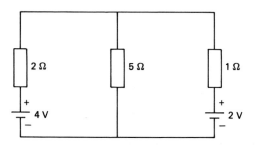

Figure 4.12

First remove the 2-V source and replace with the 1 Ω resistor as shown in Figure 4.13.

Combining the 5-Ω and the 1-Ω resistors gives the circuit of Figure 4.14.

$$I_1 = \frac{4}{2 + 0.833} = 1.412 \text{ A}$$

Figure 4.13 shows that I_1 will split in the ratio 5 : 1 where

$$I_2 = \frac{1.412 \times 1}{5 + 1} = 0.236 \text{ A}$$

$$I_3 = \frac{1.412 \times 5}{5 + 1} = 1.176 \text{ A}$$

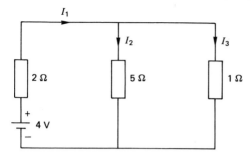

Figure 4.13

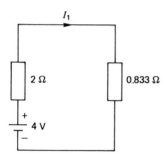

Figure 4.14

Secondly remove the 4-V source and replace with the 2-Ω resistor as shown in Figure 4.15. Combining the 2-Ω and 5-Ω resistors gives the circuit of Figure 4.16.

$$I_4 = \frac{2}{1 + 1.43} = 0.823 \text{ A}$$

Figure 4.15 shows that I_4 will split in the ratio 5 : 2 where

$$I_5 = \frac{0.823 \times 2}{5 + 2} = 0.235 \text{ A}$$

$$I_6 = \frac{0.823 \times 5}{5 + 2} = 0.588 \text{ A}$$

Superimposing Figures 4.13 and 4.15 shows that the currents in

branch 1 = $I_1 - I_6$ = 1.412 − 0.588 = 0.824 A

branch 2 = $I_2 + I_5$ = 0.236 + 0.235 = 0.471 A

branch 3 = $I_3 - I_4$ = 1.176 − 0.823 = 0.353 A

These results agree with the ones obtained using Kirchhoff's laws.

Example 7 Find the currents in each branch of the circuit in Figure 4.17 using the superposition theorem. This circuit is identical to Figure 4.8.

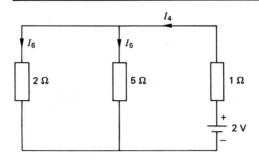

Figure 4.15

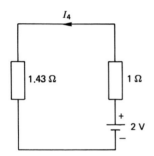

Figure 4.16

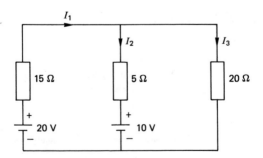

Figure 4.17

Network theorems 61

First remove the 10-V source and replace with the 5-Ω resistor as shown in Figure 4.18. Combining the 5-Ω and the 20-Ω resistors gives the circuit of Figure 4.19.

$$I_1 = \frac{20}{15 + 4} = 1.053 \text{ A}$$

Figure 4.18 shows that the I_1 will split in the ratio 5 : 20 where

$$I_2 = \frac{1.053 \times 20}{5 + 20} = 0.842 \text{ A}$$

$$I_3 = \frac{1.053 \times 5}{5 + 20} = 0.2106 \text{ A}$$

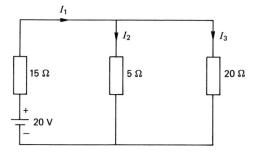

Figure 4.18

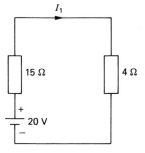

Figure 4.19

Secondly remove the 20-V source and replace with the 15-Ω resistor as shown in Figure 4.20. Combining the 15- and 20-Ω resistors gives the circuit of Figure 4.21.

$$I_4 = \frac{10}{5 + 8.57} = 0.737 \text{ A}$$

Figure 4.20 shows that I_4 will split in the ratio 20 : 15 where

$$I_5 = \frac{0.737 \times 15}{15 + 20} = 0.316 \text{ A}$$

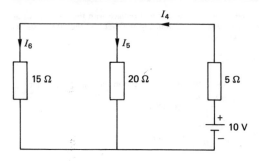

Figure 4.20

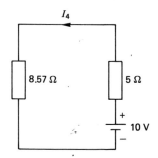

Figure 4.21

$$I_6 = \frac{0.737 \times 20}{15 + 20} = 0.421 \text{ A}$$

Superimposing Figures 4.18 and 4.20 shows that currents in

$$15 \, \Omega = I_1 - I_6 = 1.053 - 0.421 = 0.632 \text{ A}$$
$$20 \, \Omega = I_3 + I_5 = 0.2106 + 0.316 = 0.562 \text{ A}$$
$$5 \, \Omega = I_2 - I_4 = 0.842 - 0.737 = 0.105 \text{ A}$$

4.3 Thevenin's theorem

The theorem is concerned with active resistive networks, i.e. networks containing one or more sources of e.m.f., and the method of finding the current in any resistor within the network.

To find the current in a resistor connected between two points A and B in a complex circuit, proceed as follows:

1 Remove the resistor R from the circuit which is connected between two points A and B.
2 Calculate the open circuit e.m.f., V_{OC}, appearing between A and B.
3 Replace all sources of e.m.f. in the circuit, by resistors equal in value to their internal resistances.

Network theorems 63

4 Calculate the resistance, R_{AB}, appearing between A and B (with R still removed).
5 Connect R to a generator of e.m.f. V_{OC} and having internal resistance R_{AB} and calculate the current flowing.

Example 8 Find the current in the 8-Ω resistor of Figure 4.3 using Thevenin's theorem.

First remove the 8-Ω resistor, which leaves the circuit of Figure 4.22, and calculate the open circuit voltage across A and B.

$$V_{AB} = \frac{12}{(4 + 12 + 2)} \times (4 + 12) = 10.67 \text{ V}$$

Next remove the 12-V source which gives the circuit of Figure 4.23 and calculate the resistance R looking into AB.

$$R = \frac{2 \times (4 + 12)}{2 + (4 + 12)} = 1.777 \text{ Ω}$$

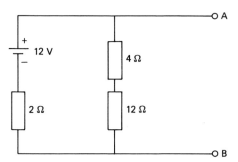

Figure 4.22

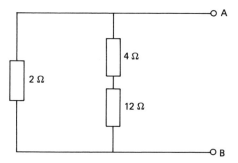

Figure 4.23

The equivalent Thevenin's circuit is shown in Figure 4.24 where

$$I = \frac{10.67}{1.777 + 8} = 1.091 \text{ A}$$

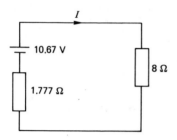

Figure 4.24

Example 9 Find the current in the 4-Ω resistor of the circuit used in Example 8.

Removing the 4-Ω resistor from the circuit of Figure 4.3 leaves the circuit of Figure 4.25.

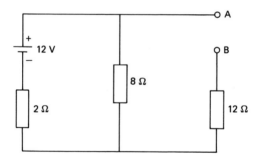

Figure 4.25

V_{AB} will equal the voltage across the 8-Ω resistor since no current is flowing in the 12-Ω resistor.

$$V_{AB} = \frac{12 \times 8}{8 + 2} = 9.6 \text{ V}$$

Removing the 12-V source gives the circuit of Figure 4.26. The resistance looking into AB is

$$R = 12 + \frac{(8 \times 2)}{(8 + 2)} = 13.6 \text{ }\Omega$$

The equivalent Thevenin's circuit is shown in Figure 4.27 where

$$I = \frac{9.6}{13.6 + 4} = 0.545 \text{ A}$$

Network theorems 65

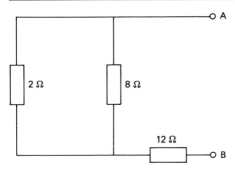

Figure 4.26

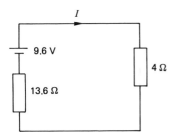

Figure 4.27

Example 10 Find the current in the 12-Ω resistor using Thevenin's theorem for the circuit of Figure 4.3.

Removing the 12-Ω resistor leaves the circuit of Figure 4.28. V_{AB} will equal the voltage across the 8-Ω resistor, since no current is flowing in the 4-Ω resistor.

$$V_{AB} = \frac{12 \times 8}{8 + 2} = 9.6 \text{ V}$$

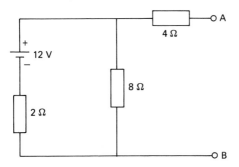

Figure 4.28

Removing the 12-V source gives the circuit of Figure 4.29. The resistance looking into AB is

$$R = \frac{4 + (8 \times 2)}{(8 + 2)} = 5.6 \, \Omega$$

The equivalent Thevenin's circuit is shown in Figure 4.30 where

$$I = \frac{9.6}{5.6 + 12} = 0.545 \, \text{A}$$

Note that the answers to Examples 9 and 10 are the same. This must be so since the 4-Ω and the 12-Ω resistors are in series in the actual circuit.

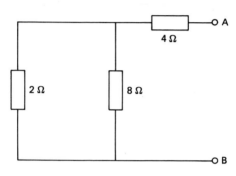

Figure 4.29

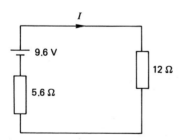

Figure 4.30

Example 11 Using Thevenin's theorem find the current in the 5-Ω resistor of the circuit in Figure 4.31. This circuit is identical to Figure 4.6.

Removing the 5-Ω resistor leaves the circuit of Figure 4.32.

$$I = \frac{4 - 2}{2 + 1} = 0.667 \, \text{A}$$

V_{AB} will equal the voltage drop across the 1-Ω resistor plus the battery voltage.

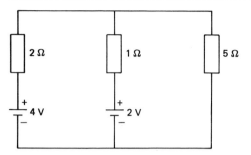

Figure 4.31

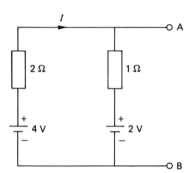

Figure 4.32

$$V_AB = (1 \times 0.667) + 2 = 2.667 \text{ V}$$

Removing the two sources gives the circuit of Figure 4.33. The resistance looking into AB is

$$R = \frac{2 \times 1}{1 + 2} = 0.667 \text{ }\Omega$$

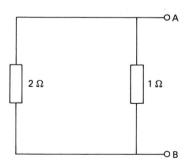

Figure 4.33

The equivalent Thevenin's circuit is shown in Figure 4.34 where

$$I = \frac{2.667}{5 + 0.667} = 0.471 \text{ A}$$

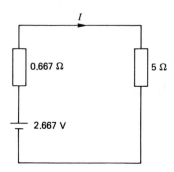

Figure 4.34

Example 12 Using Thevenin's theorem find the current in the 20-Ω resistor in Figure 4.35. The circuit is identical to Figure 4.8.

Removing the 20-Ω resistor leaves the circuit of Figure 4.36.

$$I = \frac{20 - 10}{15 + 5} = 0.5 \text{ A}$$

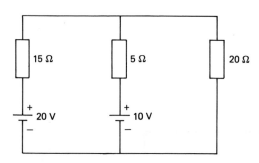

Figure 4.35

V_{AB} will equal the voltage drop across the 5-Ω resistor, plus the battery voltage (due to the polarities).

$$V_{AB} = (5 \times 0.5) + 10 = 12.5 \text{ V}$$

Removing the two sources gives the circuit of Figure 4.37. The resistance at the input to AB is

$$R = \frac{15 \times 5}{15 + 5} = 3.75 \text{ Ω}$$

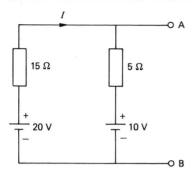

Figure 4.36

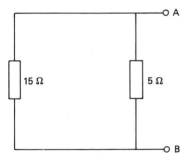

Figure 4.37

The equivalent Thevenin's circuit is shown in Figure 4.38 where

$$I = \frac{12.5}{20 + 3.75} = 0.526 \text{ A}$$

Example 13 Use Thevenin's theorem to find the current in the 8-Ω resistor in Figure 4.39. The circuit is identical to Figure 4.10. Removing the 8-Ω resistor leaves the circuit of Figure 4.40.

$$I = \frac{4 + 4}{2 + 4} = 1.333 \text{ A}$$

Note how in this circuit the battery voltages add together. V_{AB} will be equal to the voltage drop across the 4-Ω resistor minus the battery voltage (due to the polarities).

$$V_{AB} = (4 \times 1.333) - 4 = 1.332 \text{ V}$$

Removing the two sources gives the circuit of Figure 4.41. The resistance looking into AB is

$$R = \frac{4 \times 2}{4 + 2} = 1.333 \text{ }\Omega$$

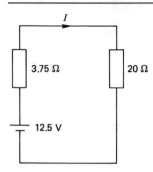

Figure 4.38

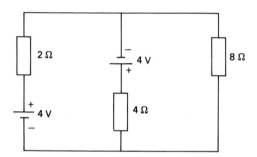

Figure 4.39

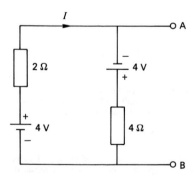

Figure 4.40

The equivalent Thevenin's circuit is shown in Figure 4.42 where

$$I = \frac{1.33}{8 + 1.333} = 0.142 \text{ A}$$

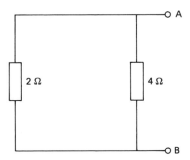

Figure 4.41

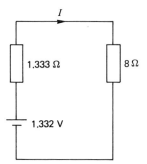

Figure 4.42

Example 14 Find the current in the 6-Ω resistor, using Thevenin's theorem in Figure 4.43. This network is identical to the Wheatstone bridge network of Figure 4.5. Removing the 6-Ω resistor leaves the circuit of Figure 4.44.

The voltage at A with reference to C is

$$V_{AC} = \frac{10 \times 5}{5 + 5} = 5 \text{ V}$$

The voltage at B with reference to C is

$$V_{BC} = \frac{10 \times 2}{3 + 2} = 4 \text{ V}$$

Then

$$V_{AB} = V_{AC} - V_{BC}$$
$$= 5 - 4 = 1 \text{ V (A positive with respect to B)}$$

Removing the 10-V source and replacing it with a short circuit gives Figure 4.45 which can be simplified and redrawn as shown in Figure 4.46.

72 Circuit calculations pocket book

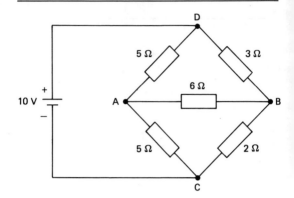

Figure 4.43

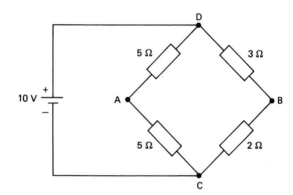

Figure 4.44

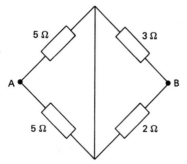

Figure 4.45

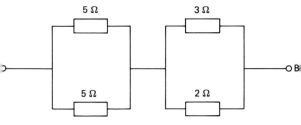

Figure 4.46

The resistance between A and B is equal to

$$R = \frac{5 \times 5}{5 + 5} + \frac{3 \times 2}{3 + 2} = 3.7 \ \Omega$$

The equivalent Thevenin's circuit is shown in Figure 4.47 where

$$I = \frac{1}{6 + 3.7} = 0.103 \ \text{A}$$

The result confirms the result in Example 2 but this method is considerably simpler.

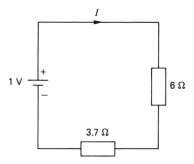

Figure 4.47

Example 15 Use Thevenin's theorem to find the current in the 3-Ω resistor of Figure 4.43. Removing the 3-Ω resistor gives the circuit of Figure 4.48 which can be redrawn as in Figure 4.49.

The 5-Ω resistor in parallel with the series 6-Ω and the 2-Ω resistors gives an equivalent resistance of

$$\frac{5 \times 8}{5 + 8} = 3.077 \ \Omega$$

$$V_{\text{CD}} = \frac{10 \times 3.077}{5 + 3.077} = 3.81 \ \text{V}$$

$$V_{\text{AC}} = 10 - 3.81 = 6.19 \ \text{V}$$

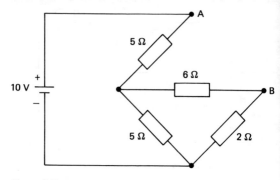

Figure 4.48

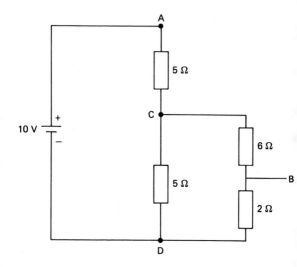

Figure 4.49

$$V_{CB} = \frac{3.81 \times 6}{6 + 2} = 2.86 \text{ V}$$

$$V_{AB} = 6.19 + 2.86 = 9.05 \text{ V}$$

To find the resistance looking in at AB first redraw the circuit as shown in Figure 4.50.

5 Ω in parallel with 5 Ω = 2.5 Ω

6 + 2.5 = 8.5 Ω

$$R = \frac{8.5 \times 2}{8.5 + 2} = 1.62$$

Network theorems 75

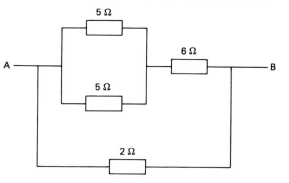

Figure 4.50

The equivalent Thevenin's circuit is shown in Figure 4.51 where

$$I = \frac{9.05}{1.62 + 3} = 1.96 \text{ A}$$

This confirms the result in Example 2.

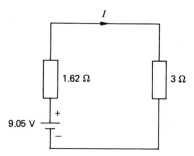

Figure 4.51

Example 16 Use Thevenin's theorem to find the current in the top 5-Ω resistor of Figure 4.43. Removing the 5-Ω resistor gives the circuit of Figure 4.52 which can be redrawn as in Figure 4.53. Adding the 6-Ω and the 5-Ω resistors gives 11-Ω in parallel with the 2-Ω resistor. The equivalent resistance is

$$\frac{11 \times 2}{11 + 2} = 1.692 \text{ Ω}$$

$$V_{CD} = \frac{10 \times 1.692}{3 + 1.692} = 3.60 \text{ V}$$

$$V_{AC} = 10 - 3.6 = 6.4 \text{ V}$$

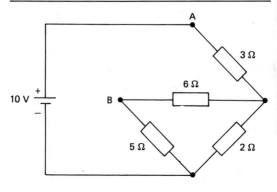

Figure 4.52

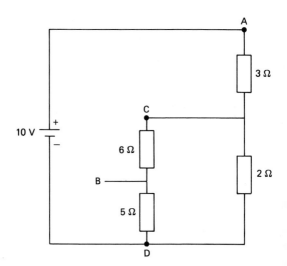

Figure 4.53

$$V_{CB} = \frac{3.60 \times 6}{6 + 5} = 1.96$$

$$V_{AB} = 6.4 + 1.96 = 8.36 \text{ V}$$

To find the resistance looking in at AB first redraw the circuit as shown in Figure 4.54.

The resistance between A and B is found by calculation.

$$\frac{3 \times 2}{3 + 2} + 6 = 7.2 \text{ }\Omega$$

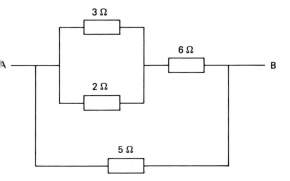

Figure 4.54

The 7.2-Ω resistor is in parallel with the 5-Ω resistor, giving

$$\frac{7.2 \times 5}{7.2 + 5} = 2.95 \ \Omega$$

The equivalent Thevenin's circuit is shown in Figure 4.55 where

$$I = \frac{8.36}{5 + 2.95} = 1.05 \ \text{A}$$

This result confirms the difference between I_1 and I_2 in Example 2.

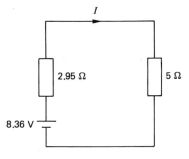

Figure 4.55

4.4 Norton's theorem

Current in a load resistor connected between two points A and B of an active network is the same as if the load resistor was connected to a constant current generator, whose generated current is equal to the short circuit current measured between A and B. This constant current generator has infinite internal resistance and is placed in parallel with the resistance of the network looking back into AB with all sources of e.m.f. replaced by their internal resistances. The procedure for solving a problem using this theorem is to:

78 Circuit calculations pocket book

1. Short circuit the branch containing the resistor.
2. Determine the short circuit current in the branch.
3. Remove all sources of e.m.f. and replace by their internal resistances.
4. Replace all active current sources, if there are any with open circuits, and replace by their internal resistances.
5. Determine the resistance looking in between A and B.
6. Determine the load current flowing, from Norton's equivalent circuit.

Examples 17 to 20, will show how this procedure works.

This theorem is similar to Thevenin's theorem. Here the generator is of the constant current type with the resistance in parallel, whilst in the case of Thevenin's theorem the equivalent generator is of the constant voltage variety with the resistance in series.

Example 17 Use Norton's theorem to find the current flowing in the 12-Ω resistance for the network shown in Figure 4.56.

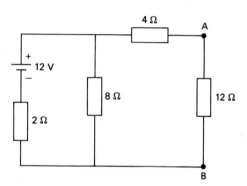

Figure 4.56

First remove the 12-Ω resistor and replace with a short circuit as shown in Figure 4.57.

Figure 4.58 is equivalent to Figure 4.57 where

$$I_{sc} = \frac{12}{2} = 6 \text{ A}$$

Removing the 12-V source in Figure 4.57, the resistance looking in at a break between A and B is

$$R = \frac{8 \times 2}{8 + 2} = 1.6 \text{ Ω}$$

Norton's equivalent circuit is shown in Figure 4.59 where $\ominus$ is the symbol for a current source. This current will divide between the two branches in the ratio 1.6 : 16 (since branch 1 = 1.6 Ω and branch 2 = 4 + 12 = 16 Ω).

$$\text{Current } I \text{ in the 12 Ω} = \frac{6 \times 16}{1.6 + 16} = 0.545 \text{ A}$$

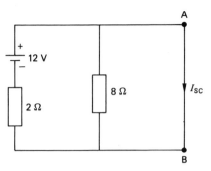

Figure 4.57

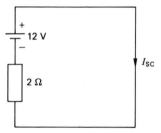

Figure 4.58

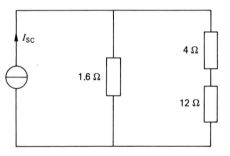

Figure 4.59

This circuit is identical to Figure 4.3 and confirms the answer shown in Figure 4.4.

Example 18 Use Norton's theorem to find the current flowing in the 5-Ω resistor for the circuit of Figure 4.60.

First remove the 5-Ω resistor and replace with a short circuit (Figure 4.61).

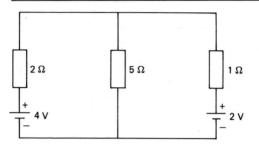

Figure 4.60

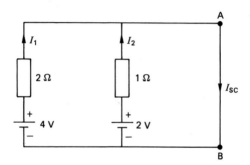

Figure 4.61

$$I_1 = \frac{4}{2} = 2 \text{ A}$$

$$I_2 = \frac{2}{1} = 2 \text{ A}$$

$$I_{sc} = I_1 + I_2 = 4 \text{ A}$$

Removing the two sources in Figure 4.61 and looking in at a break between A and B the resistance is

$$\frac{1 \times 2}{2 + 1} = 0.667 \text{ }\Omega$$

Norton's equivalent circuit is shown in Figure 4.62. The current will divide between the two branches in the ratio 0.667 : 5.

$$\text{Current } I \text{ in the 5 }\Omega = \frac{4 \times 0.667}{0.667 + 5} = 0.471 \text{ A}$$

This circuit is identical to Figure 4.6 and confirms the answer shown in Figure 4.7.

Example 19 Use Norton's theorem to find the current flowing in the 20-Ω resistor for the circuit in Figure 4.63.

Network theorems 81

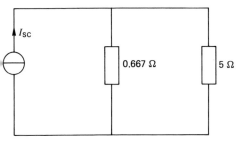

Figure 4.62

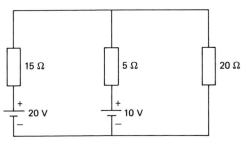

Figure 4.63

First remove the 20-Ω resistor and replace with a short circuit (Figure 4.64).

$$I_1 = \frac{20}{15} = 1.333$$

$$I_2 = \frac{10}{5} = 2$$

$$I_{sc} = I_1 + I_2 = 3.333 \text{ A}$$

Removing the two sources in Figure 4.64 and looking in at a break between A and B the resistance is

$$\frac{15 \times 5}{15 + 5} = 3.75 \text{ Ω}$$

Norton's equivalent circuit is shown in Figure 4.65. The current will divide between the two branches in the ratio 3.75 : 20.

$$\text{Current } I \text{ in the 20 Ω} = \frac{3.333 \times 3.75}{3.75 + 20} = 0.526 \text{ A}$$

The circuit is identical to Figure 4.8 and confirms the answer shown in Figure 4.9.

Example 20 Use Norton's theorem to find the current flowing in the 8-Ω resistor in the circuit of Figure 4.66.

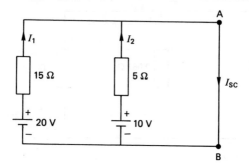

Figure 4.64

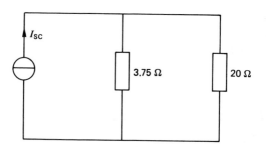

Figure 4.65

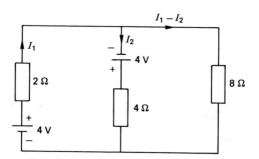

Figure 4.66

First remove the 8-Ω resistor and replace with a short circuit (Figure 4.67).

$$I_1 = \frac{4}{2} = 2 \text{ A}$$

$$I_2 = \frac{4}{4} = 1 \text{ A}$$

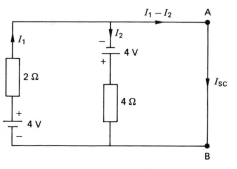

Figure 4.67

$$I_{sc} = I_1 - I_2 = 2 - 1 = 1 \text{ A}$$

Removing the two sources in Figure 4.67 and looking in at a break between A and B the resistance is

$$\frac{2 \times 4}{2 + 4} = 1.333 \text{ } \Omega$$

Norton's equivalent circuit is shown in Figure 4.68. The current will divide between the two branches in the ratio 1.333 : 8.

Current I in the 8 Ω = $\dfrac{1 \times 1.333}{8 + 1.333} = 0.142$ A

The circuit is identical to Figure 4.10 and confirms the answer shown in Figure 4.11.

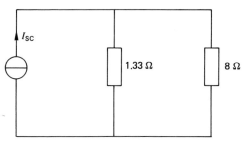

Figure 4.68

4.5 Maximum power transfer

Maximum power is transferred from a source of e.m.f. to a load, when the load resistance equals the internal resistance of the source.

Example 21 Calculate the current flowing, and the power transferred to the load resistance R_L, for the circuit of Figure 4.69 when the load is varied in steps of 1 Ω from 0 to 10 Ω.

84 Circuit calculations pocket book

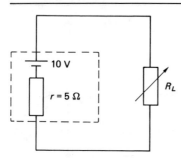

Figure 4.69

Table 4.1 Power transfer

R_L (Ω)	I(A)	P (W) = $I^2 R_L$
0	2	0
1	1.667	2.77
2	1.43	4.08
3	1.25	4.69
4	1.111	4.94
5	1.0	5.0
6	0.909	4.96
7	0.833	4.86
8	0.769	4.734
9	0.714	4.592
10	0.667	4.444

The current is calculated using Ohm's law, and the power calculated using $P = I^2 R$. The results in Table 4.1 shows that maximum power is transferred when $R_L = 5\,\Omega = r$. When $R_L = 0\,\Omega$, a short circuit condition, all the power, 20 W, is dissipated in the battery, which could cause damage.

It should also be remembered that maximum power is delivered to a load, in more complicated networks, when the load equals the

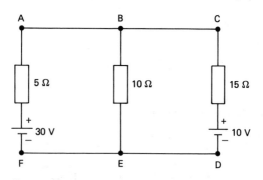

Figure 4.70

Thevenin resistance as seen by the load, 1.777 Ω, for example, in Figure 4.24. Under maximum power transfer conditions the efficiency is only 50%. Note

$$P_{max} = \frac{V^2}{4r}$$

Problems

1. Find the currents in the 5-Ω, 10-Ω and 15-Ω resistors in Figure 4.70, using Kirchhoff's laws. Attempt to solve the problem using the other three theorems – superposition, Thevenin's and Norton's.
2. Verify that the potential differences between A and F, B and E, and C and D are all equal in Figure 4.70.
3. Change the values of the three resistors and the two battery voltages in Figure 4.70, and find the values of the currents in the three branches. (The answers will be correct if the potential differences across the three branches are equal.)

5 Time

5.1 Introduction

Time is one of the parameters in equations relating to a.c. voltages which will be discussed in Chapter 6.

By definition an alternating voltage changes from one level to another level in a given time, and often it is necessary to calculate this time interval. The time can be long in counter circuits, for example, but in most circuits it is very short, where a second is considered to be a very long time! The unit of time is the second, s, and the aim is to show the submultiples used and to show the conversion of one unit into another.

5.2 Units used

The main units are the following:

1. The second, s
2. The millisecond, ms, where $1 \text{ ms} = 1/10^3 \text{ s} = 10^{-3} \text{ s}$
3. The microsecond, µs, where $1 \text{ µs} = 1/10^6 \text{ s} = 10^{-6} \text{ s}$
4. The nanosecond, ns, where $1 \text{ ns} = 1/10^9 \text{ s} = 10^{-9} \text{ s}$

The relationships can be transposed to give the following.

$$\text{Since} \quad 1 \text{ ms} = \frac{1}{10^3} \text{ s}$$

$$10^3 \text{ ms} = 1 \text{ s}$$

$$\text{Since} \quad 1 \text{ µs} = \frac{1}{10^6} \text{ s}$$

$$10^6 \text{ µs} = 1 \text{ s}$$

$$\text{Since} \quad 1 \text{ ns} = \frac{1}{10^9} \text{ s}$$

$$10^9 \text{ ns} = 1 \text{ s}$$

It then follows that

$$10^3 \text{ ms} = 10^6 \text{ µs}$$

$$1 \text{ ms} = \frac{10^6}{10^3} = 10^3 \text{ µs}$$

$$10^3 \text{ ms} = 10^9 \text{ ns}$$

$$1 \text{ ms} = \frac{10^9}{10^3} \text{ ns} = 10^6 \text{ ns}$$

$$10^6 \text{ µs} = 10^9 \text{ ns}$$

$$1 \text{ µs} = \frac{10^9}{10^6} = 10^3 \text{ ns}$$

5.3 Conversion of seconds to milliseconds, microseconds and nanoseconds

Since $1 \text{ s} = 10^3 \text{ ms}$

$0.125 \text{ s} = 0.125 \times 10^3 = 125 \text{ ms}$

Since $1 \text{ s} = 10^6 \text{ } \mu\text{s}$

$0.001 \text{ s} = 0.001 \times 10^6 = 1000 \text{ } \mu\text{s}$

$0.00025 \text{ s} = 0.00025 \text{ s} \times 10^6 = 250 \text{ } \mu\text{s}$

Since $1 \text{ s} = 10^9 \text{ ns}$

$0.000000425 \text{ s} = 0.000000425 \times 10^9 = 425 \text{ ns}$

```
10 PRINT "PROG 4:9."
20 PRINT "SECONDS TO MILLISECONDS"
30 INPUT "ENTER TIME IN SECONDS" ; S
40 LET MS=S*(10^3)
50 PRINT "ANSWER = " MS "MILLISECONDS"
60 PRINT "SECONDS TO MICROSECONDS"
70 LET US=S*(10^6)
80 PRINT "ANSWER = " US "MICROSECONDS"
90 PRINT "SECONDS TO NANOSECONDS"
100 LET NS=S*(10^9)
110 PRINT "ANSWER = " NS "NANOSECONDS"
```

Note also that

Since $1 \text{ ms} = 10^3 \text{ } \mu\text{s}$

$0.25 \text{ ms} = 0.25 \times 10^3 = 250 \text{ } \mu\text{s}$

Since $1 \text{ ms} = 10^6 \text{ ns}$

$0.00015 \text{ ms} = 0.00015 \times 10^6 = 150 \text{ ns}$

Since $1 \text{ } \mu\text{s} = 10^3 \text{ ns}$

$0.03 \text{ } \mu\text{s} = 0.03 \times 10^3 = 30 \text{ ns}$

Table 5.1 shows three different times, each expressed in four different units. Note the 10^6 relationship between columns 2 and 4, and 10^3 between columns 3 and 4.

Table 5.1 Time units

s	ms	μs	ns
0.00018	0.18	180	180×10^3
0.0075	7.5	7.5×10^3	7.5×10^6
0.04	40	40×10^3	40×10^6

5.4 Conversion of milliseconds, microseconds and nanoseconds to seconds

The examples in Section 5.3 showed that

Since $10^3 \text{ ms} = 1 \text{ s}$

$x \text{ ms} = \dfrac{x}{10^3} \text{ s}$

Since 10^6 μs = 1 s

$$x \text{ μs} = \frac{x}{10^6} \text{ s}$$

Since 10^9 ns = 1 s

$$x \text{ ns} = \frac{x}{10^9} \text{ s}$$

Since 10^3 μs = 1 ms

$$x \text{ μs} = \frac{x}{10^3} \text{ ms}$$

Since 10^6 ns = 1 ms

$$x \text{ ns} = \frac{x}{10^6} \text{ ms}$$

Since 10^3 ns = 1 μs

$$x \text{ ns} = \frac{x}{10^3} \text{ μs}$$

```
10 PRINT "PROG 50"
20 PRINT "MILLISECONDS TO SECONDS"
30 INPUT "ENTER TIME IN MILLISECONDS" ; MS
40 LET S=MS/(10^3)
50 PRINT "ANSWER = " S "SECONDS"
```

```
10 PRINT "PROG 51"
20 PRINT "MICROSECONDS TO SECONDS"
30 INPUT "ENTER TIME IN MICROSECONDS" ; US
40 LET S=US/(10^6)
50 PRINT "ANSWER = " S "SECONDS"
```

```
10 PRINT "PROG 52"
20 PRINT "NANOSECONDS TO SECONDS"
30 INPUT "ENTER TIME IN NANOSECONDS" ; NS
40 LET S=NS/(10^9)
50 PRINT "ANSWER = " S "SECONDS"
```

From these six examples the following can be written:

$$\frac{\text{ms}}{10^3} = \text{s} \quad \frac{\text{μs}}{10^6} = \text{s} \quad \frac{\text{ns}}{10^9} = \text{s}$$

$$\frac{\text{μs}}{10^3} = \text{ms} \quad \frac{\text{ns}}{10^6} = \text{ms} \quad \frac{\text{ns}}{10^3} = \text{μs}$$

Problems

1 Verify that Table 5.1 is correct by converting column 4 to microseconds, milliseconds and seconds; column 3 to milliseconds and seconds; and column 2 to seconds.
2 Express the times in seconds equivalent to 3.6×10^9 ns and 3.6×10^6 μs.
3 Write down the times in milliseconds and nanoseconds which correspond to 33 microseconds.

6 The a.c. voltage

6.1 Introduction

Unlike d.c. voltages, which have a constant amplitude, a.c. voltages vary through a cycle of values which is repeated continuously at some fixed rate.

A.c. voltages can be generated by both alternators and oscillators. The most important waveform shape is the sine wave. Others, such as triangular and square waves, are also used in electronic circuits.

A.c. voltages have parameters associated with them such as frequency, wavelength, amplitude and periodic time. This chapter shows how these quantities are calculated, the relationship between them, the units used, and how they can be derived graphically and by computer programs. Computer programs enable tedious and time-consuming problems, such as finding the r.m.s. and average values, to be solved in seconds, once the program has been written and stored.

6.2 Voltage generating

A simple machine consisting of one coil XOY, rotating between a pair of magnetic poles NS (Figure 6.1), will generate an alternating voltage which can be represented by a sine wave (Figure 6.2). Assume that the coil rotates anti-clockwise.

1. At position A the voltage is zero.
2. At position B (¼ turn) the voltage is a maximum (+ve).
3. At position C (½ turn) the voltage is zero.
4. At position D (¾ turn) the voltage is a maximum (−ve).
5. At position A (1 turn) the voltage is again zero.

It is customary to plot voltage amplitude against degrees or radians where

1 revolution = 360°
360° = 2π radians

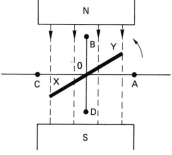

Figure 6.1

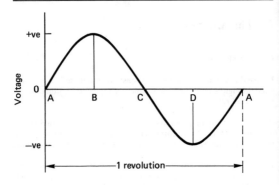

Figure 6.2

Using these units gives an easy method of calculating the voltage value at any instant, as shown in Section 6.3.

Example 1 Convert 30°, 45°, 135°, 210° into radians.

$$30° = \frac{2\pi \times 30}{360} = \frac{\pi}{6} = 0.524$$

$$45° = \frac{2\pi \times 45}{360} = \frac{\pi}{4} = 0.785$$

$$135° = \frac{2\pi \times 135}{360} = \frac{3\pi}{4} = 2.356$$

$$210° = \frac{2\pi \times 210}{360} = \frac{7\pi}{6} = 3.665$$

```
10 PRINT "PROG 53"
20 PRINT "DEGREES TO RADIANS"
30 INPUT "ENTER DEGREES" ; D
40 LET R=2*3.142*D/360
50 PRINT "ANSWER = " R "RADIANS"
```

Example 2 Express angles of 1.8, 4.0, 5.2, 5π/6 radians in degrees.

$$1.8 = \frac{360 \times 1.8}{2\pi} = 103.132°$$

$$4.0 = \frac{360 \times 4}{2\pi} = 229.183°$$

$$5.2 = \frac{360 \times 5.2}{2\pi} = 297.938°$$

$$\frac{5\pi}{6} = \frac{360 \times 5\pi}{2\pi \times 6} = \frac{60 \times 5}{2} = 150°$$

```
10 PRINT "PROG 54"
20 PRINT "RADIANS TO DEGREES"
30 INPUT "ENTER RADIANS" : R
40 LET D=(360*R)/(2*3.142)
50 PRINT "ANSWER = " D "DEGREES"
```

6.3 Instantaneous value

Figure 6.3 shows the voltage waveform plotted against degrees and radians. The instantaneous value at any point such as B can be found by finding the value of the height AY.

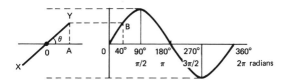

Figure 6.3

Now $AY = \sin \theta$ where θ is the angle through which the coil has moved since the start of the revolution (provided that $OY = 1$).

Assume $\theta = 40°$. Then $\sin 40° = 0.6427$.

Voltage at B = $0.6427 \times$ maximum, or 64.27% of maximum.

Example 3 Calculate the instantaneous voltages at 30° intervals, from 0° to 360°, for an e.m.f. with a maximum value of 240 V.

Column 4 could be used to plot the sine wave.

Table 6.1

θ (degrees)	θ (radians)	Sin θ	Instantaneous voltage
0	0	0	0
30	π/6	0.5	120
60	2π/6	0.866	207.8
90	π/2	1.0	240
120	2π/3	0.866	207.8
150	5π/6	0.5	120
180	π	0	0
210	7π/6	−0.5	−120
240	4π/3	−0.866	−207.8
270	3π/2	−1.0	−240
300	5π/3	−0.866	−207.8
330	11π/6	−0.5	−120
360	2π	0	0

```
10 PRINT "PROG 55"
20 PRINT "INSTANTANEOUS VALUES"
30 INPUT "ENTER MAXIMUM VALUE" : E
40 INPUT "ENTER ANGLE IN DEGREES" ; X
50 LET V=E*SIN(X*3.141593/180)
60 PRINT "INSTANTANEOUS VOLTAGE = " V "VOLTS"
```

6.4 Peak, peak to peak, and r.m.s. values

A cathode ray oscilloscope (CRO) is used to view a voltage waveform, and to measure the peak voltage (V_p) (Figure 6.4), which is the amplitude. Note that the peak to peak voltage $V_{p/p} = 2V_p$ (V_p is the maximum value). A.c. voltmeters, both analogue and digital, will read the r.m.s. (root mean squared) value. The r.m.s. value of the voltage is equal to the value of a d.c. voltage that would produce the same heating effect in a conductor. The relationship between V_p and $V_{r.m.s.}$ is given by

$$\text{r.m.s.} \times \sqrt{2} = V_p$$

$$\text{r.m.s.} \times 1.414 = V_p$$

$$\text{r.m.s.} = \frac{V_p}{1.414}$$

$$= 0.7071 \times V_p, \text{ since } \frac{1}{1.414} = 0.7071$$

This relationship will be proved graphically in Section 6.8.

Example 4 Given that the mains voltage is 240 V r.m.s., calculate the peak and peak to peak values.

$$V_p = 240 \times \sqrt{2} = 339.41 \text{ V}$$
$$V_{p/p} = 2 \times 339.41 = 678.82 \text{ V}$$

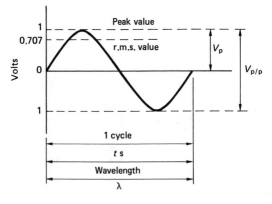

Figure 6.4

The a.c. voltage 93

```
10 PRINT "PROG 56"
20 PRINT "PEAK AND PEAK TO PEAK VALUES"
30 INPUT "ENTER RMS VOLTAGE" ; V
40 LET VP=V*SQR(2)
50 LET VPP=2*VP
60 PRINT "PEAK VOLTAGE = " VP "VOLTS"
70 PRINT "PEAK TO PEAK VOLTAGE = " VPP "VOLTS"
```

Example 5 Given that the peak to peak value of a voltage is 100 V, what is the r.m.s. value?

$$V_p = \frac{100}{2} = 50 \text{ V}$$

r.m.s. $= 0.7071 \times 50 = 35.36$ V

```
10 PRINT "PROG 57"
20 PRINT "RMS VALUE"
30 INPUT "ENTER PEAK TO PEAK VOLTAGE" ; VPP
40 LET VP=VPP/2
50 LET R=.7071*VP
60 PRINT "RMS VALUE = " R "VOLTS"
```

Example 6 A small signal measured on a digital voltmeter (DVM) reads 750 µV. If the signal is observed on a CRO, what is the amplitude in millivolts?

DVM reading = 750 µV r.m.s.

Amplitude $= V_p = 750 \times \sqrt{2} = 1060.66$ µV

Amplitude in mV $= \dfrac{1060.66}{10^3} = 1.06$ mV

```
10 PRINT "PROG 58"
20 PRINT "PEAK VOLTAGE"
30 INPUT "ENTER VOLTAGE IN MICROVOLTS" ; UV
40 LET VP=UV*SQR(2)
50 PRINT "AMPLITUDE IN MILLIVOLTS = "VP/1000 "MILLIVOLTS"
```

Example 7 A sine wave has an amplitude of 0.6 mV; what is the r.m.s. value in microvolts?

$0.7071 \times 0.6 = 0.4243$ mV r.m.s.

$= 0.4243 \times 10^3 = 424.3$ µV

```
10 PRINT "PROG 59"
20 PRINT "RMS VALUE"
30 INPUT "ENTER AMPLITUDE IN MILLIVOLTS" ; MV
40 LET UV=.7071*MV*1000
50 PRINT "RMS VALUE IN MICROVOLTS = " UV "MICROVOLTS"
```

Example 8 Measurements on two sine waves were recorded as (i) 90 V r.m.s., (ii) 120 V peak. Which one had the greater amplitude?

For (i) $V_p = 90 \times \sqrt{2} = 127.26$

(See program 56, above.)

Example 9 A mains distribution cable carries 10 kV r.m.s. The cable insulation is rated at the peak voltage. What is this rating?

$$V_p = 10\ 000 \times \sqrt{2} = 14\ 142\ \text{V} = 14.142\ \text{kV}$$

(See program 56, page 93.)

6.5 Frequency

Consider again our simple generator. One revolution corresponds to one cycle (Figure 6.3). Assume that the generator is running at 50 revolutions per second. This would produce 50 cycles per second. This is called the frequency (hertz). From this it can be seen that one cycle is produced in a time of 1/50 second. Sine waves therefore can also be plotted against time.

Frequency and time are related by

$$F = \frac{1}{t} \text{ and } t = \frac{1}{F}$$

where F is in hertz (Hz) and t, the periodic time, is in seconds (s).

Example 10 Calculate the periodic time of frequencies 250 Hz and 2500 Hz.

$$t = \frac{1}{f} = \frac{1}{250}\ \text{s} = \frac{1 \times 10^3}{250} = 4\ \text{ms}$$

$$t = \frac{1}{2500} \times 10^6 = 400\ \mu\text{s}$$

```
10 PRINT "PROG 60"
20 PRINT "PERIODIC TIME"
30 INPUT "ENTER FREQUENCY" ; F
40 LET T=1/F
50 PRINT "PERIODIC TIME = " T "SECONDS"
60 PRINT "PERIODIC TIME = " T*1000 "MILLISECONDS"
70 PRINT "PERIODIC TIME = " T*(10^6) "MICROSECONDS"
```

Example 11 Given periodic times of 1 ms and 2 μs calculate the frequencies

$$F = \frac{1}{t} = \frac{1}{10^3}\ \text{s} = 0.001\ \text{s}$$

$$F = \frac{1}{0.001} = 1000\ \text{Hz} = 1\ \text{kHz}$$

$$F = \frac{1}{t} = \text{where } t = \frac{2}{10^6}$$

$$F = 1 \div \frac{2}{10^6} = \frac{10^6}{2} = 500\ 000\ \text{Hz} = 500\ \text{kHz}$$

```
10 PRINT "PROG 61".
20 PRINT "FREQUENCY"
30 INPUT "ENTER PERIODIC TIME IN MILLISECONDS" ; MV
40 LET T=MV/1000
50 LET F=1/T
60 PRINT "FREQUENCY = " F "HERTZ"
```

```
70 INPUT "ENTER PERIODIC TIME IN MICROSECONDS" : US
80 LET T=US/(10^6)
90 LET F=1/T
100 PRINT "FREQUENCY = " F "HERTZ"
```

Example 12 Twenty cycles of a voltage sine wave occur in 100 ms. Calculate the frequency.

$$\text{Periodic time of 1 cycle} = \frac{100}{20} = 5 \text{ ms}$$

$$f = \frac{1}{t} = 1 \div \frac{5}{1000} = \frac{1000}{5} = 200 \text{ Hz}$$

(See program 61, above.)

Alternative method.

$$20 \text{ cycles occur in 100 ms} = \frac{100}{10^3} = 0.1 \text{ s}$$

$$\text{In 1 s number of cycles} = \frac{1}{0.1} \times 20 = 200 \text{ Hz}$$

Example 13 A two-pole a.c. generator is running at 300 rev/min. Calculate the frequency of the voltage generated and the periodic time.

$$300 \text{ rev/min} = \frac{300}{60} = 5 \text{ rev/s} = 5 \text{ Hz}$$

$$t = \frac{1}{f} = \frac{1}{5} \text{ s or } \frac{1 \times 1000}{5} = 200 \text{ ms}$$

(See program 60, page 94.)

Example 14 At what speed in rev/min would an a.c. generator be running if the periodic time of the waveform was 25 ms?

$$f = 1 \div \frac{25}{1000} = \frac{1000}{25} = 40 \text{ Hz}$$

$$\text{Rev/min} = 40 \times 60 = 2400$$

(See program 61, page 94.)

6.6 Wavelength

The sine wave has one more property, which is called the wavelength, λ, and which is of importance when dealing, for example, with radio waves. Radio waves are sine waves modulated by words and music.

Radio stations are identified by their wavelengths and frequencies, f. The wavelength is shown in Figure 6.4. Given that λ is in metres and f is in hertz, then

$$\lambda f = K$$

where K is a constant equal to 3×10^8 m/s and is the speed at which the waves travel from point A to point B in free space.

Example 15 Radio 4 has a wavelength of 1500 m. What is the frequency?

$$\lambda f = K$$

$$f = \frac{K}{\lambda}$$

$$= \frac{3 \times 10^8}{1500} = 200\,000 \text{ Hz}$$

$$= 200 \text{ kHz}$$

```
10 PRINT "PROG 62"
20 PRINT "FREQUENCY AND WAVELENGTH"
30 INPUT "ENTER WAVELENGTH" ; L
40 LET F=3*(10^8)/L
50 PRINT "FREQUENCY = " F "HERTZ"
```

Example 16 A radio station is transmitting at a frequency of 1 MHz. What wavelength is this on the dial?

$$1 \text{ MHz} = 10^6 \text{ Hz}$$

$$\lambda = \frac{K}{f} = \frac{3 \times 10^8}{10^6} = 300 \text{ m}$$

```
10 PRINT "PROG 63"
20 PRINT "FREQUENCY AND WAVELENGTH"
30 INPUT "ENTER FREQUENCY" ; F
40 LET L=3*(10^8)/F
50 PRINT "WAVELENGTH = " L "METRES"
```

Example 17 What is the frequency corresponding to a wave 10 000 m long?

$$10 \text{ km} = 10 \times 10^3 \text{ m}$$

$$f = \frac{3 \times 10^8}{10 \times 10^3} = 3 \times 10^4 = 30\,000 = 30 \text{ kHz}$$

(See program 62, above.)

Example 18 Infra-red rays which are sinusoidal in nature occur around a frequency of 10^{13} Hz. Calculate the wavelength.

$$\lambda f = 3 \times 10^8$$

$$\lambda = \frac{3 \times 10^8}{f}$$

$$= \frac{3 \times 10^8}{10^{13}} = \frac{3}{10^5} = 3 \times 10^{-5} \text{ m}$$

(See program 63, above.)

Example 19 What would be the frequency of a transmitter radiating a signal with a wavelength of ¼ m?

$$¼ \text{ m} = 0.25 \text{ m}$$

$$f = \frac{3 \times 10^8}{0.25} = 12 \times 10^8 \text{ Hz}$$

= 1200 MHz

= 1.2 GHz

(See program 62, page 96.)

Example 20 A sine wave has a wavelength of 1 km. Calculate the frequency.

1 km = 10^3 m

$$f = \frac{3 \times 10^8}{10^3} = 3 \times 10^5 \text{ Hz}$$

(See program 62, page 96.)

Example 21 What wavelength does a frequency of 15 MHz correspond to?

$$f = \frac{3 \times 10^8}{15 \times 10^6} = \frac{10^2}{5} = 20 \text{ m}$$

(See program 63, page 96.)

6.7 Average value for a sine wave

The average value over a complete cycle is zero. The average value over half a cycle is 0.637 V_p. This result is important in some applications, e.g. power supply rectifying circuits. The average value can be found graphically, by finding the average of a number, n, of equidistant midordinates over the half cycle. The greater the value of n, the more accurate the answer will be. Let $n = 12$. This gives 12 intervals of 15°. The first midordinate will be at 7.5°, the second at 22.5°, etc. We can then calculate the values of the midordinates, which gives us the instantaneous voltages, using the same method as in Figure 6.3.

The midordinate is the half-way point in each interval. If 18 intervals is chosen, for example, giving 18 intervals of 10°, the first midordinate would be at 5°, the second at 15°, etc., up to and including 175°.

Voltage number	Midordinate value	Instantaneous voltage value
V_1	sin 7.5	0.130
V_2	sin 22.5	0.382
V_3	sin 37.5	0.608
V_4	sin 52.5	0.793
V_5	sin 67.5	0.923
V_6	sin 82.5	0.991
V_7	sin 97.5	0.991
V_8	sin 112.5	0.923
V_9	sin 127.5	0.793
V_{10}	sin 142.5	0.608
V_{11}	sin 157.5	0.382
V_{12}	sin 172.5	0.130
		7.654

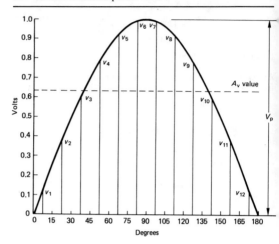

Figure 6.5

$$\text{Average voltage} = \frac{7.654}{12} = 0.637 \text{ V}$$

The results are plotted in Figure 6.5. The average value $= 0.637\, V_p$. The average value is also equal to

$$\frac{\text{The area enclosed over the half cycle}}{\text{Length of the base of the half cycle}}$$

```
10 PRINT "PROG 64"
20 PRINT "AVERAGE VALUE OF A SINE WAVE"
30 INPUT "ENTER FIRST MIDORDINATE IN DEGREES" ; S
40 LET T=S*2
50 LET N=180/T
60 LET A=SIN(S*3.141593/180)
70 PRINT A
80 LET W=0
90 LET Y=0
100 LET S=S+T
110 LET Z=SIN(S*3.141593/180)
120 PRINT Z
130 LET W=Z+W
140 LET Y=Y+1
150 IF Y<N-1 THEN GOTO 100
160 LET C=W+A
170 PRINT "TOTAL = " C
180 LET V=C/N
190 PRINT "AVERAGE = " V
```

6.8 Graphical determination of r.m.s. value of a sine wave

It was stated above that the r.m.s. voltage of a sine wave equals $0.7071\, V_p$. The name r.m.s. is derived from the fact that it is the square root of the average value of the squares of all the different values the voltage can take during a complete cycle.

$$V_{\text{r.m.s.}} = \sqrt{\left(\frac{V_1^2 + V_2^2 + V_3^2 \ldots + V_n^2}{n}\right)}$$

Using the data from the example in Section 6.7 where $n = 12$.

V_1^2	= 0.017
V_2^2	= 0.146
V_3^2	= 0.371
V_4^2	= 0.629
V_5^2	= 0.854
V_6^2	= 0.983
V_7^2	= 0.983
V_8^2	= 0.854
V_9^2	= 0.629
V_{10}^2	= 0.371
V_{11}^2	= 0.146
V_{12}^2	= 0.017
	5.998

$$\text{r.m.s. value} = \sqrt{\left(\frac{5.998}{12}\right)} = 0.707$$

$$= 0.707\ V_p$$

$$\text{The form factor for a sine wave} = \frac{\text{r.m.s. value}}{\text{peak value}}$$

$$= \frac{0.707}{0.637} = 1.11$$

```
10 PRINT "PROG 65"
20 PRINT "RMS VALUE OF A SINE WAVE"
30 INPUT "ENTER FIRST MIDORDINATE IN DEGREES" ; S
40 LET T=S*2
50 LET N=180/T
60 LET A=SIN(S*3.141593/180)^2
70 PRINT A
80 LET W=0
90 LET Y=0
100 LET S=S+T
110 LET Z=SIN(S*3.141593/180)^2
120 PRINT Z
130 LET W=Z+W
140 LET Y=Y+1
150 IF Y<N-1 THEN GOTO 100
160 LET C=W+A
170 PRINT "TOTAL = " C
180 LET V=SQR(C/N)
190 PRINT "RMS VALUE = " ; V
```

6.9 Graphical determination of average and r.m.s. values of a non-sinusoidal waveform

The methods outlined in Sections 6.7 and 6.8 can be applied to one half cycle of any waveform where both half cycles are symmetrical about the zero axis. Consider the following problem.

A triangular waveform has the following values over one half cycle. Plot the half cycle and calculate: (a) average value, (b) r.m.s. value, (c) form factor.

Time (s)	0	1	2	3	4
Voltage (V)	0	40	80	40	0

$V_1 = 10$	$V_1^2 =$	100
$V_2 = 30$	$V_2^2 =$	900
$V_3 = 50$	$V_3^2 =$	2500
$V_4 = 70$	$V_4^2 =$	4900
$V_5 = 70$	$V_5^2 =$	4900
$V_6 = 50$	$V_6^2 =$	2500
$V_7 = 30$	$V_7^2 =$	900
$V_8 = 10$	$V_8^2 =$	100
320		16 800

(a) Average value = $\dfrac{320}{8} = 40$

(b) r.m.s. value = $\sqrt{\left(\dfrac{16\,800}{8}\right)} = 45.826$

(c) Form factor = $\dfrac{45.826}{40} = 1.15$

Note that the exact value for (b) is 46.18 ($V_{max}/\sqrt{3}$). The waveform is plotted in Figure 6.6.

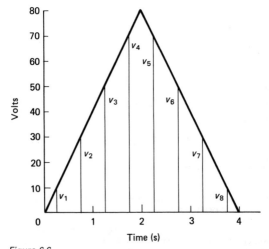

Figure 6.6

The a.c. voltage 101

```
10 PRINT "PROG 66"
20 INPUT "ENTER V1,V2,V3,V4" ; V1,V2,V3,V4
30 INPUT "ENTER V5,V6,V7,V8" ; V5,V6,V7,V8
40 LET V=V1+V2+V3+V4+V5+V6+V7+V8
50 LET Z=V/8
60 PRINT "AVERAGE VALUE = " ; Z
70 LET W=V1^2+V2^2+V3^2+V4^2+V5^2+V6^2+V7^2+V8^2
80 LET X=SQR(W/8)
90 PRINT "RMS VALUE = " X
100 LET Y=X/Z
110 PRINT "FORM FACTOR = " Y
```

Problems

1. Express the following angles in degrees: (a) 4 radians, (b) 1.5 radians, (c) 0.8 radians.
2. Express the following angles in radians: (a) 75°, (b) 180°, (c) 4.6°, (d) 300°.
3. Calculate the instantaneous voltages at 15° intervals, from 0° to 360°, for a sinusoidal voltage with a maximum value of 100 V. Plot the waveform and show both radians and degrees on the time axis.
4. In Problem 3 what are the peak to peak and r.m.s. values?
5. A sine wave has an r.m.s. value of 620 µV. What is the peak value in millivolts?
6. What should be the insulation rating of a cable carrying a mains voltage of 33 kV?
7. What is the periodic time of a 45 kHz signal? Calculate the wavelength in free space.
8. A sine wave has a periodic time of 1.5 µs. Calculate the frequency and wavelength.
9. Two radio stations are found on the FM band at frequencies of 88 and 108 MHz. What are their respective wavelengths?
10. Calculate from first principles the average and r.m.s. values of a sine wave, together with the form factor using (a) 6 midordinates, (b) 18 midordinates. Comment on the results.
11. An alternating voltage has the following values:

Time(s)	0	1	2	3	4	5	6	7	8
Voltage (V)	0	30	60	60	60	60	60	30	0

    These voltages are joined by straight lines. Calculate the average, r.m.s. and form factor.

12. A triangular waveform has the following values over one half cycle. Plot the waveform and calculate: (a) average value, (b) r.m.s. value, (c) form factor.

Time (s)	0	1	2	3	4	5	6	7	8	9	10
Voltage (V)	0	20	40	60	80	100	80	60	40	20	0

7 Capacitors

7.1 Introduction

The capacitor shown in Figure 7.1 consists of two metal plates separated by an insulator. It can store electricity, with a negative charge on one plate and a positive charge on the other. The insulator is referred to as a dielectric. Some common dielectrics are air, paper, mica, bakelite, glass, rubber and porcelain. The quantity symbol for capacitance is C, and the unit symbol is the farad, F. Multiple plate capacitors also exist. The most common is the variable capacitor used for tuning radio receivers. The behaviour of capacitors in d.c. and a.c. circuits is dealt with in Chapters 9 and 11 respectively.

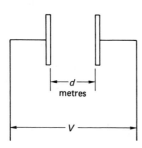

Figure 7.1

7.2 Units

The farad is an inconveniently large unit, and capacitance is usually expressed using subunits. The subunits and the relationships between them are:

1. The microfarad, μF, where $1 \text{ μF} = 1/10^6 \text{ F}$ or 10^{-6} F
2. The nanofarad, nF, where $1 \text{ nF} = 1/10^9 \text{ F}$ or 10^{-9} F
3. The picofarad, pF, where $1 \text{ pF} = 1/10^{12} \text{ F}$ or 10^{-12} F

Transposing these relationships shows that

$$10^6 \text{ μF} = 1 \text{ F}$$
$$10^9 \text{ nF} = 1 \text{ F}$$
$$10^{12} \text{ pF} = 1 \text{ F}$$

and since $10^6 \text{ μF} = 10^9 \text{ nF} = 10^{12} \text{ pF}$, then

$$1 \text{ μF} = 10^3 \text{ nF}$$
$$1 \text{ μF} = 10^6 \text{ pF}$$
$$1 \text{ nF} = 10^3 \text{ pF}$$
$$1 \text{ nF} = \frac{1}{10^3} \text{ μF or } 10^{-3} \text{ μF}$$

$$1 \text{ pF} = \frac{1}{10^6} \text{ μF or } 10^{-6} \text{ μF}$$

$$1 \text{ pF} = \frac{1}{10^3} \text{ nF or } 10^{-3} \text{ nF}$$

Example 1 Express a 0.02-μF capacitor in farads, nanofarads and picofarads.

$$0.02 \text{ μF} = 0.02 \times 10^{-6} \text{ F} = 2 \times 10^{-8} \text{ F}$$

$$0.02 \text{ μF} = 0.02 \times 10^3 \text{ nF} = 20 \text{ nF}$$

$$0.02 \text{ μF} = 0.02 \times 10^6 \text{ pF} = 20 \times 10^3 \text{ pF}$$

Further examples are shown in Table 7.1.

```
10 PRINT "PROG 67"
20 PRINT "CONVERSION OF MICROFARADS TO OTHER UNITS"
30 INPUT "ENTER MICROFARADS" ; UF
40 LET F=UF/10^6
50 LET NF=UF*10^3
60 LET PF=UF*10^6
70 PRINT UF "MICROFARADS = " F "FARADS
80 PRINT UF "MICROFARADS = " NF "NANOFARADS"
90 PRINT UF "MICROFARADS = " PF "PICOFARADS"
```

Table 7.1 Capacitor values expressed in different units

F	μF	nF	pF
10^{-4}	100	100×10^3	100×10^6
10^{-6}	1	10^3	10^6
0.01×10^{-6}	0.01	10	10^4
0.0001×10^{-6}	0.0001	0.1	100
0.000002×10^{-6}	0.000002	0.002	2

7.3 Capacitor formulas

(a) Capacitance

The capacitance of a parallel plate capacitor with vacuum or air as the dielectric is given by the formula

$$C = \frac{\epsilon_o A}{d}$$

where C is in farads
ϵ_o = the permittivity of free space and has the value 8.854×10^{-12} farad per metre, F/m
A = the area of a plate in square metres
d = distance between plates in metres, shown in Figure 7.1,

i.e.

$$C = \frac{8.854 \times 10^{-12} (\text{F/m}) \times A \text{ (m}^2\text{)}}{d \text{ (m)}}$$

The value of the capacitance can be increased in two ways:

1 By increasing the number of plates or plate areas.
2 By using a dielectric with a higher permittivity compared to air. Given that air has a relative permittivity of 1, mica, for example, can have a relative permittivity of 6.

The full formula for a multi-plate capacitor is therefore

$$C = \frac{\epsilon_0 \epsilon_r (n-1) A}{d}$$

where ϵ_r = the relative permittivity of the dielectric
n = the number of plates

It is essential to express both the area of the plates and the distance between them in metres, as the worked examples will show.

Example 2 A capacitor has six metal plates, separated by mica having a thickness of 0.4 mm, and a relative permittivity of 6. The area of one side of each plate is 300 cm². Calculate the capacitance.

$\epsilon_r = 6, n - 1 = 5$

$$A = \frac{300}{10^4} \text{ m}^2 = 0.03 \text{ m}^2$$

$$d = \frac{0.4}{10^3} \text{ m} = 0.0004 \text{ m}$$

$$C = \frac{8.854 \times 10^{-12} \times 6 \times 5 \times 0.03}{0.0004} \text{ F}$$

$$= 0.0199 \times 10^{-6} \text{ F}$$

$$= 0.0199 \times 10^{-6} \times 10^6 \text{ µF} = 0.0199 \text{ µF}$$

```
10 PRINT "PROG 68"
20 PRINT "CAPACITANCE"
30 INPUT "ENTER RELATIVE PERMITTIVITY" ; ER
40 INPUT "ENTER THE NUMBER OF PLATES" ; N
50 INPUT "ENTER PLATE AREA IN SQR CMS" ; A
60 INPUT "ENTER DISTANCE BETWEEN PLATES IN MM" ; D
70 LET N=N-1
80 LET A=A/10^4
90 LET D=D/10^3
100 LET C=8.854*ER*N*A/(D*(10^12))
110 PRINT "CAPACITANCE = " C*(10^6) "MICROFARADS"
```

Example 3 Find the separation between the plates of a four-metal-plate capacitor, if the dielectric has a relative permittivity of 7. The area of each plate is 900 mm², and the capacitance is 0.1 µF.

$\epsilon_r = 7, n - 1 = 3$

$$A = \frac{900}{10^6} \text{ m}^2 = 0.0009 \text{ m}^2$$

$$C = 0.1 \times 10^{-6} \text{ F}$$

$$d = \frac{\epsilon_0 \epsilon_r (n-1) A}{C}$$

$$d = \frac{8.854 \times 10^{-12} \times 7 \times 3 \times 0.0009}{0.1 \times 10^{-6}}$$

$$= 1.67 \times 10^{-6} \text{ m} = 1.67 \times 10^{-6} \times 10^3 \text{ mm}$$

$$= 1.67 \times 10^{-3} \text{ mm} = 0.00167 \text{ mm}$$

Capacitors 105

```
10 PRINT "PROG 69"
20 PRINT "DISTANCE BETWEEN PLATES"
30 INPUT "ENTER RELATIVE PERMITTIVITY" ; ER
40 INPUT "ENTER THE NUMBER OF PLATES" ; N
50 INPUT "ENTER AREA IN SQR MMS" ; A
60 INPUT "ENTER CAPACITANCE IN MICROFARADS" ; C
70 LET N=N-1
80 LET A=A/10^6
90 LET C=C/10^6
100 LET D=8.854*ER*N*A/(C*(10^12))
110 PRINT "DISTANCE BETWEEN PLATES = " D *(10^3) "MILLIMETRES"
```

Example 4 A capacitor has the following parameters: $C = 500$ pF, $\epsilon_r = 6$, $n = 8$, $d = 1$ mm. Calculate the area of the plates.

$$n - 1 = 7$$

$$C = 500 \times 10^{-12} \text{ F}$$

$$d = \frac{1}{10^3} \text{ m} = 0.001 \text{ m}$$

$$A = \frac{Cd}{\epsilon_o \epsilon_r (n - 1)}$$

$$= \frac{500 \times 10^{-12} \times 0.001}{8.854 \times 10^{-12} \times 6 \times 7} = 1.34 \times 10^{-3} \text{ m}^2$$

$$= 1.34 \times 10^{-3} \times 10^4 \text{ cm}^2 = 13.44 \text{ cm}^2$$

```
10 PRINT "PROG 70"
20 PRINT "AREA OF PLATES"
30 INPUT "ENTER CAPACITANCE IN PICOFARADS" ; C
40 INPUT "ENTER DISTANCE BETWEEN PLATES IN MMS" ; D
50 INPUT "ENTER RELATIVE PERMITTIVITY" ; ER
60 INPUT "ENTER NUMBER OF PLATES" : N
70 LET C=C/10^12
80 LET D=D/10^3
90 LET N=N-1
100 LET A=C*D*(10^12)/(8.854*ER*N)
110 PRINT "AREA = " A*(10^4) "SQUARE CENTIMETRES"
```

Example 5 Find the number of plates in a capacitor having the following parameters: $C = 509.9905$ pF $= 509.9905 \times 10^{-12}$ F, $d = 0.001$ m, $\epsilon_r = 6$, $A = 1.2 \times 10^{-3}$ m^2.

$$n - 1 = \frac{Cd}{\epsilon_o \epsilon_r A}$$

$$= \frac{509.9905 \times 10^{-12} \times 0.001}{8.854 \times 10^{-12} \times 6 \times 1.2 \times 10^{-3}} = 8$$

Since $n - 1 = 8$, $n = 8 + 1 = 9$ plates

```
10 PRINT "PROG 71"
20 PRINT "NUMBER OF PLATES"
30 INPUT "ENTER CAPACITANCE IN PICOFARADS" ; C
40 INPUT "ENTER DISTANCE BETWEEN PLATES IN M" ; D
50 INPUT "ENTER RELATIVE PERMITTIVITY" ; ER
60 INPUT "ENTER AREA OF PLATES IN SQR M" ; A
70 LET C=C/10^12
80 LET N=C*D*(10^12)/(8.854*ER*A)
90 LET N=N+1
100 PRINT "NUMBER OF PLATES = " N
```

106 Circuit calculations pocket book

Example 6 A 0.01-μF capacitor has six plates. The area of each plate is 250 cm² and the distance between each one is 0.5 mm. Calculate the relative permittivity.

$$C = 0.01 \times 10^{-6} \text{ F}$$

$$d = \frac{0.5}{10^3} = 0.0005 \text{ m}$$

$$n - 1 = 5$$

$$A = \frac{250}{10^4} = 0.025 \text{ m}^2$$

$$\epsilon_r = \frac{Cd}{\epsilon_o(n-1)A}$$

$$= \frac{0.01 \times 10^{-6} \times 0.0005}{8.854 \times 10^{-12} \times 5 \times 0.025} = 4.5$$

```
10 PRINT "PROG 72"
20 PRINT "RELATIVE PERMITTIVITY"
30 INPUT "ENTER CAPACITANCE IN MICROFARADS" ; C
40 INPUT "ENTER DISTANCE BETWEEN PLATES IN MM" ; D
50 INPUT "ENTER NUMBER OF PLATES" ; N
60 INPUT "ENTER AREA OF PLATES IN SQR CM" ; A
70 LET C=C/10^6
80 LET D=D/10^3
90 LET N=N-1
100 LET A=A/10^4
110 LET ER=C*D*(10^12)/(8.854*N*A)
120 PRINT "RELATIVE PERMITTIVITY = " ER
```

(b) Charge on plates

A capacitor connected to a constant voltage supply accepts charge until the voltage across the plates equals the supply voltage. Capacitance is a measure of how much charge can be stored for each volt across the capacitor.

Charge $Q = CV$
where Q is in coulombs, C
       C is in farads,   F
       V is in volts,   V

Example 7 A 680 pF capacitor has 6 V across the plates. Calculate the charge stored.

$$Q = CV$$

$$= 680 \times 10^{-12} \times 6 = 4.08 \times 10^{-9} \text{ C}$$

$$= 4.08 \times 10^{-9} \times 10^9 \text{ nC} = 4.08 \text{ nC}$$

```
10 PRINT "PROG 73"
20 PRINT "CAPACITOR CHARGE"
30 INPUT "ENTER CAPACITANCE IN PICOFARADS" ; C
40 INPUT "ENTER VOLTAGE IN VOLTS" ; V
50 LET C=C/10^12
60 LET Q=C*V
70 PRINT "CHARGE = " Q*10^9 " NANOCOULOMBS"
```

Example 8 A 0.1 μF capacitor stores a charge of 0.02×10^{-3} C. What is the voltage across the terminals?

$$V = \frac{Q}{C} = \frac{0.02 \times 10^{-3}}{0.1 \times 10^{-6}} = \frac{0.02 \times 10^3}{0.1}$$
$$= 200 \text{ V}$$

```
10 PRINT "PROG 74"
20 PRINT "CAPACITOR VOLTAGE"
30 INPUT "ENTER CAPACITANCE IN MICROFARADS" ; C
40 INPUT "ENTER CHARGE IN COULOMBS" ; Q
50 LET C=C/10^6
60 LET V=Q/C
70 PRINT "VOLTAGE = " V "VOLTS"
```

Example 9 Calculate the value of a capacitor which stores a charge of 2 mC when charged to 400 V.

$$C = \frac{Q}{V} = \frac{2 \times 10^{-3}}{400} \text{ F} = \frac{2 \times 10^{-3} \times 10^6}{400} \text{ } \mu\text{F}$$
$$= 5 \text{ } \mu\text{F}$$

```
10 PRINT "PROG 75"
20 PRINT "CAPACITOR VALUE"
30 INPUT "ENTER CHARGE IN MILLICOULOMBS" ; Q
40 INPUT "ENTER VOLTAGE IN VOLTS" ; V
50 LET Q=Q/10^3
60 LET C=Q/V
70 PRINT "VALUE = " C*10^6 "MICROFARADS"
```

(c) Energy stored

When charge is moved into a capacitor work is done, and hence energy is expended. This energy is stored between the plates in the form of an electric field.

Energy stored, $W = \frac{1}{2}CV^2$

where W is in joules, J
C is in farads, F
V is in volts, V

Example 10 Calculate the energy stored in a 0.1 µF capacitor with an applied voltage of 250 V.

$$W = \frac{1}{2} \times 0.1 \times 10^{-6} \times (250)^2$$
$$= 3.125 \times 10^{-3} \text{ J}$$
$$= 3.125 \times 10^{-3} \times 10^3 = 3.125 \text{ mJ}$$

```
10 PRINT "PROG 76"
20 PRINT "ENERGY STORED"
30 INPUT "ENTER CAPACITANCE IN MICROFARADS" ; C
40 INPUT "ENTER VOLTAGE IN VOLTS" ; V
50 LET C=C/10^6
60 LET W=.5*C*V^2
70 PRINT "ENERGY STORED = " W*10^3 "MILLIJOULES"
```

Example 11 The energy stored in a 0.25 µF capacitor is 8×10^{-4} J. What is the terminal voltage?

$$V = \sqrt{\left(\frac{W}{0.5C}\right)}$$

$$= \sqrt{\left(\frac{8 \times 10^{-4}}{0.5 \times 0.25 \times 10^{-6}}\right)}$$
$$= 80 \text{ V}$$

```
10 PRINT "PROG 77"
20 PRINT "TERMINAL VOLTAGE"
30 INPUT "ENTER CAPACITANCE IN MICROFARADS" ; C
40 INPUT "ENTER ENERGY STORED IN JOULES" ; W
50 LET C=C/10^6
60 LET V=SQR(W/(.5*C))
70 PRINT "TERMINAL VOLTAGE = " V "VOLTS"
```

Example 12 Calculate the value of a capacitor that will store 10 J with an applied voltage of 100 V.

$$C = \frac{W}{0.5V^2}$$

$$= \frac{10}{0.5 \times (100)^2} = 2 \times 10^{-3} \text{ F}$$

$$= 2 \times 10^{-3} \times 10^{-6} \text{ μF} = 2000 \text{ μF}$$

```
10 PRINT "PROG 78"
20 PRINT "CAPACITOR VALUE"
30 INPUT "ENTER ENERGY STORED IN JOULES" ; W
40 INPUT "ENTER VOLTAGE IN VOLTS" ; V
50 LET C=W/(.5*V^2)
60 PRINT "CAPACITOR VALUE = " C*10^6 "MICROFARADS"
```

(d) Electric field strength

The electric field strength between two plates is the potential drop per unit length. It is the potential gradient between the plates.

$$\text{Electric field strength, } E = \frac{V}{d} V$$

where V = the voltage between plates
d = the distance between plates in metres

Example 13 Two plates are separated by a dielectric 0.5 mm thick. If a voltage of 250 V is maintained across the capacitor, calculate the electric field strength.

$$E = \frac{V}{d}$$

$$= \frac{250}{0.5 \times 10^{-3}} = \frac{250 \times 10^3}{0.5}$$

$$= 500 \times 10^3 \text{ V/m} = 500 \text{ kV/m}$$

```
10 PRINT "PROG 79"
20 PRINT "ELECTRIC FIELD STRENGTH"
30 INPUT "ENTER DIELECTRIC THICKNESS IN MM" ; D
40 INPUT "ENTER TERMINAL VOLTAGE IN VOLTS" ; V
50 D=D/10^3
60 LET E=V/D
70 PRINT "FIELD STRENGTH = " E "VOLTS/METRE"
```

Capacitors 109

Example 14 For a capacitor, given that the applied voltage is 100 V and E is 100 kV/m, calculate the spacing between the two plates.

$$d = \frac{V}{E}$$

$$= \frac{100}{100\,000} = 0.001 \text{ m}$$

$$= 0.001 \times 10^3 = 1 \text{ mm}$$

```
10 PRINT "PROG 80"
20 PRINT "DIELECTRIC THICKNESS"
30 INPUT "ENTER TERMINAL VOLTAGE IN VOLTS" ; V
40 INPUT "ENTER FIELD STRENGTH IN VOLTS/METRE" ; E
50 LET D=V/E
60 PRINT "DIELECTRIC THICKNESS = " D*10^3 "MILLIMETRES"
```

Example 15 A capacitor has a dielectric 0.25 mm thick, and a field strength of 250 kV/m. Calculate the applied voltage.

$$V = dE$$

$$= 0.25 \times 10^{-3} \times 250 \times 10^3$$

$$= 62.5 \text{ V}$$

```
10 PRINT "PROG 81"
20 PRINT "TERMINAL VOLTAGE"
30 INPUT "ENTER DIELECTRIC THICKNESS IN MM" ; D
40 INPUT "ENTER FIELD STRENGTH IN VOLTS/METRE" ; E
50 LET D=D/10^3
60 LET V=D*E
70 PRINT "TERMINAL VOLTAGE = " V "VOLTS"
```

(e) Electric flux density
One line of electric flux is assumed to emanate from a positive charge of one coulomb. If the plates carry Q coulombs then the flux density, D, between the plates will be given by

$$D = \frac{Q}{A} \text{ C/m}^2$$

where Q is in coulombs, C
A is in square metres

Example 16 A capacitor consists of two metal plates 40×80 mm and holds a charge of 0.1 μC. Calculate the electric flux density.

$$D = \frac{0.1 \times 10^{-6}}{40 \times 10^{-3} \times 80 \times 10^{-3}}$$

$$= \frac{0.1 \times 10^3 \times 10^3}{40 \times 80 \times 10^6}$$

$$= 3.125 \times 10^{-5} \text{ C/m}^2$$

$$= 3.125 \times 10^{-5} \times 10^6 \text{ μC/m}^2 = 31.25 \text{ μC/m}^2$$

```
10 PRINT "PROG 82"
20 PRINT "ELECTRIC FLUX DENSITY"
30 INPUT "ENTER CHARGE IN MICROCOULOMBS" ; Q
```

```
40 INPUT "ENTER PLATE LENGTH IN MM" ; X
50 INPUT "ENTER PLATE WIDTH IN MM" ; Y
60 LET Q=Q/10^6
70 LET A=X*Y/(10^6)
80 LET D=Q/A
90 PRINT "FLUX DENSITY = " D*10^6 "MICROCOULOMBS/SQRM"
```

Example 17 Given the following parameters for a two-plate capacitor, calculate the area of a plate.

$$D = 100 \ \mu C/m^2$$

$$Q = 0.2 \ \mu C$$

$$A = \frac{Q}{D} = \frac{0.2 \times 10^{-6} \ C}{100 \times 10^{-6} \ C/m^2}$$

$$= 2 \times 10^{-3} \ m^2$$

$$= 2 \times 10^{-3} \times 10^6 \ mm^2 = 2000 \ mm^2$$

```
10 PRINT "PROG 83"
20 PRINT "PLATE AREA"
30 INPUT "ENTER FLUX DENSITY IN MICROCOULOMBS/SQRM" ; D
40 INPUT "ENTER CHARGE IN MICROCOULOMBS" ; Q
50 LET D=D/10^6
60 LET Q=Q/10^6
70 LET A=Q/D
80 PRINT "PLATE AREA = " A*10^6 "SQRMM"
```

Example 18 A capacitor has two circular plates with diameters of 4 mm. If the flux density is 20 $\mu C/m^2$, calculate the charge.

$$Q = DA$$

$$\text{where } A = \pi \left(\frac{d}{2}\right)^2 = \pi \left(\frac{4}{2}\right)^2 = 12.567 \ mm^2$$

$$= 12.567 \times 10^{-6} \ m^2$$

$$\text{Then } Q = 20 \times 10^{-6} \times 12.567 \times 10^{-6} \ C$$

$$= 2.5134 \times 10^{-10} \ C$$

$$= 2.5134 \times 10^{-10} \times 10^6 \ \mu C$$

$$= 2.5134 \times 10^{-4} \ \mu C$$

$$= 251 \ pC$$

```
10 PRINT "PROG 84"
20 PRINT "CHARGE"
30 INPUT "ENTER FLUX DENSITY IN MICROCOULOMBS/SQRM" ; D
40 INPUT "ENTER PLATE RADIUS IN MM" ; R
50 LET D=D/10^6
60 LET A=3.141593*(R^2)/(10^6)
70 LET Q=D*A
80 PRINT "CHARGE = " Q*10^6 "MICROCOULOMBS"
```

7.4 Capacitor circuits

(a) Series

Example 19 Figure 7.2 shows two capacitors C_1 and C_2, connected in series. The method of finding the combined capacitance C is

Capacitors 111

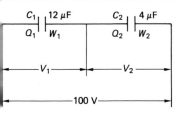

Figure 7.2

identical to the method of finding the combined value of two resistors connected in parallel, i.e.

$$C = \frac{C_1 C_2}{C_1 + C_2} = \frac{12 \times 4}{12 + 4} = 3 \ \mu F$$

The voltage drop across each capacitor is

$$V_1 = \frac{VC_2}{C_1 + C_2} = \frac{100 \times 4}{16} = 25 \ V$$

and

$$V_2 = \frac{VC_1}{C_1 + C_2} = \frac{100 \times 12}{16} = 75 \ V$$

Since the same charging current flows through both capacitors, each capacitor will carry the same charge.

$$Q_1 = C_1 V_1 = 12 \times 10^{-6} \times 25 = 300 \times 10^{-6} \ C$$
$$= 300 \times 10^{-6} \times 10^6 \ \mu C = 300 \ \mu C$$
$$Q_2 = C_2 V_2 = 4 \times 10^{-6} \times 75 = 300 \times 10^{-6} \ C$$
$$= 300 \times 10^{-6} \times 10^6 \ \mu C = 300 \ \mu C$$

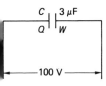

Figure 7.3

Figure 7.3 is equivalent to Figure 7.2 and

$$Q = CV = 3 \times 10^{-6} \times 100 = 300 \times 10^{-6} \ C$$
$$= 300 \times 10^{-6} \times 10^6 \ \mu C = 300 \ \mu C$$

i.e.

$$Q = Q_1 = Q_2$$

The energy stored in C_1 in joules, J, is

$$W_1 = \tfrac{1}{2} C_1 (V_1)^2 = \tfrac{1}{2} \times 12 \times 10^{-6} \times (25)^2$$

$$= 3750 \times 10^{-6} \text{ J} = 3750 \times 10^{-6} \times 10^{6} \text{ μJ}$$
$$= 3750 \text{ μJ} = 3.75 \text{ mJ}$$

The energy stored in C_2 in joules, J, is

$$W_2 = \tfrac{1}{2}C_2(V_2)^2 = \tfrac{1}{2} \times 4 \times 10^{-6} \times (75)^2$$
$$= 11\,250 \times 10^{-6} \text{ J} = 11\,250 \times 10^{-6} \times 10^{6} \text{ μJ}$$
$$= 11\,250 \text{ μJ} = 11.25 \text{ mJ}$$

The energy stored in C, Figure 7.3, in joules, J, is

$$W = \tfrac{1}{2}C(V)^2 = \tfrac{1}{2} \times 3 \times 10^{-6} \times (100)^2$$
$$= 15\,000 \times 10^{-6} \text{ J} = 15\,000 \times 10^{-6} \times 10^{-6} \text{ μJ}$$
$$= 15\,000 \text{ μJ} = 15 \text{ mJ}$$

Note $W = W_1 + W_2$.

There is an alternative method of calculating V_1 and V_2. First find the value of Q, and since $Q = Q_1 = Q_2$ for a series circuit, then

$$V_1 = \frac{Q}{C_1} = \frac{300 \times 10^{-6}}{12 \times 10^{-6}} = 25 \text{ V}$$

$$V_2 = \frac{Q}{C_2} = \frac{300 \times 10^{-6}}{4 \times 10^{-6}} = 75 \text{ V}$$

```
10 PRINT "PROG 85"
20 PRINT "SERIES CIRCUIT"
30 INPUT "ENTER C1 IN MICROFARADS" ; C1
40 INPUT "ENTER C2 IN MICROFARADS" ; C2
50 INPUT "ENTER INPUT VOLTAGE IN VOLTS" ; V
60 LET C=C1*C2/(C1+C2)
70 PRINT "C = " C "UF"
80 LET V1=V*C2/(C1+C2)
90 PRINT "V1 = " V1 "V"
100 LET V2=V*C1/(C1+C2)
110 PRINT "V2 = " V2 "V"
120 LET Q1=C1*V1*10^6/(10^6)
130 PRINT "Q1 = " Q1 "UC"
140 LET Q2=C2*V2*10^6/10^6
150 PRINT "Q2 = " Q2 "UC"
160 LET Q=C*V*10^6/(10^6)
170 PRINT "Q = " Q "UC"
180 LET W1=.5*C1*(V1^2*10^6)/(10^6)
190 PRINT "W1 = " W1 "UJ"
200 LET W2=.5*C2*(V2^2*10^6)/(10^6)
210 PRINT "W2 = " W2 "UJ"
220 LET W=.5*C*(V^2*10^6)/(10^6)
230 PRINT "W = " W "UJ"
```

Example 20 Figure 7.4 shows three capacitors in series. The combined capacitance is calculated using the same method as for three resistors in parallel.

$$\frac{1}{C} = \frac{1}{20} + \frac{1}{20} + \frac{1}{40} = 0.125$$

Then

$$C = \frac{1}{0.125} = 8 \text{ μF}$$

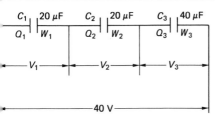

Figure 7.4

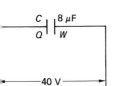

Figure 7.5

The equivalent circuit is shown in Figure 7.5.

$$Q = CV = 8 \times 10^{-6} \times 40 \text{ C}$$
$$= 8 \times 10^{-6} \times 10^6 \text{ μC} = 320 \text{ μC}$$

Since in a series circuit $Q = Q_1 = Q_2 = Q_3$, then

$$V_1 = \frac{Q_1}{C_1} = \frac{320 \text{ μC}}{20 \text{ μF}} = 16 \text{ V}$$

$$V_2 = \frac{Q_2}{C_2} = \frac{320 \text{ μC}}{20 \text{ μF}} = 16 \text{ V}$$

$$V_3 = \frac{Q_3}{C_3} = \frac{320 \text{ μC}}{40 \text{ μF}} = 8 \text{ V}$$

Energy stored in C_1

$$W_1 = \tfrac{1}{2} C_1 (V_1)^2 = \tfrac{1}{2} \times 20 \times 10^{-6} \times (16)^2 \text{ J}$$
$$= 2560 \times 10^{-6} \times 10^6 \text{ μJ} = 2560 \text{ μJ} = 2.56 \text{ mJ}$$

Energy stored in C_2 = energy stored in C_1 (since $C_2 = C_1$)

$$W_2 = 2560 \text{ μJ} = 2.56 \text{ mJ}$$

Energy stored in C_3

$$W_3 = \tfrac{1}{2} C_3 (V_3)^2 = \tfrac{1}{2} \times 40 \times 10^{-6} \times (8)^2 \text{ J}$$
$$= 1280 \times 10^{-6} \times 10^{-6} \text{ μJ} = 1280 \text{ μJ} = 1.28 \text{ mJ}$$

Energy stored in C, the equivalent capacitance, is

$$W = \tfrac{1}{2} C(V)^2 = \tfrac{1}{2} \times 8 \times 10^{-6} \times (40)^2 \text{ J}$$
$$= 6400 \times 10^{-6} \times 10^6 \text{ μJ} = 6400 \text{ μJ} = 6.4 \text{ mJ}$$

Note $W = W_1 + W_2 + W_3$.

```
10 PRINT "PROG 86"
20 PRINT "SERIES CIRCUIT"
30 INPUT "ENTER C1 IN MICROFARADS" ; C1
40 INPUT "ENTER C2 IN MICROFARADS" ; C2
50 INPUT "ENTER C3 IN MICROFARADS" ; C3
60 INPUT "ENTER INPUT VOLTAGE IN VOLTS" ; V
70 LET X=1/C1+1/C2+1/C3
80 LET C=1/X
90 PRINT "C = " C "UF"
100 LET Q=C*V*10^6/10^6
110 PRINT "Q = " Q "UC"
120 LET V1=Q/C1
130 PRINT "V1 = " V1 "V"
140 LET V2=Q/C2
150 PRINT "V2 = " V2 "V"
160 LET V3=Q/C3
170 PRINT "V3 = " V3 "V"
180 LET W1=.5*C1*(V1^2*10^6)/(10^6)
190 PRINT "W1 = " W1 "UJ"
200 LET W2=.5*C2*(V2^2*10^6)/(10^6)
210 PRINT "W2 = " W2 "UJ"
220 LET W3=.5*C3*(V3^2*10^6)/(10^6)
230 PRINT "W3 = " W3 "UJ"
240 LET W=.5*C*(V^2*10^6)/(10^6)
250 PRINT "W = " W "UJ"
```

(b) Parallel

Example 21 Figure 7.6 shows two capacitors C_1 and C_2 connected in parallel. The method of finding the combined capacitance C identical to the method of finding the combined value of two resistors in series, i.e.

$$C = C_1 + C_2 = 8 + 4 = 12 \; \mu F$$

Total charge stored, Q, is

$$Q = CV = 12 \times 10^{-6} \times 10 = 120 \times 10^{-6} \; C$$
$$= 120 \times 10^{-6} \times 10^6 \; \mu C = 120 \; \mu C$$
$$Q_1 = C_1 V = 8 \times 10^{-6} \times 10 = 80 \times 10^{-6} \; C$$
$$= 80 \times 10^{-6} \times 10^6 \; \mu C = 80 \; \mu C$$
$$Q_2 = C_2 V = 4 \times 10^{-6} \times 10 = 40 \times 10^{-6} \; C$$
$$= 40 \times 10^{-6} \times 10^6 \; \mu C = 40 \; \mu C$$

Note $Q = Q_1 + Q_2$.

$$W = \text{total energy stored} = \tfrac{1}{2} C(V)^2$$
$$= \tfrac{1}{2} \times 12 \times 10^{-6} \times (10)^2 = 600 \times 10^{-6} \; J$$

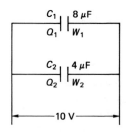

Figure 7.6

$$= 600 \times 10^{-6} \times 10^6 \, \mu J = 600 \, \mu J$$

W_1 = energy stored in $C_1 = \frac{1}{2}C_1(V)^2$

$$= \tfrac{1}{2} \times 8 \times 10^{-6} \times (100)^2 = 400 \times 10^{-6} \, J$$

$$= 400 \times 10^{-6} \times 10^6 \, \mu J = 400 \, \mu J$$

W_2 = energy stored in $C_2 = \frac{1}{2}C_2(V)^2$

$$= \tfrac{1}{2} \times 4 \times 10^{-6} \times (100)^2 = 200 \times 10^{-6} \, J$$

$$= 200 \times 10^{-6} \times 10^6 \, \mu J = 200 \, \mu J$$

Note $W = W_1 + W_2$.

```
10 PRINT "PROG 87"
20 PRINT "PARALLEL CIRCUIT"
30 INPUT "ENTER C1 IN MICROFARADS" ; C1
40 INPUT "ENTER C2 IN MICROFARADS" ; C2
50 INPUT "ENTER VOLTAGE IN VOLTS" ; V
60 LET C=C1+C2
70 PRINT "C = " C "UF"
80 LET Q=C*V*10^6/10^6
90 PRINT "Q = " Q "UC"
100 LET Q1=C1*V*10^6/10^6
110 PRINT "Q1 = " Q1 "UC"
120 LET Q2=C2*V*10^6/10^6
130 PRINT "Q2 = " Q2 "UC"
140 LET W=.5*C*(V^2*10^6)/(10^6)
150 PRINT "W = " W "UJ"
160 LET W1=.5*C1*(V^2*10^6)/(10^6)
170 PRINT "W1 = " W1 "UJ"
180 LET W2=.5*C2*(V^2*10^6)/(10^6)
190 PRINT "W2 = " W2 "UJ"
```

Example 22 Figure 7.7 shows three capacitors connected in parallel. The method of finding the combined capacitance C is identical to the method of finding the combined value of three resistors in series, i.e.

$$C = C_1 + C_2 + C_3 = 12 \, \mu F$$

Total charge stored

$$Q = CV$$
$$= 12 \times 10^{-6} \times 25 \times 10^6 = 300 \, \mu C$$

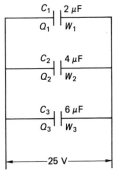

Figure 7.7

$Q_1 = 2 \times 10^{-6} \times 25 \times 10^6 = 50 \ \mu C$

$Q_2 = 4 \times 10^{-6} \times 25 \times 10^6 = 100 \ \mu C$

$Q_3 = 6 \times 10^{-6} \times 25 \times 10^6 = 150 \ \mu C$

Note $Q = Q_1 + Q_2 + Q_3$ since $I = I_1 + I_2 + I_3$.

W = total energy stored

$= \tfrac{1}{2} C (V)^2$

$= \tfrac{1}{2} \times 12 \times 10^{-6} \times (25)^2 \times 10^6 = 3750 \ \mu J = 3.75 \ mJ$

$W_1 = \tfrac{1}{2} \times 2 \times 10^{-6} \times (25)^2 \times 10^6 = 625 \ \mu J = 0.625 \ mJ$

$W_2 = \tfrac{1}{2} \times 4 \times 10^{-6} \times (25)^2 \times 10^6 = 1250 \ \mu J = 1.25 \ mJ$

$W_3 = \tfrac{1}{2} \times 6 \times 10^{-6} \times (25)^2 \times 10^6 = 1875 \ \mu J = 1.875 \ mJ$

Note $W = W_1 + W_2 + W_3$.

```
10 PRINT "PROG 88"
20 PRINT "PARALLEL CIRCUIT"
30 INPUT "ENTER C1 IN MICROFARADS" ; C1
40 INPUT "ENTER C2 IN MICROFARADS" ; C2
50 INPUT "ENTER C3 IN MICROFARADS" ; C3
60 INPUT "ENTER VOLTAGE IN VOLTS" ; V
70 LET C=C1+C2+C3
80 PRINT "C = " C "UF"
90 LET Q=C*V
100 PRINT "Q = " Q "UC"
110 LET Q1=C1*V
120 PRINT "Q1 = " Q1 "UC"
130 LET Q2=C2*V
140 PRINT "Q2 = " Q2 "UC"
150 LET Q3=C3*V
160 PRINT "Q3 = " Q3 "UC"
170 LET W=.5*C*(V^2)
180 PRINT "W = " W "UJ"
190 LET W1=.5*C1*(V^2)
200 PRINT "W1 = " W1 "UJ"
210 LET W2=.5*C2*(V^2)
220 PRINT "W2 = " W2 "UJ"
230 LET W3=.5*C3*(V^2)
240 PRINT "W3 = " W3 "UJ"
```

(c) Series–parallel

Example 23 Figure 7.8 shows a series–parallel arrangement. C_2 and C_3 can be combined to give C_5 where

$C_5 = 4 + 2 = 6 \ \mu F$

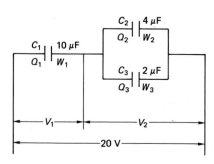

Figure 7.8

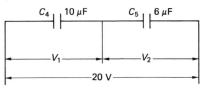

Figure 7.9

This gives the equivalent circuit shown in Figure 7.9.

$$C = \frac{10 \times 6}{10 + 6} = 3.75 \ \mu F \text{ (shown in Figure 7.10)}$$

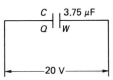

Figure 7.10

The total charge in the circuit, therefore, is

$$Q = CV = 3.7 \ \mu F \times 20 = 75 \ \mu C$$

With reference to Figure 7.9 the voltage drops are V_1 and V_2.

$$V_1 = \frac{20 \times 6}{10 + 6} = 7.5 \text{ and } V_2 = \frac{20 \times 10}{10 + 6} = 12.5$$

$$Q_1 = C_1 V_1 = 10 \ \mu F \times 7.5 = 75 \ \mu C$$

$$Q_2 = C_2 V_2 = 4 \ \mu F \times 12.5 = 50 \ \mu C$$

$$Q_3 = C_3 V_2 = 2 \ \mu F \times 12.5 = 25 \ \mu C$$

Note that $Q = Q_1$ since total current I flows through both, and

$$Q_2 + Q_3 = Q$$

The total energy stored in the circuit, W, is

$$W = \tfrac{1}{2} C(V)^2 = \tfrac{1}{2} \times 3.75 \ \mu F \times (20)^2 = 750 \ \mu J$$

Energy stored in C_1

$$W_1 = \tfrac{1}{2} \times 10 \ \mu F \times (7.5)^2 = 281.25 \ \mu J$$

Energy stored in C_2

$$W_2 = \tfrac{1}{2} \times 4 \ \mu F \times (12.5)^2 = 312.5 \ \mu J$$

Energy stored in C_3

$$W_3 = \tfrac{1}{2} \times 2 \ \mu F \times (12.5)^2 = 156.25 \ \mu J$$

Note the total energy = energy stored in C_1 + energy stored in C_2 + energy stored in C_3.

```
10 PRINT "PROG 89"
20 PRINT "SERIES-PARALLEL"
30 INPUT "ENTER C1 IN MICROFARADS" ; C1
40 INPUT "ENTER C2 IN MICROFARADS" ; C2
50 INPUT "ENTER C3 IN MICROFARADS" ; C3
60 INPUT "ENTER VOLTAGE IN VOLTS" : V
70 LET C5=C2+C3
80 PRINT C5 "UF"
90 LET C=(C1*C5)/(C1+C5)
100 PRINT C "UF"
110 LET V1=V*C5/(C1+C5)
120 PRINT V1 "V"
130 LET V2=V*C1/(C1+C5)
140 PRINT V2 "V"
150 LET Q=C*V
160 PRINT Q "UC"
170 LET Q1=C1*V1
180 PRINT Q1 "UC"
190 LET Q2=C2*V2
200 PRINT Q2 "UC"
210 LET Q3=C3*V2
220 PRINT Q3 "UC"
230 LET W=.5*C*V^2
240 PRINT W "UJ"
250 LET W1=.5*C1*V1^2
260 PRINT W1 "UJ"
270 LET W2=.5*C2*V2^2
280 PRINT W2 "UJ"
290 LET W3=.5*C3*V2^2
300 PRINT W3 "UJ"
```

Example 24 The circuit given in Figure 7.11 can be simplified to that shown in Figure 7.12 by first combining C_1 and C_2 to give C_5 where

$$C_5 = 6 + 4 = 10 \ \mu F$$

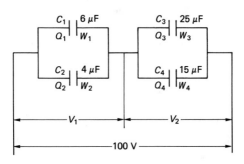

Figure 7.11

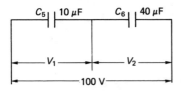

Figure 7.12

and then combining C_3 and C_4 to give C_6 where

$$C_6 = 25 + 15 = 40 \ \mu F$$

The circuit in Figure 7.12 can be simplified further to give the equivalent circuit capacitance C, shown in Figure 7.13 where

$$C = \frac{10 \times 40}{10 + 40} = 8 \ \mu F$$

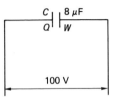

Figure 7.13

Referring to Figure 7.12 the voltage drops are

$$V_1 = \frac{100 \times 40}{10 + 40} = 80 \text{ V} \quad \text{and} \quad V_2 = \frac{100 \times 10}{10 + 40} = 20 \text{ V}$$

$$Q = CV = 8 \ \mu F \times 100 = 800 \ \mu C$$
$$Q_1 = C_1 V_1 = 6 \ \mu F \times 80 = 480 \ \mu C$$
$$Q_2 = C_2 V_1 = 4 \ \mu F \times 80 = 320 \ \mu C$$
$$Q_3 = C_3 V_2 = 25 \ \mu F \times 20 = 500 \ \mu C$$
$$Q_4 = C_4 V_2 = 15 \ \mu F \times 20 = 300 \ \mu C$$

Note that $Q_1 + Q_2 = Q$ and $Q_3 + Q_4 = Q$.

The total energy stored in the circuit equals the energy stored in C, the equivalent capacitance

$$W = \tfrac{1}{2} C(V)^2 = \tfrac{1}{2} \times 8 \ \mu F \times (100)^2 = 40\ 000 \ \mu J = 40 \text{ mJ}$$

Energy stored in C_1

$$W_1 = \tfrac{1}{2} C_1 (V_1)^2 = \tfrac{1}{2} \times 6 \ \mu F \times (80)^2 = 19\ 200 \ \mu J = 19.2 \text{ mJ}$$

Energy stored in C_2

$$W_2 = \tfrac{1}{2} C_2 (V_1)^2 = \tfrac{1}{2} \times 4 \ \mu F \times (80)^2 = 12\ 800 \ \mu J = 12.8 \text{ mJ}$$

Energy stored in C_3

$$W_3 = \tfrac{1}{2} C_3 (V_2)^2 = \tfrac{1}{2} \times 25 \ \mu F \times (20)^2 = 5000 \ \mu J = 5 \text{ mJ}$$

Energy stored in C_4

$$W_4 = \tfrac{1}{2} C_4 (V_2)^2 = \tfrac{1}{2} \times 15 \ \mu F \times (20)^2 = 3000 \ \mu J = 3 \text{ mJ}$$

Note total energy = energy in C_1 + energy in C_2 + energy in C_3 + energy in C_4.

```
10 PRINT "PROG 90"
20 PRINT "SERIES-PARALLEL"
30 INPUT "ENTER C1 IN MICROFARADS" ; C1
40 INPUT "ENTER C2 IN MICROFARADS" ; C2
```

```
50 INPUT "ENTER C3 IN MICROFARADS" ; C3
60 INPUT "ENTER C4 IN MICROFARADS" ; C4
70 INPUT "ENTER VOLTAGE IN VOLTS" ; V
80 LET C5=C1+C2
90 PRINT "C5 = " C5 "UF"
100 LET C6=C3+C4
110 PRINT "C6 = " C6 "UF"
120 LET C=(C5*C6)/(C5+C6)
130 PRINT "C = " C "UF"
140 LET V1=V*C6/(C5+C6)
150 PRINT "V1 = " V1 "V"
160 LET V2=V*C5/(C5+C6)
170 PRINT "V2 = " V2 "V"
180 LET Q=C*V
190 PRINT "Q = " Q "UC"
200 LET Q1=C1*V1
210 PRINT "Q1 = " Q1 "UC"
220 LET Q2=C2*V1
230 PRINT "Q2 = " Q2 "UC"
240 LET Q3=C3*V2
250 PRINT "Q3 = " Q3 "UC"
260 LET Q4=C4*V2
270 PRINT "Q4 = " Q4 "UC"
280 LET W=.5*C*V^2
290 PRINT "W = " W "UJ"
300 LET W1=.5*C1*V1^2
310 PRINT "W1 = " W1 "UJ"
320 LET W2=.5*C2*V1^2
330 PRINT "W2 = " W2 "UJ"
340 LET W3=.5*C3*V2^2
350 PRINT "W3 = " W3 "UJ"
360 LET W4=.5*C4*V2^2
370 PRINT "W4 = " W4 "UJ"
```

Example 25 To calculate the parameters of the circuit in Figure 7.14, first combine C_2 and C_3 to give C_5, and the equivalent circuit of Figure 7.15.

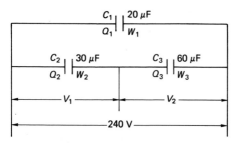

Figure 7.14

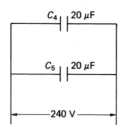

Figure 7.15

$$C_5 = \frac{30 \times 60}{30 + 60} = 20 \ \mu F$$

Figure 7.15 can be simplified further to give the equivalent circuit capacitance C, shown in Figure 7.16 where

$$C = 20 + 20 = 40 \ \mu F$$

$$V_1 = \frac{60 \times 240}{30 + 60} = 160 \ V \text{ and } V_2 = \frac{30 \times 240}{30 + 60} = 80 \ V$$

$$Q = CV = 40 \ \mu F \times 240 = 9600 \ \mu C = 9.6 \ mC$$

$$Q_1 = C_1 V = 20 \ \mu F \times 240 = 4800 \ \mu C = 4.8 \ mC$$

$$Q_2 = C_2 V_1 = 30 \ \mu F \times 160 = 4800 \ \mu C = 4.8 \ mC$$

$$Q_3 = C_3 V_2 = 60 \ \mu F \times \ 80 = 4800 \ \mu C = 4.8 \ mC$$

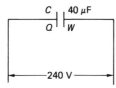

Figure 7.16

Note that since

$$C_4 = C_5$$

then

$$Q_1 = Q_2 = Q_3$$

and

$$Q_1, Q_2 \text{ and } Q_3 = \frac{Q}{2}$$

$$W = \tfrac{1}{2} C (V)^2 = \tfrac{1}{2} \times 40 \ \mu F \times (240)^2 = 1 \ 152 \ 000 \ \mu J = 1.152 \ J$$

$$W_1 = \tfrac{1}{2} C_1 (V)^2 = \tfrac{1}{2} \times 20 \ \mu F \times (240)^2 = 576 \ 000 \ \mu J = 0.576 \ J$$

$$W_2 = \tfrac{1}{2} C_2 (V_1)^2 = \tfrac{1}{2} \times 30 \ \mu F \times (160)^2 = 384 \ 000 \ \mu J = 0.384 \ J$$

$$W_3 = \tfrac{1}{2} C_3 (V_2)^2 = \tfrac{1}{2} \times 60 \ \mu F \times \ (80)^2 = 192 \ 000 \ \mu J = 0.192 \ J$$

Note $W_1 + W_2 + W_3 = W$.

```
10 PRINT "PROG 91"
20 PRINT "SERIES-PARALLEL"
30 INPUT "ENTER C1 IN MICROFARADS" ; C1
40 INPUT "ENTER C2 IN MICROFARADS" ; C2
50 INPUT "ENTER C3 IN MICROFARADS" ; C3
60 INPUT "ENTER VOLTAGE IN VOLTS " ; V
70 LET C5=C2*C3/(C2+C3)
80 PRINT "C5 = " C5 "UF"
90 LET C=C1+C5
100 PRINT "C = " C "UF"
110 LET V1=C3*V/(C2+C3)
120 PRINT "V1 = " V1 "V"
130 LET V2=C2*V/(C2+C3)
```

```
140 PRINT "V2 = " V2 "V"
150 LET Q=C*V
160 PRINT "Q = " Q "UC"
170 LET Q1=C1*V
180 PRINT "Q1 = " Q1 "UC"
190 LET Q2=C2*V1
200 PRINT "Q2 = " Q2 "UC"
210 LET Q3=C3*V2
220 PRINT "Q3 = " Q3 "UC"
230 LET W=.5*C*V^2
240 PRINT "W = " W "UJ"
250 LET W1=.5*C1*V^2
260 PRINT "W1 = " W1 "UJ"
270 LET W2=.5*C2*V1^2
280 PRINT "W2 = " W2 "UJ"
290 LET W3=.5*C3*V2^2
300 PRINT "W3 = " W3 "UJ"
```

7.5 Reactance

The behaviour of capacitors in a.c. circuits will be dealt with in Chapter 11. The relationship between the applied voltage and the current flowing resembles Ohm's law. The opposition to the current is called the capacitive reactance X_C, and the unit symbol is the ohm, Ω

$$X_C = \frac{1}{2\pi f C} \text{ (for a sine wave only)}$$

where C is in farads, F
f is in hertz, Hz
X_C is in ohms, Ω

Example 26 Calculate the reactance of a 10 µF capacitor at a frequency of 50 Hz.

$$X_C = \frac{1}{2 \times \pi \times 50 \times 10 \times 10^{-6}} = \frac{10^6}{2 \times \pi \times 50 \times 10}$$
$$= 318.3 \; \Omega$$

```
10 PRINT "PROG 92"
20 PRINT "CAPACITOR REACTANCE"
30 INPUT "ENTER CAPACITANCE IN MICROFARADS" ; C
40 INPUT "ENTER FREQUENCY" ; F
50 LET XC=10^6/(2*3.141593*F*C)
60 PRINT "XC = " XC "OHMS"
```

Example 27 A capacitor has a reactance of 100 Ω at a frequency of 1 kHz. What is the value?

$$C = \frac{1}{2 \times \pi \times f \times X_C} = \frac{1}{2 \times \pi \times 1000 \times 100} \text{ F}$$
$$= \frac{1 \times 10^6}{2 \times \pi \times 1000 \times 100} \text{ µF} = 1.59 \text{ µF}$$

```
10 PRINT "PROG 93"
20 PRINT "CAPACITOR VALUE"
30 INPUT "ENTER REACTANCE" ; XC
40 INPUT "ENTER FREQUENCY" ; F
50 LET C=10^6/(2*3.141593*F*XC)
60 PRINT "C = " C "UF"
```

Capacitors 123

Example 28 At what frequency is the reactance of a 10-pF capacitor equal to 1 MΩ?

$$f = \frac{1}{2\pi C X_C} = \frac{1}{2 \times \pi \times 10 \times 10^{-12} \times 10^6}$$

$$= \frac{1 \times 10^{12}}{2 \times \pi \times 10 \times 10^6} = 15\,915 \text{ Hz}$$

```
10 PRINT "PROG 94"
20 PRINT "FREQUENCY"
30 INPUT "ENTER CAPACITANCE IN PICOFARADS" ; C
40 INPUT "ENTER REACTANCE IN OHMS" ; XC
50 LET F=10^12/(2*3.141593*C*XC)
60 PRINT "F = " F "HERTZ"
```

Problems

1. Convert 0.001 microfarads to farads, nanofarads and picofarads.
2. A capacitor has four plates, separated by mica 0.5 mm thick. The relative permittivity is 8 and the area of each plate is 200 cm². Calculate the capacitance in microfarads.
3. Find the distance between the plates of a five-plate capacitor, if the dielectric has a relative permittivity of 6. The area of each plate is 500 mm² and the capacitance is 0.25 μF.
4. A 470 pF capacitor has six plates. The plate separation distance is 2 mm and the relative permittivity is 6. Calculate the area of the plates in square centimetres.
5. A 150 pF capacitor has 120 V across the plates. Calculate the charge stored.
6. If a 12-μF capacitor stores 0.02 C, what is the voltage across the plates?
7. Calculate the value of a capacitor which stores 20 mC with an applied voltage of 50 V.
8. Calculate the energy stored in a 2.2 μF capacitor if the applied voltage is 200 V.
9. A 25-μF capacitor stores 0.001 J. What is the terminal voltage?
10. Calculate the value of a capacitor that will store 16 J with an applied voltage of 240 V.
11. In a two-plate capacitor the dielectric is 0.2 mm thick. If a voltage of 150 V is applied, calculate the field strength.
12. If the field strength in a capacitor is 120 000 V/m with a terminal voltage of 90 V, calculate the thickness of the dielectric.
13. If a capacitor has a dielectric 1 mm thick, and a field strength of 130 000 V/m, calculate the terminal voltage.
14. A capacitor has plates of area 25 × 50 mm. If it holds a charge of 0.5 μC, calculate the flux density.
15. If a capacitor has a flux density of 60 μC/m² and a charge of 0.3 μC, calculate the area of the plate in square millimetres.
16. A capacitor has two circular plates with diameters of 8 mm. If the flux density is 18 μC/m², calculate the charge.
17. A circuit consists of a 3-μF capacitor in series with a 5-μF capacitor. If the applied voltage is 25 V calculate: (a) the combined capacitance, (b) the voltage drop across each one, (c) the charge in each one, (d) the energy stored in each one.
18. Three capacitors of 100, 400 and 80 μF are connected in series. If the applied voltage is 20 V, calculate: (a) the energy stored in

each capacitor, (b) the voltage drops, (c) the energy stored in each one.

19   Two capacitors C_1 and C_2 are connected in parallel. C_1 is 24 µF and C_2 is 8 µF, and the applied voltage is 6 V. Calculate: (a) the combined capacitance, (b) the charge stored in C_1 and C_2, (c) the energy stored in C_1 and C_2.

20   Three capacitors C_1, C_2 and C_3 are connected in parallel. The values are 2.5, 3.5 and 4 µF. If 12 V is applied to the combination, calculate: (a) the combined capacitance, (b) the charge stored in C_1, C_2 and C_3, (c) the energy stored in C_1, C_2 and C_3.

21   If C_1, C_2 and C_3 in Figure 7.8 have values of 10, 5 and 5 µF, and the applied voltage is 10 V, calculate: (a) V_1 and V_2, (b) the charge stored in C_1, C_2 and C_3, (c) the energy stored in C_1, C_2 and C_3.

22   C_1, C_2, C_3 and C_4 in Figure 7.11 have values of 3, 7, 8 and 7 µF. The applied voltage is 25 V. Calculate: (a) V_1 and V_2, (b) Q_1, Q_2, Q_3 and Q_4, (c) W_1, W_2, W_3 and W_4.

23   If C_1, C_2 and C_3 in Figure 7.14 have values of 14, 10 and 15 µF and the applied voltage is 200 V calculate: (a) V_1 and V_2, (b) Q_1, Q_2 and Q_3, (c) W_1, W_2 and W_3.

24   Calculate the reactance of a 16 µF capacitor at 50 Hz.

25   A capacitor has a reactance of 20 Ω at a frequency of 750 Hz. Calculate the value in microfarads.

26   At what frequency is the reactance of a 68-pF capacitor equal to 470 kΩ?

8 Inductors

8.1 Introduction

Inductors consist of coils of insulated wire wound on bobbins or formers.

Inductance is the property of the coil which tends to prevent a change of current in a circuit. When current is flowing in a coil of wire, magnetic flux is produced which cuts the coil, and when this current is changing a voltage is induced in the coil. The induced voltage exactly balances the applied voltage which is producing the current.

Inductance, L, is measured in henrys, which have unit symbol H, and a closed circuit has an inductance of one henry if an e.m.f. of one volt is induced when the current is varying at one ampere per second.

8.2 Units

Inductance is expressed in henrys, and the subunits and the relationships between them are:

The millihenry where $1 \text{ mH} = \dfrac{1}{10^3}$ H or 10^{-3} H

The microhenry where $1 \text{ μH} = \dfrac{1}{10^6}$ H or 10^{-6} H

Transposing these relationships shows that

10^3 mH = 1 H

10^6 μH = 1 H

and since

10^3 mH = 10^6 μH

then

$1 \text{ mH} = 10^3 \text{ μH}$

and

$1 \text{ μH} = 10^{-3} \text{ mH}$

Example 1 Express a 0.25-H inductor in millihenrys and microhenrys.

$0.25 \text{ H} = 0.25 \times 10^3 = 250 \text{ mH}$

$0.25 \text{ H} = 0.25 \times 10^6 = 25 \times 10^4 \text{ μH}$ or $2.5 \times 10^5 \text{ μH}$

```
10 PRINT "PROG 95"
20 PRINT "CONVERSION OF HENRYS TO OTHER UNITS"
30 INPUT "ENTER HENRYS" ; H
40 LET MH=H*10^3
50 LET UH=H*10^6
60 PRINT H "HENRYS = " MH "MILLIHENRYS"
70 PRINT H "HENRYS = " UH "MICROHENRYS"
```

Example 2 An inductor has a value of 820 μH. Express this in (i) millihenrys and (ii) henrys

(i) $1\ \mu\text{H} = 1 \times 10^{-3}\ \text{mH}$

$820\ \mu\text{H} = 1 \times 10^{-3} \times 820 = 0.82\ \text{mH}$

(ii) $1\ \mu\text{H} = 1 \times 10^{-6}\ \text{H}$

$820\ \mu\text{H} = 1 \times 10^{-6} \times 820 = 8.2 \times 10^{-4}\ \text{H}$

```
10 PRINT "PROG 96"
20 PRINT "CONVERSION OF MICROHENRYS TO OTHER UNITS"
30 INPUT "ENTER MICROHENRYS" ; UH
40 LET MH=UH/10^3
50 LET H=UH/10^6
60 PRINT UH "MICROHENRYS = " MH "MILLIHENRYS"
70 PRINT UH "MICROHENRYS = " H "HENRYS"
```

8.3 Inductor formulas

(a) Induced voltage
The average voltage induced in a coil is given by the formula

$e = N\ d\Phi/dt$

where $d\Phi/dt$ = rate of change of magnetic flux with time
Φ (Greek letter phi) = flux in webers, Wb
t = time in seconds
N = number of turns on the coil

Example 3 Calculate the e.m.f. induced in a coil of 300 turns if a flux of 80 mWb acting through the coil is reversed in 0.05 s.

$e = N\ d\Phi/dt$

$d\Phi = 2 \times 80 \times 10^{-3}\ \text{Wb} = 160 \times 10^{-3}\ \text{Wb}$

$$e = \frac{300 \times 160 \times 10^{-3}}{0.05} = 960\ \text{V}$$

The factor 2 is due to the fact that the flux decreases from 80 mWb to zero, and then increases to 80 mWb in the reverse direction.

```
10 PRINT "PROG 97"
20 PRINT "INDUCED E.M.F"
30 INPUT "ENTER NUMBER OF TURNS" : N
40 INPUT "ENTER FLUX IN MILLIWEBERS" ; PHI
50 INPUT "ENTER TIME IN SECONDS" ; S
60 LET E=N*2*PHI/(S*10^3)
70 PRINT "E = " E "VOLTS"
```

Example 4 A voltage of 6 V is induced in a coil when a flux of 100 mWb is reversed in 1.5 s. Calculate the number of turns.

$$N = \frac{e\ dt}{2\ d\Phi} = \frac{6 \times 1.5}{2 \times 100 \times 10^{-3}}$$

$$= \frac{6 \times 1.5 \times 10^3}{2 \times 100}$$

$= 45\ \text{turns}$

```
10 PRINT "PROG 98"
20 PRINT "NUMBER OF TURNS"
30 INPUT "ENTER INDUCED VOLTAGE IN VOLTS" ; E
40 INPUT "ENTER TIME IN SECONDS" ; S
50 INPUT "ENTER FLUX IN MILLIWEBERS" ; PHI
60 LET N=E*S*(10^3)/(2*PHI)
70 PRINT "N = " N "TURNS"
```

Example 5 A voltage of 240 V is induced in a coil having 1200 turns when the flux is reversed in 0.5 s. Calculate the value of the flux in milliwebers.

$$\Phi = \frac{e \times t \times 10^3}{2N} \quad (10^3 \text{ gives answer in mWb})$$

$$= \frac{240 \times 0.5 \times 10^3}{2 \times 1200}$$

$$= 50 \text{ mWb}$$

```
10 PRINT "PROG 99"
20 PRINT "FLUX"
30 INPUT "ENTER INDUCED VOLTAGE IN VOLTS" ; E
40 INPUT "ENTER TIME IN SECONDS" ; S
50 INPUT "ENTER NUMBER OF TURNS" ; N
60 LET PHI=E*S*(10^3)/(2*N)
70 PRINT "PHI = " PHI "MILLIWEBERS"
```

An alternative formula for the induced voltage is obtained by considering that in an air-cored coil the flux linking with the coil is proportional to the current. The rate of change of flux, $d\Phi/dt$, is proportional to the rate of change of this current dI/dt, thus

$$e = \text{a constant} \times dI/dt$$

This constant is the self-inductance, L, and therefore

$$e = L \, dI/dt$$

Example 6 Calculate the e.m.f. induced in a coil of inductance 10 H if the current through it increases from 4 A to 10 A in 100 ms.

$$e = L \, dI/dt$$

$$= 10 \times \frac{6}{100 \times 10^{-3}} \text{ where } dI = 10 - 4 = 6$$

$$= 600 \text{ V}$$

```
10 PRINT "PROG 100"
20 PRINT "INDUCED E.M.F"
30 INPUT "ENTER INDUCTANCE IN HENRYS" ; L
40 INPUT "ENTER INITIAL CURRENT" ; I1
50 INPUT "ENTER FINAL CURRENT" ; I2
60 INPUT "ENTER TIME IN MILLISECONDS" ; MS
70 LET E=(L/MS)*(I2-I1)*(10^3)
80 PRINT "E = " E "VOLTS"
```

Example 7 Calculate the e.m.f. induced in a 3-H coil if a current of 6 A is reversed in 20 ms.

$$e = \frac{L \times 2 \times I}{t}$$

$$= \frac{3 \times 2 \times 6}{20 \times 10^{-3}} = 1800 \text{ V}$$

Note the factor 2 in this example due to the reversal of current.

```
10 PRINT "PROG 101"
20 PRINT "INDUCED E.M.F"
30 INPUT "ENTER INDUCTANCE IN HENRYS" ; L
40 INPUT "ENTER CURRENT IN AMPS" ; I
50 INPUT "ENTER TIME IN MILLISECONDS" ; MS
60 LET E=(L/MS)*2*I*(10^3)
70 PRINT "E = " E "VOLTS"
```

Example 8 A voltage of 400 V is induced in a coil with an inductance of 0.5 H when a current is reversed in 25 ms. Calculate the value of the current.

$$I = \frac{et}{2L}$$

$$= \frac{400 \times 25 \times 10^{-3}}{2 \times 0.5} \text{ A}$$

$$= 10 \text{ A}$$

```
10 PRINT "PROG 102"
20 PRINT "CURRENT"
30 INPUT "ENTER INDUCED E.M.F IN VOLTS" ; E
40 INPUT "ENTER TIME IN MILLISECONDS" ; MS
50 INPUT "ENTER INDUCTANCE IN HENRYS" ; L
60 LET I=E/(2*10^3*L)*MS
70 PRINT "I = " I "AMPS"
```

Example 9 When a current of 20 A is reversed in a coil in 200 ms, a voltage of 2000 V is induced. Calculate the value of the inductance in henrys.

$$L = \frac{et}{2I}$$

$$= \frac{2000 \times 200 \times 10^{-3}}{2 \times 20} \text{ H}$$

$$= 10 \text{ H}$$

```
10 PRINT "PROG 103"
20 PRINT "INDUCTANCE"
30 INPUT "ENTER INDUCED E.M.F IN VOLTS" ; E
40 INPUT "ENTER TIME IN MILLISECONDS" ; MS
50 INPUT "ENTER CURRENT IN AMPS" ; I
60 LET L=E/(2*10^3*I)*MS
70 PRINT "L = " L "HENRYS"
```

(b) Self-inductance

Equating the two formulas for the induced voltage, e, shown in Section 8.3(a) gives

$$L \, dI/dt = N \, d\Phi/dt$$

and this gives

$$L = N \, (d\Phi/dt) \, dt/dI$$

$$= N \, d\Phi/dI \text{ H}$$

Example 10 A coil of 200 turns is wound on a non-magnetic former. A current of 5 A produces a flux of 400 μWb. Calculate the inductance of the coil.

$$L = N \, d\Phi/dI = \frac{200 \times 400 \times 10^{-6}}{5} \text{ H}$$

$$= 0.016 \text{ H}$$

$$= 0.016 \times 10^3 \text{ mH}$$

$$= 16 \text{ mH}$$

Calculate the value of the induced e.m.f. if the current is reversed in 40 ms.

$$e = N \, d\Phi/dt = \frac{200 \times 2 \times 400 \times 10^{-6}}{40 \times 10^{-3}}$$

$$= \frac{200 \times 2 \times 400}{40 \times 10^3} = 4 \text{ V}$$

Alternatively

$$e = L \, dI/dt$$

$$= \frac{16 \times 10^{-3} \times 2 \times 5}{40 \times 10^{-3}}$$

$$= 4 \text{ V}$$

```
10 PRINT "PROG 104"
20 PRINT "INDUCTANCE AND INDUCED E.M.F"
30 INPUT "ENTER NUMBER OF TURNS" ; N
40 INPUT "ENTER FLUX IN MICROWEBERS" ; PHI
50 INPUT "ENTER CURRENT IN AMPS" ; I
60 INPUT "ENTER TIME IN MILLISECONDS" ; MS
70 LET L=N*PHI/(I*10^6)
80 PRINT "L = " L "HENRYS"
90 LET E=N*2*PHI*10^3/(MS*10^6)
100 PRINT "E = " E "VOLTS"
110 LET E=(L/MS)*2*I*10^3
120 PRINT "E BY ALTERNATE METHOD = " E "VOLTS"
```

The inductance of a coil will also depend on the permeability of the core material. A coil wound on a bobbin with an iron core will produce more flux than a coil wound on a hollow air core bobbin.

Considering first the air core inductor, the inductance can be calculated from

$$L = \frac{\mu_o N^2 A}{l}$$

where μ_o = permeability of free space with unit symbol henry per metre, and is equal to $4\pi \times 10^{-7}$ H/m
N = number of turns
A = cross-sectional area in square metres
l = length in metres

Example 11 Find the inductance of a coil that has 250 turns on an air-cored cylinder, 5 cm long and 0.3 cm in diameter.

$$L = \frac{4\pi \times 10^{-7} \times 250^2 \times [\pi(0.15 \times 10^{-2})^2]}{5 \times 10^{-2}} \text{ H}$$

$$= \frac{4\pi \times 10^{-7} \times 250^2 \times \pi \times 2.25 \times 10^{-6}}{5 \times 10^{-2}} \text{ H}$$

$$= \frac{4\pi \times 250^2 \times \pi \times 2.25 \times 10^2}{5 \times 10^7 \times 10^6} \text{ H}$$

$$= \frac{11.1}{10^6} \text{ H} = \frac{11.1 \times 10^6}{10^6} \text{ μH} = 11.1 \text{ μH}$$

(Since ratio length/diameter is greater than 16, no correction factor to the inductance figure would be necessary.)

```
10 PRINT "PROG 105"
20 PRINT "COIL INDUCTANCE"
30 INPUT "ENTER NUMBER OF TURNS" ; N
40 INPUT "ENTER LENGTH IN CM" ; X
50 INPUT "ENTER DIAMETER IN CM" ; D
60 LET R=D/2
70 LET L=4*3.141593/(10^7*X)*N^2*R^2/10^4*3.141593*100
80 PRINT "L = " L*10^6 "MICROHENRYS"
```

If a magnetic material is introduced into the core, with a relative permeability μ_r, the inductance formula becomes

$$L = \frac{\mu_0 \mu_r N^2 A}{l}$$

Example 12 Find the inductance of a coil having 60 turns wound on a toroidal ferromagnetic bobbin, 2.5 cm long and 4 mm in diameter. The core has a relative permeability of 8000.

$$L = \frac{4\pi \times 10^{-7} \times 8000 \times 60^2 \times [\pi(2 \times 10^{-3})^2]}{2.5 \times 10^{-2}} \text{ H}$$

$$= \frac{4\pi \times 10^{-7} \times 8000 \times 60^2 \times \pi \times 4 \times 10^{-6}}{2.5 \times 10^{-2}} \text{ H}$$

$$= \frac{4\pi \times 8000 \times 60^2 \times \pi \times 4 \times 10^2}{2.5 \times 10^7 \times 10^6} \text{ H} = 0.018 \text{ H}$$

(In practice, since ratio l/d is only 6, a correction factor would have to be applied to the answer.)

```
10 PRINT "PROG 106"
20 PRINT "COIL INDUCTANCE"
30 INPUT "ENTER RELATIVE PERMEABILITY" ; UR
40 INPUT "ENTER NUMBER OF TURNS" ; N
50 INPUT "ENTER LENGTH IN CM" ; X
60 INPUT "ENTER DIAMETER IN MM" ;D
70 LET R=D/2
80 LET L=UR*4*3.141593/(10^7*X)*N^2*R^2/10^6*3.141593*100
90 PRINT "L = " L*10^3 "MICROHENRYS"
```

(c) *Energy stored*

When current is flowing in a coil, energy is supplied from the voltage source, and this energy is stored in the magentic field. The stored energy, W, is measured in joules, unit symbol J, where

$$W = \tfrac{1}{2}LI^2 \text{ J}$$

L is in henrys and I is in amperes.

Inductors **131**

Example 13 Find the energy stored in a coil with inductance 10 H if the current is 1.2 A.

$$W = \tfrac{1}{2} \times 10 \times (1.2)^2 = 7.2 \text{ J}$$

```
10 PRINT "PROG 107"
20 PRINT "ENERGY STORED"
30 INPUT "ENTER INDUCTANCE IN HENRYS" ; L
40 INPUT "ENTER CURRENT IN AMPS" ; I
50 LET W=.5*L*I^2
60 PRINT "W = " W "JOULES"
```

Example 14 If 3.2 J of energy is stored in a coil when the current is 4 A, calculate the inductance.

$$L = \frac{W}{0.5I^2} \text{ H}$$

$$= \frac{3.2}{0.5 \times 4^2} = 0.4 \text{ H}$$

```
10 PRINT "PROG 108"
20 PRINT "COIL INDUCTANCE"
30 INPUT "ENTER ENERGY STORED IN JOULES" ; W
40 INPUT "ENTER CURRENT IN AMPS" ; I
50 LET L=W/(.5*I^2)
60 PRINT "L = " L "HENRYS"
```

Example 15 If a coil with an inductance of 0.75 H stores 15 J calculate the current flowing.

$$I^2 = \frac{W}{0.5L}$$

$$I = \sqrt{\left(\frac{W}{0.5L}\right)}$$

$$= \sqrt{\left(\frac{15}{0.5 \times 0.075}\right)} = 20 \text{ A}$$

```
10 PRINT "PROG 109"
20 PRINT "CURRENT"
30 INPUT "ENTER ENERGY STORED IN JOULES" ; W
40 INPUT "ENTER INDUCTANCE IN HENRYS" ; L
50 LET I=SQR(W/(.5*L))
60 PRINT "I = " I "AMPS"
```

(d) Mutual inductance

When two coils are positioned such that the flux produced by the current flowing in one links with the other, mutual inductance, M, exists between them.

If the current in the first coil changes, an e.m.f. will be induced in the second. Conversely if the current changes in the second coil, an e.m.f. will be induced in the first. The mutual inductance is the same in both cases.

The average induced e.m.f. is given by

$$e_1 = M \, dI_2/dt$$

where M is in henrys
I is in amps
t is in seconds
e_1 = e.m.f. in 1st coil
I_2 = current in 2nd coil

Example 16 The mutual inductance between two coils is 0.5 H. Calculate the average voltage induced in one coil when the current in the other coil is increased from 1 A to 10 A in 150 mS.

$$dI = 10 - 1 = 9 \text{ A}$$

$$e = \frac{0.5 \times 9}{150 \times 10^{-3}} = \frac{0.5 \times 9 \times 10^3}{150} = 30 \text{ V}$$

```
10 PRINT "PROG 110"
20 PRINT "INDUCED E.M.F"
30 INPUT "ENTER MUTUAL INDUCTANCE IN HENRYS" ; L
40 INPUT "ENTER INITIAL CURRENT IN AMPS" ; I1
50 INPUT "ENTER FINAL CURRENT IN AMPS" ; I2
60 INPUT "ENTER TIME IN MILLISECONDS" ; MS
70 LET I=I2-I1
80 LET E=L*I*(10^3/MS)
90 PRINT "E = " E "VOLTS"
```

Example 17 Four hundred volts is induced in a coil which has an inductance of 2 H when the current increases from 0 A to 10 A. Calculate the time for the increase in milliseconds.

$$dt = \frac{L \, dI}{e}$$

$$= \frac{2 \times (10 - 0) \times 10^3}{400} \text{ ms}$$

$$= 50 \text{ ms}$$

```
10 PRINT "PROG 111"
20 PRINT "TIME"
30 INPUT "ENTER MUTUAL INDUCTANCE IN HENRYS" ; L
40 INPUT "ENTER INITIAL CURRENT IN AMPS" ; I1
50 INPUT "ENTER FINAL CURRENT IN AMPS" ; I2
60 INPUT "ENTER INDUCED E.M.F" ; E
70 LET I=I2-I1
80 LET T=L*I*10^3/E
90 PRINT "T = " T "MILLISECONDS"
```

8.4 Inductor circuits

(a) Series

Example 18 Figure 8.1 shows three inductors connected in series. Calculate the total inductance, the current growth, the voltages V_1, V_2 and V_3, and the energy stored after 4 s in each inductor. The treatment of inductors in series is similar to the treatment of resistors in series, and therefore

$$L = L_1 + L_2 + L_3$$
$$= 2.4 + 3.2 + 4.4 = 10 \text{ H}$$

Inductors   133

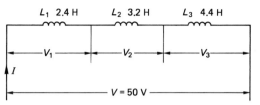

Figure 8.1

In an inductive circuit of negligible resistance the current increases at the rate of

$$\frac{V}{L} \text{ A/s}$$
$$= \frac{50}{10} = 5 \text{ A/s}$$
$$V_1 = 5 \times 2.4 = 12.0 \text{ V}$$
$$V_2 = 5 \times 3.2 = 16.0 \text{ V}$$
$$V_3 = 5 \times 4.4 = 22.0 \text{ V}$$

Note $V_1 + V_2 + V_3 = V$

After 4 s the circuit current is

$$I = 5 \times 4 = 20 \text{ A}$$

Energy stored in

$$L_1 = W_1 = \tfrac{1}{2} \times 2.4 \times (20)^2 = 480 \text{ J}$$
$$L_2 = W_2 = \tfrac{1}{2} \times 3.2 \times (20)^2 = 640 \text{ J}$$
$$L_3 = W_3 = \tfrac{1}{2} \times 4.4 \times (20)^2 = 880 \text{ J}$$

Total energy stored is

$$W = W_1 + W_2 + W_3$$
$$= 480 + 640 + 880 = 2000 \text{ J}$$

Note that

$$W = \tfrac{1}{2} L I^2$$
$$= 0.5 \times 10 \times 10^2 = 2000 \text{ J}$$

```
10 PRINT "PROG 112"
20 PRINT "INDUCTOR SERIES CIRCUIT"
30 INPUT "ENTER INDUCTANCE L1 IN HENRYS" ; L1
40 INPUT "ENTER INDUCTANCE L2 IN HENRYS" ; L2
50 INPUT "ENTER INDUCTANCE L3 IN HENRYS" ; L3
60 INPUT "ENTER VOLTAGE IN VOLTS" ; V
70 INPUT "ENTER TIME IN SECONDS" ; T
80 LET L=L1+L2+L3
90 PRINT "TOTAL INDUCTANCE L = " L "HENRYS"
100 LET IG=V/L
110 PRINT "CURRENT GROWTH = " IG "AMPS/SEC"
120 LET V1=IG*L1
130 PRINT "V1 = " V1 "VOLTS"
140 LET V2=IG*L2
150 PRINT "V2 = " V2 "VOLTS"
160 LET V3=IG*L3
```

```
170 PRINT "V3 = " V3 "VOLTS"
180 LET I=IG*T
190 PRINT "CURRENT AFTER" T "SECS =" I "AMPS"
200 LET W1=.5*L1*I^2
210 PRINT "W1 = " W1 "JOULES"
220 LET W2=.5*L2*I^2
230 PRINT "W2 = " W2 "JOULES"
240 LET W3=.5*L3*I^2
250 PRINT "W3 = " W3 "JOULES"
260 LET W=.5*L*I^2
270 PRINT "W = " W "JOULES"
```

(b) Parallel

Example 19 Figure 8.2 shows three inductors connected in parallel. Calculate the total inductance, the rate of current growth in each inductor, the total rate of current growth, and the energy stored in each inductor after 1 s.

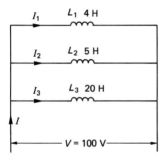

Figure 8.2

The treatment of inductors in parallel is similar to that of resistors in parallel.

$$\frac{1}{L} = \frac{1}{L_1} + \frac{1}{L_2} + \frac{1}{L_3}$$

$$= \frac{1}{4} + \frac{1}{5} + \frac{1}{20}$$

$$= 0.25 + 0.2 + 0.05 = 0.5$$

$$L = \frac{1}{0.5} = 2 \text{ H}$$

Rate of current growth in

$$L_1 = \frac{V}{L_1} = \frac{100}{4} = 25 \text{ A/s}$$

$$L_2 = \frac{V}{L_2} = \frac{100}{5} = 20 \text{ A/s}$$

$$L_3 = \frac{V}{L_3} = \frac{100}{20} = 5 \text{ A/s}$$

Total rate of growth in

$$L = \frac{V}{L} = \frac{100}{2} = 50 \text{ A/s}$$

Note total rate = 25 + 20 + 5 = 50 A/s.

```
10 PRINT "PROG 113"
20 PRINT "INDUCTOR PARALLEL CIRCUIT"
30 INPUT "ENTER INDUCTANCE L1 IN HENRYS" ; L1
40 INPUT "ENTER INDUCTANCE L2 IN HENRYS" ; L2
50 INPUT "ENTER INDUCTANCE L3 IN HENRYS" ; L3
60 INPUT "ENTER VOLTAGE IN VOLTS" ; V
70 INPUT "ENTER TIME IN SECONDS" ; T
80 LET X=1/L1+1/L2+1/L3
90 LET L=1/X
100 PRINT "EQUIVALENT INDUCTANCE = " L "HENRYS"
110 LET IG1=V/L1
120 PRINT "CURRENT GROWTH IN L1 = " IG1 "AMPS/SEC"
130 LET IG2=V/L2
140 PRINT "CURRENT GROWTH IN L2 = " IG2 "AMPS/SEC"
150 LET IG3=V/L3
160 PRINT "CURRENT GROWTH IN L3 = " IG3 "AMPS/SEC"
170 LET IT=V/L
180 PRINT "TOTAL CURRENT GROWTH = " IT "AMPS/SEC"
190 LET I1=IG1*T
200 PRINT "CURRENT IN L1 AFTER" T "SECS = " I1 "AMPS"
210 LET I2=IG2*T
220 PRINT "CURRENT IN L2 AFTER" T "SECS = " I2 "AMPS"
230 LET I3=IG3*T
240 PRINT "CURRENT IN L3 AFTER" T "SECS = " I3 "AMPS"
250 LET I=IT*T
260 PRINT "TOTAL CURRENT AFTER" T "SECS = " I "AMPS"
270 LET W1=.5*L1*IG1^2
280 PRINT "W1 = " W1 "JOULES"
290 LET W2=.5*L2*IG2^2
300 PRINT "W2 = " W2 "JOULES"
330 LET W3=.5*L3*IG3^2
340 PRINT "W3 = " W3 "JOULES"
350 LET WT=.5*L*IT^2
360 PRINT "TOTAL ENERGY STORED = " WT "JOULES"
```

The energy stored after 1 s in

$$L_1 = W_1 = \tfrac{1}{2} \times 4 \times (25)^2 = 1250 \text{ J}$$
$$L_2 = W_2 = \tfrac{1}{2} \times 5 \times (20)^2 = 1000 \text{ J}$$
$$L_3 = W_3 = \tfrac{1}{2} \times 20 \times (5)^2 = 250 \text{ J}$$
$$W = \text{total energy stored} = \tfrac{1}{2} \times 2 \times (50)^2 = 2500 \text{ J}$$

Note that total energy stored is

$$W = W_1 + W_2 + W_3 = 1250 + 1000 + 250 = 2500 \text{ J}$$

8.5 Reactance (sinusoidal voltages only)

The behaviour of inductors in d.c. and a.c. circuits will be dealt with in Chapters 9 and 11 respectively. The relationship between the applied voltage and the current flowing resembles Ohm's law. The opposition to the current is called inductive reactance, X_L and the unit symbol is the ohm, Ω, where

$$X_L = 2\pi f L$$

where L is in henrys, H
f is in hertz, Hz
X_L is in ohms, Ω

Example 20 Calculate the reactance of a 100-mH inductor at 50 Hz.

$$X_L = 2 \times \pi \times 50 \times 100 \times 10^{-3} \ \Omega$$
$$= 31.42 \ \Omega$$

```
10 PRINT "PROG 114"
20 PRINT "INDUCTIVE REACTANCE"
30 INPUT "ENTER INDUCTANCE IN MILLIHENRYS" ; L
40 INPUT "ENTER FREQUENCY" ; F
50 LET XL=2*3.141593*F*L/10^3
60 PRINT "XL = " XL "OHMS"
```

Example 21 Calculate the frequency at which a 2.5 µH coil has a reactance of 2.3562 Ω.

$$f = \frac{X_L}{2 \times \pi \times L \times 10^{-6}}$$
$$= \frac{10^6 \times 2.3562}{2 \times \pi \times 2.5} = 150\ 000 \ \text{Hz} = 150 \ \text{kHz}$$

```
10 PRINT "PROG 115"
20 PRINT "FREQUENCY"
30 INPUT "ENTER INDUCTANCE IN MICROHENRYS" ; L
40 INPUT "ENTER REACTANCE" ; XL
50 LET F=XL*10^6/(2*3.141593*L)
60 PRINT "F = " F "HZ"
```

Example 22 Calculate the inductance of a coil which has a reactance of 3.141593 Ω at a frequency of 25 Hz.

$$L = \frac{X_L}{2\pi f} = \frac{3.141593}{2 \times \pi \times 25} = 0.02 \ \text{H} = 20 \ \text{mH}$$

```
10 PRINT "PROG 116"
20 PRINT "INDUCTANCE"
30 INPUT "ENTER REACTANCE": XL
40 INPUT "ENTER FREQUENCY" ; F
50 LET L=XL/(2*3.141593*F)
60 PRINT "L = " L "HENRYS"
```

Problems

1 Convert 0.006 henrys to millihenrys and microhenrys.
2 Convert 960 microhenrys to millihenrys and henrys.
3 Calculate the e.m.f. induced in a coil of 450 turns if a flux of 75 mWb acting through the coil is reversed in 0.02 s.
4 Calculate the number of turns in a coil if 10 V is induced when a flux of 40 mWb is reversed in 2 s.
5 A voltage of 700 V is induced in a coil of 1000 turns when the flux is reversed in 0.35 s. Find the value of the flux in milliwebers.
6 Calculate the voltage induced in a 12.5 H inductor if the current flowing through it increases from 2 A to 8 A in 30 ms.
7 Calculate the e.m.f. induced in a coil which has an inductance of 2.75 H if a current of 12.5 A is reversed in 25 ms.
8 A voltage of 350 V is induced in a coil which has an inductance of 2.2 H, when a current is reversed in 15 ms. Find the value of the current in amps.

9 A voltage of 1600 V is induced when a current of 4 A is reversed in a coil in 80 ms. Calculate the inductance of the coil in henrys.

10 A coil of 180 turns is wound on a non-magnetic former. A current of 2.5 A produces a flux of 550 μWb. Calculate the inductance of the coil in henrys. Find also the value of the induced e.m.f. if the current is reversed in 20 ms.

11 An air-cored bobbin has 400 turns. If the bobbin is 6 cm long and has a diameter of 0.4 cm, find the inductance of the coil in microhenrys.

12 Calculate the inductance of a coil having 140 turns wound on a ferromagnetic bobbin 8 cm long and 6 mm in diameter. The core has a relative permeability of 10 000. Express the answer in millihenrys.

13 Find the energy stored in joules in a coil having an inductance of 0.15 H if the current flowing is 6 A.

14 Calculate the inductance of a coil which stores 6.5 J when the current flowing is 6.5 A.

15 A 0.055-H coil stores 11 J. Calculate the current flowing.

16 The mutual inductance between two coils is 0.8 H. Calculate the average voltage induced in one coil when the current in the other coil is increased from 2 A to 8 A in 200 ms.

17 Three hundred and twenty volts is induced in a coil which has a mutual inductance of 4 H when the current in the second coil increases from 1 A to 9 A. Calculate the time for the increase in milliseconds.

18 Figure 8.3 shows a series inductor circuit. Calculate: (a) total inductance, (b) the current growth, (c) the voltage drops V_1, V_2 and V_3, (c) the current after 5 s, (e) the energy stored in each inductor after 5 s, (f) the total energy stored after 5 s.

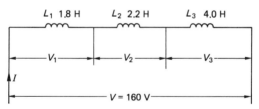

Figure 8.3

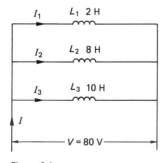

Figure 8.4

19 Figure 8.4 shows a parallel inductor circuit. Calculate: (a) the equivalent inductance, (b) the current growth in L_1, L_2 and L_3, (c) the total current growth, (d) the current in L_1, L_2 and L_3 after 12 s, (e) the total current after 12 s, (f) the energy stored in L_1, L_2 and L_3 after 1 s, (g) the total energy stored after 1 s.
20 Calculate the reactance of a 65-mH inductor at a frequency of 250 Hz.
21 At what frequency will a 35-µH coil have a reactance of 5 Ω?
22 Find the inductance of a coil that has a reactance of 100 Ω at a frequency of 100 Hz.

9 D.c. transients

9.1 Introduction

Chapter 7 showed that when a capacitor is connected to a d.c. voltage, it becomes charged until its terminal voltage equals the applied voltage. Charging cannot take place instantaneously since every circuit has a resistance.

The voltage in the capacitor builds up at a rate depending on the value of the series resistance. Chapter 8 showed that when a d.c. voltage is applied to a conductor an e.m.f. is induced in the inductance in opposition to the applied voltage. The induced voltage retards the growth of the current, and the rate of increase depends on both the value of the inductance and the value of its resistance. The inductance and the resistance are generally represented in a circuit as two separate components.

The changing values of voltages and currents in both circuits during the growth and decay periods, before a steady state is reached, are called transients.

9.2 C–R circuit

Example 1 Figure 9.1 shows a circuit to which is applied a step voltage of 100 V. The instantaneous voltages and current are denoted by v_R, v_C and i. During the charging and discharging cycles the following sequence of events will occur.

(a) At zero time, the instant of switching on, the capacitor C is uncharged and all the voltage appears across R, giving the following circuit conditions:

$v_C = 0$

$v_R = V = 100$

$i = I = \dfrac{100}{10^6} \times 10^6 = 100 \ \mu A$

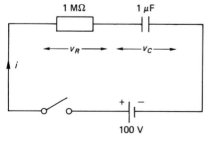

Figure 9.1

(b) The capacitor charges and at any instant
$$v_C = V(1 - e^{-t/CR})$$
$$v_R = Ve^{-t/CR}$$
$$i = Ie^{-t/CR}$$

where t is in seconds, C is in farads, R is in ohms and $e = 2.7182$ (the base for natural logarithms).

To calculate v_C at different times t, it is first necessary to calculate the product CR, which is the time constant in seconds.

$$CR = 1 \times 10^{-6} \times 10^6 = 1 \text{ s}$$

At $t = 0$

$$e^{-t/CR} = \frac{1}{e^{t/CR}} = \frac{1}{e^0} = 1$$

After 1 s

$$\frac{1}{e^{t/CR}} = \frac{1}{e} = 0.368$$

After 2 s

$$\frac{1}{e^{t/CR}} = \frac{1}{e^2} = 0.135$$

After 3 s

$$\frac{1}{e^{t/CR}} = \frac{1}{e^3} = 0.0498$$

After 4 s

$$\frac{1}{e^{t/CR}} = \frac{1}{e^4} = 0.0183$$

After 5 s

$$\frac{1}{e^{t/CR}} = \frac{1}{e^5} = 0.0067$$

For these six time intervals

$$v_C = 100(1 - 1) = 0$$
$$v_C = 100(1 - 0.368) = 63.2$$
$$v_C = 100(1 - 0.135) = 86.5$$
$$v_C = 100(1 - 0.0498) = 95.02$$
$$v_C = 100(1 - 0.0183) = 98.17$$
$$v_C = 100(1 - 0.0067) = 99.33$$

These results are shown in Table 9.1, and plotted in Figures 9.2(a) and 9.2(b).

The values of v_R are obtained by multiplying column 2 by 100, since

$$v_R = V \frac{1}{e^{t/CR}}$$

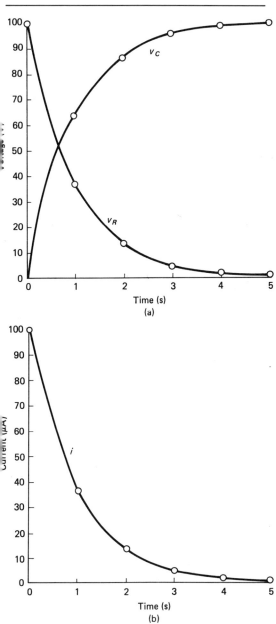

Figure 9.2

142 Circuit calculations pocket book

Table 9.1 Voltage and current transients during charge

1	2	3	4	5	6
$t(s)$	$\dfrac{1}{e^{t/CR}}$	$1 - \dfrac{1}{e^{t/CR}}$	v_C	v_R	$i\ (\mu A)$
0	1.0	0	0	100	100
1	0.368	0.632	63.2	36.8	36.8
2	0.135	0.865	86.5	13.5	13.5
3	0.0498	0.9502	95.02	4.98	4.98
4	0.0183	0.9817	98.17	1.83	1.83
5	0.0067	0.9933	99.33	0.67	0.67

The values of i are obtained by multiplying column 2 by 100 since

$$i = I \frac{1}{e^{t/CR}}$$

```
10  PRINT "PROG 117"
20  PRINT "VOLTAGE GROWTH"
30  INPUT "ENTER APPLIED VOLTAGE" ; V
40  INPUT "ENTER RESISTANCE IN OHMS" ; R
50  INPUT "ENTER CAPACITANCE IN MICROFARADS" ; C
60  INPUT "ENTER START TIME IN SECONDS" : T
70  INPUT "ENTER END TIME IN SECONDS" ; B
80  INPUT "ENTER TIME INTERVAL BETWEEN CALCULATIONS" ; D
90  LET TC=C*R/10^6
100 LET X=T/TC
110 LET Y=2.7183^X
120 LET VC=V*(1-(1/Y))
130 PRINT "VC AT " T "SECS = " VC "VOLTS"
140 LET VR=V-VC
150 PRINT "VR AT " T "SECS = " VR "VOLTS"
160 LET I=V/R
170 LET IC=I/Y
180 PRINT "CURRENT AT" T "SECS = " IC*10^6 "MICROAMPS"
190 LET T=T+D
200 IF T<=B THEN GOTO 100
```

Notes:

1 The calculations for $t = 0$ confirm the statements made in (a).
2 The product CR in seconds, known as the 'time constant', is often denoted by the Greek letter 'tau' and is the time taken for the capacitor to reach 0.632 or 63.2% of its final voltage. The time constant is also the time that would be taken to charge the capacitor, if the initial rate of charge was maintained.
3 The results show that the capacitor is almost fully charged after five time constants. Theoretically it takes an infinite time for the capacitor voltage to equal the applied voltage.
4 Columns 4 and 5 show that as v increases, $v_C + v_R = V$ at all times.

(c) If the charged capacitor is now disconnected from the supply, and is short-circuited through the resistance R (see Figure 9.3), the capacitor voltage and hence the current will fall. The current flows in the opposite direction to the charging current.

(d) At the beginning of the discharge cycle, i.e. when $t = 0$, the circuit conditions are

$v_C = V = 100$

$v_R = -v_C = -100$ (due to reversal of current)

d.c. transients **143**

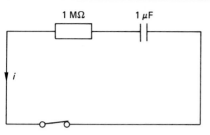

Figure 9.3

$$i = I = \frac{100 \times 10^6}{10^6} = 100 \text{ μA}$$

(e) The capacitor discharges and at any instant

$$v_C = V e^{-t/CR}$$
$$v_R = -V e^{-t/CR}$$
$$i = -I e^{-t/CR}$$

The calculations for v_C are the same as for v_R during the charging cycle, and the calculations for v_R and i are the same, each with a negative sign due to the reversal of the voltage and hence the current. The results are reproduced in Table 9.2 and plotted in Figure 9.4(a) and 9.4(b).

Table 9.2 Capacitor discharge voltages and current

t	v_C	v_R	i (μA)
0	100	−100	−100
1	36.8	−36.8	−36.8
2	13.5	−13.5	−13.5
3	4.98	−4.98	−4.98
4	1.83	−1.83	−1.83
5	0.67	−0.67	−0.67

```
10  PRINT "PROG 118"
20  PRINT "VOLTAGE DECAY"
30  INPUT "ENTER CAPACITOR VOLTAGE" ; V
40  INPUT "ENTER RESISTANCE IN OHMS" ; R
50  INPUT "ENTER CAPACITANCE IN MICROFARADS" ; C
60  INPUT "ENTER START TIME IN SECONDS" ; T
70  INPUT "ENTER END TIME IN SECONDS" ; B
80  INPUT "ENTER TIME INTERVAL BETWEEN CALCULATIONS" ; D
90  LET TC=C*R/10^6
100 LET X=T/TC
110 LET Y=2.7183^X
120 LET VC=V/Y
130 PRINT "VC AT" T "SECS = " VC "VOLTS"
140 LET VR=-VC
150 PRINT "VR AT" T "SECS = " VR "VOLTS"
160 LET I=V/R
170 LET IC=-I/Y
180 PRINT "I AT" T "SECS = " IC*10^6 "MICROAMPS"
190 LET T=T+D
200 IF T<=B THEN GOTO 100
```

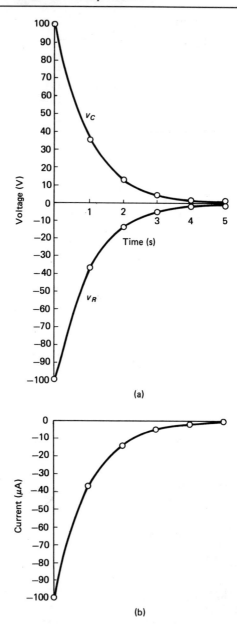

Figure 9.4

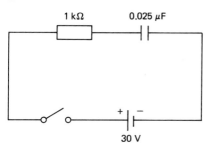

Figure 9.5

Notes:

1. The calculations for $t = 0$ confirm the statement made in (d).
2. The capacitor voltage falls by a factor of 0.632 or 63.2% of its initial value during one time constant, leaving 36.8 V (36.8%) of the initial voltage.
3. The capacitor is almost fully discharged after five time constants.
4. $v_C = -v_R$ at all times.

Example 2 Draw up a table of the capacitor and resistance voltages, and the current during the charge and discharge transients at times $t = 0$, $t = CR$, $t = 2CR$, $t = 3CR$, $t = 4CR$ and $t = 5CR$ for the circuit in Figure 9.5.

Time constant $CR = 0.025 \times 10^{-6} \times 10^3 \times 10^6 \, \mu s = 25 \, \mu s$

The results are shown in Tables 9.3 and 9.4. Substituting for CR and t in $1/e^{t/CR}$ and $1 - 1/e^{t/CR}$ gives the results shown in columns 2 and 3. These results are identical to those in Table 9.1.

The values of v_C are obtained by multiplying column 3 by $V = 30$ V

The values of v_R are obtained by multiplying column 2 by $V = 30$ V

The values of i are obtained by multiplying column 2 by I, where

$$I = \frac{V}{R} = \frac{(30)}{10^3} \times 10^3 \, \text{mA} = 30 \, \text{mA}$$

Table 9.3 Voltages and current transients during charging

1	2	3	4	5	6
t (μs)	$\dfrac{1}{e^{t/CR}}$	$1 - \dfrac{1}{e^{t/CR}}$	v_C	v_R	i (mA)
0	1	0	0	30	30
25	0.368	0.632	18.96	11.04	11.04
50	0.135	0.865	25.95	4.05	4.05
75	0.0498	0.9502	28.506	1.494	1.494
100	0.0183	0.9817	29.451	0.549	0.549
125	0.0067	0.9933	29.799	0.201	0.201

```
10 PRINT "PROG 119"
20 PRINT "VOLTAGE GROWTH"
30 INPUT "ENTER APPLIED VOLTAGE" ; V
40 INPUT "ENTER RESISTANCE IN OHMS" ; R
50 INPUT "ENTER CAPACITANCE IN MICROFARADS" : C
60 INPUT "ENTER START TIME IN MICROSECONDS" ;T
70 INPUT "ENTER END TIME IN MICROSECONDS" ; B
80 INPUT "ENTER TIME INTERVAL BETWEEN CALCULATIONS" ; D
90 LET TC=C*R*10^6/10^6
100 LET X=T/TC
110 LET Y=2.7183^X
120 LET VC=V*(1-(1/Y))
130 PRINT "VC AT" T "MICROSECS = " VC "VOLTS"
140 LET VR=V-VC
150 PRINT "VR AT" T "MICROSECS = " VR "VOLTS"
160 LET I=V/R
170 LET IC=I/Y
180 PRINT "I AT" T "MICROSECS = " IC*10^3 "MILLIAMPS"
190 LET T=T+D
200 IF T<=B THEN GOTO 100
```

Notes:

1 The values shown in columns 2 and 3 will always be the same irrespective of the values of C and R, for multiples of the time constant.

2 After one time constant

$$v_C = 18.96 \text{ V}$$

and

$$\frac{18.96}{30} \times 100 = 63.2\% \text{ of } V$$

This will always be true irrespective of circuit values.

Table 9.4 Voltage and current transients during discharge

1	2	3	4
t (μs)	v_C	v_R	i (mA)
0	30	−30	−30
25	11.04	−11.04	−11.04
50	4.05	−4.05	−4.05
75	1.494	−1.494	−1.494
100	0.549	−0.549	−0.549
125	0.201	−0.201	−0.201

The values of v_C are the same as for v_R during the charging transient as shown in Example 1.

The values of v_R and i are the same, each one negative due to the reversal of the voltage polarity across R.

Note that after $1CR$, $v_C = 11.04$, and

$$v_C = 36\% \text{ of } V \left(\frac{11.04}{30} \times 100 = 36.8\% \right)$$

This will always be true irrespective of circuit values.

d.c. transients 147

```
10 PRINT "PROG 120"
20 PRINT "VOLTAGE DECAY"
30 INPUT "ENTER CAPACITOR VOLTAGE" ; V
40 INPUT "ENTER RESISTANCE IN OHMS" ; R
50 INPUT "ENTER CAPACITANCE IN MICROFARADS" ; C
60 INPUT "ENTER START TIME IN MICROSECONDS" : T
70 INPUT "ENTER END TIME IN MICROSECONDS" ; B
80 INPUT "ENTER TIME INTERVAL BETWEEN CALCULATIONS" ; D
90 LET TC=C*R*10^6/10^6
100 LET X=T/TC
110 LET Y=2.7183^X
120 LET VC=V/Y
130 PRINT "VC AT" T "MICROSECS = " VC "VOLTS"
140 LET VR=-VC
150 PRINT "VR AT" T "MICROSECS = " VR "VOLTS"
160 LET I=V/R
170 LET IC=-I/Y
180 PRINT "I AT" T "MICROSECS = " IC*10^3 "MILLIAMPS"
190 LET T=T+D
200 IF T<=B THEN GOTO 100
```

Example 3 Repeat Example 2 for the circuit shown in Figure 6.

Time constant $CR = 220 \times 10^{-6} \times 150 \times 10^3$ ms

$$= 33 \text{ ms}$$

$$I = \frac{V}{R} = \frac{60}{150} \times 10^3 = 400 \text{ mA}$$

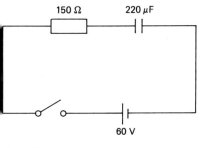

Figure 9.6

Table 9.5 Transients during charging

1	2	3	4	5	6
t (ms)	$\dfrac{1}{e^{t/CR}}$	$1 - \dfrac{1}{e^{t/CR}}$	v_C	v_R	i (mA)
0	1	0	0	60	400
33	0.368	0.632	37.92	22.08	147.2
66	0.135	0.865	51.9	8.1	54
99	0.0498	0.9502	57.012	2.988	19.92
132	0.0183	0.9817	58.902	1.098	7.32
165	0.0067	0.9933	59.598	0.402	2.68

148 Circuit calculations pocket book

```
10 PRINT "PROG 121"
20 PRINT "VOLTAGE GROWTH"
30 INPUT "ENTER APPLIED VOLTAGE" ; V
40 INPUT "ENTER RESISTANCE IN OHMS" ; R
50 INPUT "ENTER CAPACITANCE IN MICROFARADS" ; C
60 INPUT "ENTER START TIME IN MILLISECONDS" ;T
70 INPUT "ENTER END TIME IN MILLISECONDS" ; B
80 INPUT "ENTER TIME INTERVAL BETWEEN CALCULATIONS" ; D
90 LET TC=C*R*10^3/10^6
100 LET X=T/TC
110 LET Y=2.7183^X
120 LET VC=V*(1-(1/Y))
130 PRINT "VC AT" T "MILLISECS = " VC "VOLTS"
140 LET VR=V-VC
150 PRINT "VR AT" T "MILLISECS = " VR "VOLTS"
160 LET I=V/R
170 LET IC=I/Y
180 PRINT "I AT" T "MILLISECS = " IC*10^3 "MILLIAMPS"
190 LET T=T+D
200 IF T<=B THEN GOTO 100
```

Table 9.6 Transients during discharge

1 t (ms)	2 v_C	3 v_R	4 i (mA)
0	60	−60	−400
33	22.08	−22.08	−147.2
66	8.1	−8.1	−54
99	2.988	−2.988	−19.92
132	1.098	−1.098	−7.32
165	0.402	−0.402	−2.68

```
10 PRINT "PROG 122"
20 PRINT "VOLTAGE DECAY"
30 INPUT "ENTER CAPACITOR VOLTAGE" ; V
40 INPUT "ENTER RESISTANCE IN OHMS" ; R
50 INPUT "ENTER CAPACITANCE IN MICROFARADS" ; C
60 INPUT "ENTER START TIME IN MILLISECONDS" ; T
70 INPUT "ENTER END TIME IN MILLISECONDS" ; B
80 INPUT "ENTER TIME INTERVAL BETWEEN CALCULATIONS" ; D
90 LET TC=C*R*10^3/10^6
100 LET X=T/TC
110 LET Y=2.7183^X
120 LET VC=V/Y
130 PRINT "VC AT" T "MILLISECS = " VC "VOLTS"
140 LET VR=-VC
150 PRINT "VR AT" T "MILLISECS = " VR "VOLTS"
160 LET I=V/R
170 LET IC=-I/Y
180 PRINT "I AT" T "MILLISECS = " IC*10^3 "MILLIAMPS"
190 LET T=T+D
200 IF T<=B THEN GOTO 100
```

Example 4 Consider again the circuit shown in Figure 9.1, and determine the time for the capacitor voltage to reach 50 V. What is the current flowing at this time?

(i) $v_C = V(1 - e^{-t/CR})$

$50 = 100(1 - e^{-t/CR})$

$50 = 100 - 100e^{-t/CR}$

$100e^{-t/CR} = 100 - 50 = 50$

$e^{-t/CR} = \dfrac{50}{100} = 0.5$

$\dfrac{1}{e^{t/CR}} = 0.5$

Since $CR = 1$ then

$$\frac{1}{e^t} = 0.5$$

Transposing the formula gives

$$e^t = \frac{1}{0.5} = 2$$

$$t = \log_e 2 = 0.6931 \text{ s}$$

(ii) $i = 100e^{-t/CR}$ since $I = 100$ μA

$$= 100 \times \frac{1}{e^{t/CR}}$$

$$= 100 \times \frac{1}{e^{0.6931}} \text{ since } CR = 1$$

$$= 100 \times \tfrac{1}{2} = 50 \text{ μA}$$

```
10 PRINT "PROG 123"
20 PRINT "TIME AND CURRENT"
30 INPUT "ENTER APPLIED VOLTAGE" ; V
40 INPUT "ENTER RESISTANCE IN OHMS" ; R
50 INPUT "ENTER CAPACITANCE IN MICROFARADS" ; C
60 INPUT "ENTER CAPACITOR VOLTAGE" ; VC
70 LET TC=C*R/10^6
80 LET T=TC*LOG(V/(V-VC))
90 PRINT "T = " T "SECONDS"
100 LET X=T/TC
110 LET Y=2.7183^X
120 LET I=V*(10^6/R)
130 LET IC=I/Y
140 PRINT "CURRENT AT" T "SECS = " IC "MICROAMPS"
```

Example 5 If the capacitor in Figure 9.1 is initially charged to 100 V, calculate: (i) the time taken for the voltage to drop to 20, and (ii) the current flowing at that time.

(i) $v_C = Ve^{-t/CR}$

$20 = 100e^{-t}$ since $V = 100$ and $CR = 1$ s

$$e^{-t} = \frac{20}{100}$$

$$\frac{1}{e^t} = 0.2$$

$$e^t = \frac{1}{0.2} = 5$$

$$t = \log_e 5 = 1.6 \text{ s}$$

(ii) $i = Ie^{-t/CR} = Ie^{-t}$ since $CR = 1$ s

$= 100e^{-t}$ since $I = 100$ μA

$$= 100 \times \frac{1}{e^t} = 100 \times \frac{1}{e^{1.6}} = \frac{100}{5} = 20 \text{ μA}$$

```
10 PRINT "PROG 124"
20 PRINT "TIME AND CURRENT"
30 INPUT "ENTER INITIAL VOLTAGE" ; V
40 INPUT "ENTER RESISTANCE IN OHMS" ; R
50 INPUT "ENTER CAPACITANCE IN MICROFARADS" ; C
60 INPUT "ENTER CAPACITOR VOLTAGE" ; VC
70 LET TC=C*R/10^6
80 LET T=TC*LOG(V/VC)
90 PRINT "T = " T "SECONDS"
100 LET X=T/TC
110 LET Y=2.7183^X
120 LET I=V*(10^6/R)
130 LET IC=I/Y
140 PRINT "CURRENT AT" T "SECS = " IC "MICROAMPS"
```

9.3 *L–R* circuit

Example 6 Figure 9.7 shows a circuit to which is applied a step voltage of 20 V. The instantaneous voltages and current are denoted by v_L, v_R and i. During the current growth and decay transients the following sequence of events will occur.

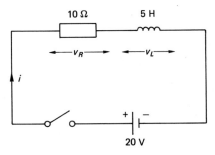

Figure 9.7

(a) A zero time, the instant of switching on, the induced voltage v_L will be a maximum, in opposition to the applied voltage as shown in Figure 9.7, and the circuit conditions are

$i = 0$ A

$v_L = 20$ V

$v_R = 0$ V

(b) The current increases until

$$i = I = \frac{V}{R} = \frac{20}{10} = 2 \text{ A}$$

At any instant

$i = I(1 - e^{-tR/L})$

$v_L = Ve^{-tR/L}$

$v_R = V(1 - e^{-tR/L})$

where t is in seconds, L is in henrys, R is in ohms, and e = 2.7183 (the base of natural logarithms).

To calculate the values of i at different times t, first calculate L/R, the time constant in seconds.

$$\frac{L}{R} = \frac{5}{10} = 0.5 \text{ s}$$

Then $2CR = 1$ s, $3CR = 1.5$ s, $4CR = 2.0$ s and $5CR = 2.5$ s.

At $t = 0$

$$\frac{1}{e^{\frac{-tR}{L}}} = \frac{1}{e^{tR/L}} = \frac{1}{e^0} = 1$$

At $t = 0.5$

$$\frac{1}{e^{0.5 \times (10/5)}} = \frac{1}{e} = 0.368$$

At $t = 1.0$

$$\frac{1}{e^{1 \times 2}} = \frac{1}{e^2} = 0.135$$

At $t = 1.5$

$$\frac{1}{e^{1.5 \times 2}} = \frac{1}{e^3} = 0.0498$$

At $t = 2.0$

$$\frac{1}{e^{2 \times 2}} = \frac{1}{e^4} = 0.0183$$

At $t = 2.5$

$$\frac{1}{e^{2.5 \times 2}} = \frac{1}{e^5} = 0.0067$$

The remaining results are shown in Table 9.7 and plotted in Figures 9.8(a) and 9.8(b).

Table 9.7 Current growth transients

1	2	3	4	5	6
t (s)	$\dfrac{1}{e^{tR/L}}$	$1 - e^{tR/L}$	i (A)	v_L	v_R
0	1	0	0	20	0
0.5	0.368	0.632	1.264	7.36	12.64
1.0	0.135	0.865	1.73	2.7	17.3
1.5	0.0498	0.9502	1.9004	0.996	19.004
2.0	0.0183	0.9817	1.9634	0.366	19.634
2.5	0.0067	0.9933	1.9866	0.134	19.866

Columns 2 and 3 give the same results as for the C–R circuits.
Column 4 is obtained by multiplying column 3 by $I = 2$ A.
Column 5 is obtained by multiplying column 2 by $V = 20$ V.
Column 6 is obtained by multiplying column 3 by $V = 20$ V.

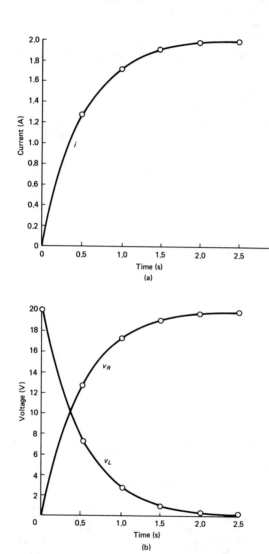

Figure 9.8

```
10 PRINT "PROG 125"
20 PRINT "CURRENT GROWTH"
30 INPUT "ENTER APPLIED VOLTAGE" ; V
40 INPUT "RESISTANCE IN OHMS" ; R
50 INPUT "ENTER INDUCTANCE IN HENRYS" ; L
60 INPUT "ENTER START TIME IN SECONDS" ; T
70 INPUT "ENTER END TIME IN SECONDS" ; B
80 INPUT "ENTER TIME INTERVAL BETWEEN CALCULATIONS" ;D
90 LET TC=L/R
100 LET X=T/TC
110 LET Y=2.7183^X
120 LET I=V/R
130 LET IL=I*(1-(1/Y))
140 PRINT "CURRENT AT" T "SECS = " IL "AMPS"
150 LET EL=V/Y
160 PRINT "EL AT" T "SECS = " EL "VOLTS"
170 LET ER=V-EL
180 PRINT "ER AT" T "SECS = " ER "VOLTS"
190 LET T=T+D
200 IF T<=B THEN GOTO 100
```

Notes:

1 The results confirm the statements made in (a).
2 The time constant L/R is the time taken for the current to reach 63.2% of its final value, since

$$\frac{1.264}{2} \times 100 = 63.2\%$$

It is also the time that the current would take to reach its maximum value, I, if the initial rate of growth was maintained.

3 The maximum current is almost reached in a time of $5CR$.
4 Columns 5 and 6 show that $v_L + v_R = V$ at all times.

(c) If the supply is now disconnected and a short circuit placed across the inductor, the induced e.m.f. will be a maximum opposing the change in current. Its polarity will therefore be reversed as shown in Figure 9.9. The current flows in the same direction as in the growth transient.

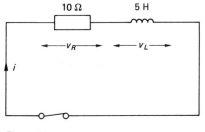

Figure 9.9

(d) At the beginning of the decay transient, i.e. when $t = 0$, the circuit conditions are

$i = I = 2$ A

$v_L = -V = -20$ V

$v_R = V = 20$ V

(e) The current decays and at any instant

$$i = I e^{-tR/L}$$
$$v_L = -V e^{-tR/L}$$
$$v_R = V e^{-tR/L}$$

The results are shown in Table 9.8 and plotted in Figures 9.10(a) and 9.10(b).

Table 9.8 Currrent decay transients

1	2	3	4	5
t (s)	$\dfrac{1}{e^{tR/L}}$	i (A)	v_L	v_R
0	1	2	−20	20
0.5	0.368	0.736	−7.36	7.36
1.0	0.135	0.27	−2.7	2.7
1.5	0.0498	0.0996	−0.996	0.996
2.0	0.0183	0.0366	−0.366	0.366
2.5	0.0067	0.0134	−0.134	0.134

```
10 PRINT "PROG 126"
20 PRINT "CURRENT DECAY"
30 INPUT "ENTER APPLIED VOLTAGE" ; V
40 INPUT "ENTER RESISTANCE IN OHMS" ; R
50 INPUT "ENTER INDUCTANCE IN HENRIES" ; L
60 INPUT "ENTER START TIME IN SECONDS" ; T
70 INPUT "ENTER END TIME IN SECONDS" ; B
80 INPUT "ENTER TIME INTERVAL BETWEEN CALCULATIONS" ; D
90 LET TC=L/R
100 LET X=T/TC
110 LET Y=2.7183^X
120 LET I=V/R
130 LET IL=I/Y
140 PRINT "CURRENT AT" T "SECS = " IL "AMPS"
150 LET ER=V/Y
160 PRINT "ER AT" T "SECS = " ER "VOLTS"
170 LET EL=-V/Y
180 PRINT "EL AT" T "SECS = " EL "VOLTS"
190 LET T=T+D
200 IF T<=B THEN GOTO 100
```

Notes:

1 The calculations for $t = 0$ confirm the statements made in (d).
2 The current falls by a factor of 0.632 or 63.2% of its initial value after one time constant, leaving

$$\frac{0.736}{2} \times 100 = 36.8\%$$

3 The current has almost decayed to zero after $5CR$.
4 $v_R = -v_L$ at all times.

Example 7 A coil of inductance 4 H and resistance 40 Ω is suddenly connected to a 100-V supply. Determine: (i) The time constant, (ii) the time for the current to reach 0.5 A, (iii) the current after 0.4 s.

(i) Time constant $= \dfrac{L}{R} = \dfrac{4}{40} = 0.1$ s

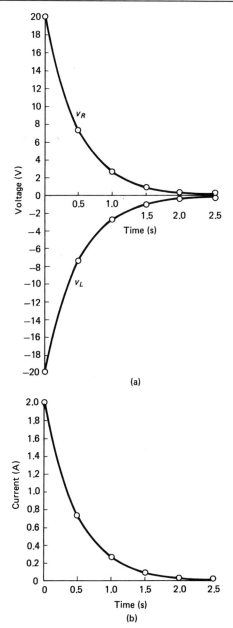

Figure 9.10

(ii)
$$I = \frac{V}{R} = \frac{100}{40} = 2.5 \text{ A}$$
$$i = I(1 - e^{-tR/L})$$
$$0.5 = 2.5(1 - e^{-tR/L})$$
$$0.5 = 2.5 - 2.5e^{-tR/L}$$
$$2.5e^{-tR/L} = 2.5 - 0.5 = 2.0$$
$$e^{-tR/L} = \frac{2.0}{2.5} = 0.8$$
$$\frac{1}{e^{tR/L}} = 0.8$$

Transposing
$$e^{tR/L} = \frac{1}{0.8} = 1.25$$
$$e^{10t} = 1.25 \text{ since } \frac{R}{L} = 10$$
$$10t = \log_e 1.25 = 0.223$$
$$t = 0.0223 \text{ s}$$

(iii) $i = 2.5(1 - e^{-0.4 \times (40/4)})$
$= 2.5(1 - e^{-4})$
$= 2.5\left(1 - \frac{1}{e^4}\right) = 2.5 \times 0.982 = 2.45 \text{ A}$

```
10 PRINT "PROG 127"
20 PRINT "TIME AND CURRENT"
30 INPUT "ENTER APPLIED VOLTAGE" ; V
40 INPUT "RESISTANCE IN OHMS" ; R
50 INPUT "ENTER INDUCTANCE IN HENRYS" ; L
60 INPUT "INDUCTOR CURRENT IN AMPS" ; IL
70 LET TC=L/R
80 PRINT "TIME CONSTANT = " TC "SECONDS"
90 LET I=V/R
100 LET T=TC*LOG(I/(I-IL))
110 PRINT "TIME TO REACH" IL "AMPS = " T "SECS"
120 INPUT "ENTER TIME IN SECONDS" ; T
130 LET X=T/TC
140 LET Y=2.718^X
150 LET IL=I-I/Y
160 PRINT "CURRENT AFTER" T "SECONDS = " IL "AMPS"
```

Example 8 A 2-H inductor, with a resistance of 10 Ω, is driven from a 50-V d.c. supply. The steady state has been reached, the supply disconnected, and a short circuit placed across the inductor. Calculate: (i) the time for the current to drop to half of its steady-state value, and (ii) the voltage across the coil at that time, when a short circuit is placed across the network.

(i) $I = \frac{V}{R} = \frac{50}{10} = 5$

$i = Ie^{-tR/L}$

$\frac{5}{2} = 5e^{-t \times 10/2}$

$$2.5 = 5e^{-5t}$$

$$2.5 = \frac{5}{e^{5t}}$$

$$e^{5t} = \frac{5}{2.5} = 2$$

$$5t = \log_e 2 = 0.693$$

$$t = \frac{0.693}{5} = 0.1386 \text{ s}$$

(ii) $v_L = -Ve^{-tR/L}$

$$= -50 \times \frac{1}{0.1386 \times 5}$$

$$= -25 \text{ V}$$

```
10 PRINT "PROG 128"
20 PRINT "TIME AND VOLTAGE"
30 INPUT "ENTER APPLIED VOLTAGE" ; V
40 INPUT "RESISTANCE IN OHMS" ; R
50 INPUT "ENTER INDUCTANCE IN HENRYS" ; L
60 LET TC=L/R
70 LET I=V/R
80 LET IL=I/2
90 LET T=TC*LOG(I/IL)
110 PRINT "TIME TO REACH" IL "AMPS = " T "SECS"
120 LET X=T/TC
130 LET Y=2.7183^X
140 LET EL=-V/Y
150 PRINT "COIL VOLTAGE AT" T "SECONDS = " EL "VOLTS"
```

Example 9 Figure 9.11 shows a 2-H coil and resistance 2 Ω connected in parallel with a 10-Ω resistor across a 200-V supply; steady conditions have been reached. If the supply is suddenly disconnected, determine: (i) initial current, (ii) initial voltage across the 10-Ω resistor, (iii) current after 0.5 s, (iv) voltage after 0.5 s across R.

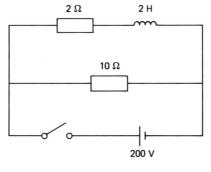

Figure 9.11

(i) Initially the current I through the coil is

$$I = \frac{V}{R} = \frac{200}{2} = 100 \text{ A}$$

(ii) The 100 A current will flow through the 10-Ω resistor and the initial voltage across the 10-Ω resistor

$$= 100 \times 10 = 1000 \text{ V}$$

(iii)

$$i = Ie^{-tR/L}$$

$$= 100 \times \frac{1}{0.5 \times (12/2)}$$

(note $R = 10 + 2 \, \Omega$)

$$= 100 \times \frac{1}{e^3} = 4.978 \text{ A}$$

(iv)

$$v_R = Ve^{-tR/L}$$

$$= 1000 \times \frac{1}{e^3} = 49.78 \text{ V}$$

```
10 PRINT "PROG 129"
20 PRINT "CURRENT AND VOLTAGE"
30 INPUT "ENTER APPLIED VOLTAGE" ; V
40 INPUT "ENTER RESISTOR VALUE" ; R1
50 INPUT "ENTER COIL RESISTANCE" ; R2
60 INPUT "ENTER COIL INDUCTANCE" ; L
70 LET I=V/R2
80 PRINT "INITIAL CURRENT = " I "AMPS"
90 LET VR=I*R1
110 PRINT "RESISTOR VOLTAGE = " VR "VOLTS"
120 INPUT "ENTER TIME IN SECONDS" ; T
130 LET R=R1+R2
140 LET TC=L/R
150 LET X=T/TC
160 LET Y=2.7183^X
170 LET IL=I/Y
180 PRINT "CURRENT AFTER" T "SECONDS = " IL "AMPS"
190 LET ER=VR/Y
200 PRINT "VOLTAGE AFTER " T " SECONDS = " ER "VOLTS"
```

Problems

1. A 1.5-μF capacitor is connected in series with a 1.0-MΩ resistor. If 240 V d.c. is suddenly applied to the circuit, calculate: (a) v_C at 2 and 4 s, (b) v_R at 2 and 4 s, (c) the current at 1, 3 and 5 s.
2. When the capacitor in Problem 1 is fully charged, the supply voltage is removed and is replaced by a short circuit. Calculate: (a) v_C at 2 and 5 s, (b) v_R at 3 and 4 s, (c) the current at 1, 3 and 5 s.
3. A 0.03-μF capacitor is connected in series with a 1000-Ω resistor. If 50 V d.c. is suddenly applied to the network, calculate: (a) v_C at 30 and 120 μs, (b) v_R at 30 and 120 μs, (c) the current at 30, 60 and 120 μs.
4. When the capacitor in Problem 3 is fully charged, the supply voltage is removed and is replaced by a short circuit. Calculate:

(a) v_C at 30 and 120 μs, (b) v_R at 30 and 120 μs, (c) the current at 30, 60 and 120 μs.

5 A 200-μF capacitor is connected in series with a 200-Ω resistor. If 40 V d.c. is suddenly connected to the circuit, calculate: (a) v_C at 80 and 200 ms, (b) v_R at 40 and 120 ms, (c) the current at 200 ms.

6 When the capacitor in Problem 5 is fully charged, the supply voltage is removed and is replaced by a short circuit. Calculate: (a) v_C at 80 and 200 ms, (b) v_R at 40 and 120 ms, (c) the current at 80 and 200 ms.

7 A 2-μF capacitor is connected in series with a 500-kΩ resistor. If 200 V d.c. is applied to the circuit, calculate: (a) the time for the capacitor voltage to reach 150 V, (b) the current flowing at this time.

8 A 4-μF capacitor has been charged to 240 V. If the capacitor is then discharged through a 1.0-MΩ resistor, calculate: (a) the time for the capacitor voltage to drop to 120 V, (b) the current flowing at this time.

9 A 20-H inductor is connected in series with a 100-Ω resistor. If 50 V is connected to the circuit, calculate: (a) v_L at 0.4 and 1.0 s, (b) v_R at 0.4 and 1.0 s, (c) the current at 0.2, 0.4 and 0.6 s.

10 Assume that the current in Problem 9 has reached a maximum. The supply voltage is then disconnected and replaced with a short circuit across the inductor. Calculate: (a) v_L at 0.2 and 0.6 s, (b) v_R at 0.2 and 0.6 s, (c) the current at 0.2, 0.4 and 0.6 s.

11 A coil with an inductance of 10 H and a resistance of 50Ω is suddenly connected to 120 V d.c. Calculate: (a) the time constant, (b) the time for the current to reach 1.0 A, (c) the current after 1.0 s.

12 A 2-H coil with a resistance of 12Ω is connected to a 24-V d.c. supply. The steady state has been reached and the supply is suddenly disconnected and replaced by a short circuit. Calculate: (a) the time for the current to drop to 1.0 A, (b) the coil voltage at this time.

13 A coil having an inductance of 4 H and a resistance of 1.0 Ω is connected in parallel with a 12-Ω resistor. The circuit is driven from a 250-V d.c. supply. If the steady conditions have been reached, and the supply is suddenly disconnected, calculate: (a) the initial current, (b) the initial voltage across the 12-Ω resistor, (c) the current after 2 s, (d) the voltage across the 12-Ω resistor after 2 s.

10 Electromagnetism

10.1 Force on a conductor, F

A current-carrying conductor placed in a magnetic field has a force exerted on it where

Force = BlI

where force, F is in newtons
B = magnetic flux density in teslas, T
l = conductor length in metres, m
I = current in amperes, A

Example 1 A conductor carrying 50 A is at right angles to a magnetic field having a density of 0.5 T. Calculate the force on the conductor in newtons per metre length.

$F = BlI$

$= 0.5 \times 1 \times 50 = 25$ N

```
10 PRINT "PROG 130"
20 PRINT "FORCE ON A CONDUCTOR"
30 INPUT "ENTER FLUX DENSITY IN TESLAS" ; B
40 INPUT "ENTER LENGTH IN METRES" ; L
50 INPUT "ENTER CURRENT IN AMPS" ; I
60 LET F=B*L*I
70 PRINT "FORCE = " F "NEWTONS"
```

Example 2 If the force on a conductor 100 mm long is 2 N, and the current in the conductor is 0.5 A, calculate the flux density.

$$B = \frac{F}{lI} = \frac{2}{0.1 \times 0.5} = 40 \text{ T}$$

```
10 PRINT "PROG 131"
20 PRINT "FLUX DENSITY"
30 INPUT "ENTER FORCE IN NEWTONS" ; F
40 INPUT "ENTER LENGTH IN MILLIMETRES" ; L
50 INPUT "ENTER CURRENT IN AMPS" ; I
60 LET B=F*10^3/(L*I)
70 PRINT "FLUX DENSITY = " B "TESLAS"
```

Example 3 Calculate the length of a conductor which has a force of 1 N exerted on it when carrying a current of 8 A in a magnetic field of 5 T.

$$l = \frac{F}{BI} = \frac{1}{5 \times 8} = 0.025 \text{ m} = 25 \text{ mm}$$

```
10 PRINT "PROG 132"
20 PRINT "CONDUCTOR LENGTH"
30 INPUT "ENTER FORCE IN NEWTONS" ; F
40 INPUT "ENTER FLUX DENSITY IN TESLAS" ; B
50 INPUT "ENTER CURRENT IN AMPS" ; I
60 LET L=F/(B*I)
70 PRINT "CONDUCTOR LENGTH = " L*10^3 "MILLIMETRES"
```

Example 4 If $F = 12$ N, $B = 3$ T, and $l = 5$ m, calculate I.

$$I = \frac{F}{Bl} = \frac{12}{3 \times 5} = 0.8 \text{ A}$$

```
10 PRINT "PROG 133"
20 PRINT "CURRENT"
30 INPUT "ENTER FORCE IN NEWTONS" : F
40 INPUT "ENTER FLUX DENSITY IN TESLAS" ; B
50 INPUT "ENTER LENGTH IN METRES" ; L
60 LET I=F/(B*L)
70 PRINT "CURRENT = " I "AMPS"
```

10.2 E.m.f. generated, E

With uniform flux density B teslas

When a conductor cuts or is cut by a magnetic field, an e.m.f. is generated in the conductor. The magnitude of the generated e.m.f. depends on the rate at which the conductor cuts or is cut by the magnetic field. The required formula is

$$E = Blv$$

where E = electromotive force in volts, V
B = magnetic flux density in teslas, T
l = conductor length in metres, m
v = velocity in metres per second, m/s

Example 5 Calculate the magnitude of the e.m.f. induced in a conductor 100 mm long, moving with a velocity of 10 m/s, at right angles to a magnetic field having a flux density of 0.5 T.

$$E = Blv = 0.5 \times 0.1 \times 10 = 0.5 \text{ V}$$

```
10 PRINT "PROG 134"
20 PRINT "INDUCED E.M.F"
30 INPUT "ENTER FLUX DENSITY IN TESLAS" ; B
40 INPUT "ENTER LENGTH IN MILLIMETRES" ; L
50 INPUT "ENTER VELOCITY IN METRES/SEC" ; V
60 LET E=B*(L/10^3)*V
70 PRINT "E.M.F = " E "VOLTS"
```

Example 6 Two volts is induced in a conductor 50 mm long, moving at a velocity of 5 m/s at right angles to a magnetic field. What is the flux density?

$$B = \frac{E}{lv} = \frac{2}{0.05 \times 5} = 8 \text{ T}$$

```
10 PRINT "PROG 135"
20 PRINT "FLUX DENSITY"
30 INPUT "ENTER E.M.F IN VOLTS" ; E
40 INPUT "ENTER LENGTH IN MILLIMETRES" ;L
50 INPUT "ENTER VELOCITY IN METRES/SEC" ; V
60 LET B=E*10^3/(L*V)
70 PRINT "FLUX DENSITY = " B "TESLAS"
```

Example 7 What is the length of conductor if the voltage induced is 100 mV in a magnetic field of 0.5 T. The velocity is 2 m/s.

$$l = \frac{E}{Bv} = \frac{0.1}{0.5 \times 2} = 0.1 \text{ m} = 10 \text{ cm}$$

```
10 PRINT "PROG 136"
20 PRINT "CONDUCTOR LENGTH"
30 INPUT "ENTER E.M.F IN MILLIVOLTS" ; E
40 INPUT "ENTER FLUX DENSITY IN TESLAS" ; B
50 INPUT "ENTER VELOCITY IN METRES/SEC" ; V
60 LET L=E/(B*V*10^3)
70 PRINT "LENGTH = " L "METRES"
```

Example 8 Calculate the velocity of a 2-m conductor in a magnetic field of 6 T, if the voltage induced is 6 V.

$$v = \frac{E}{Bl} = \frac{6}{6 \times 2} = 0.5 \text{ m/s}$$

```
10 PRINT "PROG 137"
20 PRINT "VELOCITY"
30 INPUT "ENTER E.M.F IN VOLTS" ; E
40 INPUT "ENTER FLUX DENSITY IN TESLAS" ; B
50 INPUT "ENTER LENGTH IN METRES"; L
60 LET V=E/(B*L)
70 PRINT "VELOCITY = " V "METRES/SEC"
```

With total flux Φ webers

The problems so far have involved uniform flux densities of B teslas. The tesla is defined as the density of a magnetic field such that a conductor carrying one ampere at right angles to the field has a force of one newton per metre acting on it. In a magnetic field the total flux, measured in webers, Wb, and represented by the Greek letter Φ (phi), depends on the cross-sectional area where

Φ (Wb) = B (teslas) × A (metres)

The weber can be defined in two ways:

1 The amount of flux, when cut at a uniform rate by a conductor in one second, generates an e.m.f. of one volt.
2 The magnetic flux linking one turn induces in it an e.m.f. of one volt when the flux is reduced to zero at a uniform rate in one second.

The induced voltage in a coil therefore depends on the total flux, the number of turns, and the time for the field to be reversed. This is shown in the next example.

Example 9 A magnetic flux of 600 µWb acts through a coil of 1500 turns and is reversed in 0.5 s. Calculate the average value of the e.m.f. induced.

Average e.m.f. = number of turns × rate of change of flux with time

$$= \frac{1500 \times 2 \times 600 \times 10^{-6}}{0.5} = 3.6 \text{ V}$$

The factor 2 is due to the fact that the flux decreases from 600 µWb to zero and then increases to 600 µWb in the reverse direction.

Note: This type of problem has already been dealt with in Section 8.3.

```
10 PRINT "PROG 138"
20 PRINT "INDUCED E.M.F"
30 INPUT "ENTER FLUX IN MICROWEBERS" ; PHI
40 INPUT "ENTER NUMBER OF TURNS" ; N
50 INPUT "ENTER TIME IN SECONDS" ; T
60 LET E=N*(PHI/10^6)*2/T
70 PRINT "E.M.F = " E "VOLTS"
```

10.3 Magnetomotive force, F

A magnetic circuit consisting of a coil wound on either a magnetic or non-magnetic former can be compared with the electric circuit. In the electric circuit:

$$\text{Current } (I) = \frac{\text{e.m.f. } (E)}{\text{resistance } (R)}$$

In the magnetic circuit:

$$\text{Flux } (\Phi) = \frac{\text{magnetomotive force, m.m.f. } (F)}{\text{Reluctance } (R_m)}$$

Magnetomotive force is measured in amperes, A, and is produced by the current in the magnetizing coil where:

m.m.f. = NI amperes
N = number of turns
I = magnetizing current in amperes

Increasing the number of turns and the current increases the flux. This can be compared with increasing the voltage in an electric circuit, which in turn increases the current.

10.4 Magnetic field strength, H

The m.m.f. produces a magnetic field of strength H throughout the circuit. If the mean length of the magnetic circuit is 1 m then

$$H = \frac{NI}{1} \text{ A}$$

The product NI is referred to as the number of ampere turns.

Example 10 A coil of 500 turns is wound on a wooden ring having a mean circumference of 200 mm. If the current in the coil is 2 A calculate the magnetic field strength.

Mean circumference = 200 mm = 0.2 m

$$H = \frac{500 \times 2}{0.2} = 5000 \text{ A/m}$$

```
10 PRINT "PROG 139"
20 PRINT "MAGNETIC FIELD STRENGTH"
30 INPUT "ENTER NUMBER OF TURNS" ; N
40 INPUT "ENTER CURRENT IN AMPS" ; I
50 INPUT "ENTER LENGTH IN MILLIMETRES" ; L
60 LET H=N*I*10^3/L
70 PRINT "FIELD STRENGTH = " H "AMPS/METRE"
```

Example 11 A coil 150 mm in length has a field strength of 1000 A/m when the current flowing is 15 A. Calculate the number of turns.

$$N = \frac{Hl}{I} = \frac{1000 \times 0.150}{15} = 10 \text{ turns}$$

```
10 PRINT "PROG 140"
20 PRINT "NUMBER OF TURNS"
30 INPUT "ENTER FIELD STRENGTH IN AMPS/METRE" ; H
40 INPUT "ENTER CURRENT IN AMPS" ; I
50 INPUT "ENTER LENGTH IN MILLIMETRES" ; L
60 LET N=H*L/(I*10^3)
70 PRINT "NUMBER OF TURNS = " N
```

Example 12 What is the current required to produce a field strength of 500 A/m in a coil of 200 turns and a mean length of 500 mm?

$$I = \frac{Hl}{N} = \frac{500 \times 0.5}{200} = 1.25 \text{ A}$$

```
10 PRINT "PROG 141"
20 PRINT "CURRENT"
30 INPUT "ENTER FIELD STRENGTH IN AMPS/METRE" ; H
40 INPUT "ENTER NUMBER OF TURNS" ; N
50 INPUT "ENTER LENGTH IN MILLIMETRES" ; L
60 LET I=H*L/(N*10^3)
70 PRINT "CURRENT = " I "AMPS"
```

Example 13 Calculate the mean circumference of a ring having 150 turns if a current of 5 A produces a magnetic field with a strength of 2500 A/m.

$$l = \frac{NI}{H} = \frac{150 \times 5}{2500} = 0.3 \text{ m} = 300 \text{ mm}$$

```
10 PRINT "PROG 142"
20 PRINT "CIRCUMFERENCE"
30 INPUT "ENTER FIELD STRENGTH IN AMPS/METRE" ; H
40 INPUT "ENTER NUMBER OF TURNS" ; N
50 INPUT "ENTER CURRENT IN AMPS" ; I
60 LET L=N*I/H
70 PRINT "CIRCUMFERENCE = " L*10^3 "MILLIMETRES"
```

10.5 Permeability

The ratio of the flux density B to the magnetic field strength H in vacuum is called the permeability of free space. It has the symbol μ_o and is measured in henrys per metre, H/m.

$$\mu_o = 4\pi \times 10^{-7} \text{ H/m}$$

The value of μ_o is almost exactly the same for air and for all other non-magnetic materials.

$$\mu_o = \frac{B}{H} \text{ and } H = \frac{B}{4\pi \times 10^{-7}} \text{ H/m and } B = \mu_o H$$

10.6 Total flux

It was mentioned in Section 10.2 that:

Total flux (Wb) = flux density B (T) × cross-sectional area A (m^2)

It therefore follows that:

Total flux (Wb) = $\mu_o HA$

Electromagnetism 165

Example 14 A coil of 100 turns is wound on a non-magnetic former having a mean circumference of 500 mm and a cross-sectional area of 600 mm^2. If the current flowing in the coil is 10 A, calculate: (i) the magnetic field strength (H), (ii) the flux density (B), (iii) the total flux Φ.

(i) $l = 500$ mm $= 0.5$ m

$$H = \frac{NI}{l} = \frac{100 \times 10}{0.5} = 2000 \text{ A/m}$$

(ii) $B = \mu_o H = 4\pi \times 10^{-7} \times 2000 = 2.51 \times 10^{-3}$ T

$\qquad\qquad\qquad\qquad\qquad\qquad\qquad\quad = 2.51$ mT

(iii) Area $= 600$ mm$^2 = 600 \times 10^{-6}$ m^2

$\quad\Phi = \mu_o H A$

$\quad\quad = BA$

$\quad\quad = 2.51 \times 10^{-3} \times 600 \times 10^{-6}$ Wb

$\quad\quad = 2.51 \times 10^{-3} \times 600 \times 10^{-6} \times 10^6$ µWb

$\quad\quad = 1.506$ µWb

```
10 PRINT "PROG 143"
20 PRINT "H,B,PHI"
30 INPUT "ENTER NUMBER OF TURNS" ; N
40 INPUT "ENTER CURRENT IN AMPS" ; I
50 INPUT "ENTER LENGTH IN MILLIMETRES" ; L
60 INPUT "ENTER AREA IN SQR MILLIMETRES" ; A
70 LET H=N*I*10^3/L
80 PRINT "FIELD STRENGTH = " H "AMPS/METRE"
90 LET B=4*3.141593*H/10^7
100 PRINT "FLUX DENSITY = " B*10^3 "MILLITESLAS"
110 LET PHI=B*A/10^6
120 PRINT "TOTAL FLUX = " PHI*10^6 "MICROWEBERS"
```

Example 15 A coil wound on a wooden bobbin has a magnetic field strength of 2500 A/m. If the total flux is 25 µWb, calculate the cross-sectional area of the bobbin. Take μ_o as $4\pi \times 10^{-7}$.

$\Phi = \mu_o H A$

$A = \dfrac{\Phi}{\mu_o H}$

$\quad = \dfrac{25 \times 10^{-6}}{4\pi \times 10^{-7} \times 2500}$ m^2 since 25 µWb $= 25 \times 10^{-6}$ Wb

$\quad = 7.96 \times 10^{-3}$ m^2

$\quad = 7.96 \times 10^{-3} \times 10^6$ mm^2

$\quad = 7960$ mm^2

```
10 PRINT "PROG 144"
20 PRINT "AREA"
30 INPUT "ENTER FIELD STRENGTH IN AMPS/METRE" ; H
40 INPUT "ENTER TOTAL FLUX IN MICROWEBERS" ; PHI
50 LET A=PHI*10^7/(4*3.141593*10^6*H)
60 PRINT "AREA = " A*10^6 "SQR MILLIMETRES"
```

Example 16 A coil wound on a plastic former has a total flux of 15 µWb. If the cross-sectional area is 4000 mm^2, calculate the field strength. $\mu_o = 4\pi \times 10^{-7}$.

$$\Phi = \mu_o H A$$

$$H = \frac{\Phi}{\mu_o A}$$

$$= \frac{15 \times 10^{-6}}{4\pi \times 10^{-7} \times 4000 \times 10^{-6}} = 2984 \text{ A/m}$$

```
10 PRINT "PROG 145"
20 PRINT "FIELD STRENGTH"
30 INPUT "ENTER TOTAL FLUX IN MICROWEBERS" ; PHI
40 INPUT "ENTER AREA IN SQR MILLIMETRES" ;A
50 LET H=PHI*10^7*10^6/(4*3.141593*10^6*A)
60 PRINT "FIELD STRENGTH = " H "AMPS/METRE"
```

Example 17 What is the flux density in Examples 15 and 16?

$$B = \mu_o H = 4\pi \times 10^{-7} \times 2500 \text{ T}$$

$$= 3.14 \times 10^{-3} \text{ T}$$

$$= 3.14 \text{ mT}$$

$$B = \mu_o H = 4\pi \times 10^{-7} \times 2984$$

$$= 3.75 \times 10^{-3} \text{ T}$$

$$= 3.75 \text{ mT}$$

```
10 PRINT "PROG 146"
20 PRINT "FLUX DENSITY"
30 INPUT "ENTER FIELD STRENGTH IN AMPS/METRE" ; H
40 LET B=4*3.141593*H/10^7
50 PRINT "FLUX DENSITY = " ; B*10^3 "MILLITESLAS"
```

Example 18 A coil wound uniformly on a non-magnetic ring has an air gap 4 mm long and a cross-sectional area of 20 cm². What is the number of ampere turns required to establish a total flux of 10 mWb in the air gap?

$$A = 20 \times 10^{-4} \text{ m}^2$$

Flux density $B = \Phi/A = \dfrac{10 \times 10^{-3}}{20 \times 10^{-4}} = 5 \text{ T}$

Magnetic field strength $H = \dfrac{B}{\mu_o}$

$$= \frac{5}{4\pi \times 10^{-7}} = 3.98 \times 10^6 \text{ A/m}$$

Since $H = \dfrac{NI}{l}$, NI (the ampere turns) $= Hl$

where l = length of the gap $= 4 \times 10^{-3}$ m

$$NI = 3.98 \times 10^6 \times 4 \times 10^{-3}$$

$$= 15\ 920$$

```
10 PRINT "PROG 147"
20 PRINT "AMPERE TURNS"
30 INPUT "ENTER AIR GAP IN MILLIMETRES" ; L
40 INPUT "ENTER AREA IN SQR CENTIMETRES" ; A
50 INPUT "ENTER TOTAL FLUX IN MILLIWEBERS" ; PHI
60 LET B=PHI*10^4/(A*10^3)
70 LET H=B*10^7/(4*3.141593)
80 LET NI=H*L/10^3
90 PRINT "AMPERE TURNS = " NI
```

10.7 Relative and absolute permeabilities

Using a magnetic core, such as steel inserted into a coil, increases the flux produced by a given m.m.f. The ratio of the flux density in a magnetic material to the flux density in a non-magnetic material produced by the same magnetic field strength is called the relative permeability, denoted by the symbol μ_r. For a magnetic core, therefore,

$$B = \mu_0 \mu_r H$$

The product $\mu_0 \mu_r$ is called the absolute permeability, μ, and

$$B = \mu H$$

$$\mu = \frac{B}{H}$$

Since μ_r varies for different values of the magnetic field strength (H), it is convenient to represent the relationship between the flux density (B) and the magnetic field strength (H) in the form of a graph, as in Figure 10.1.

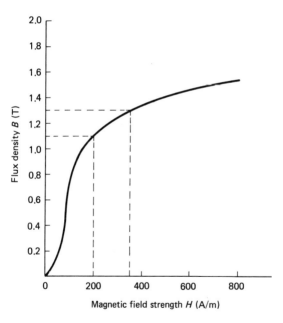

Example 19 An iron ring having a uniform cross-sectional area of 500 mm² and a mean length of 0.2 m has a *B–H* characteristic as shown in Figure 10.1. Find the current required in a coil of 400 turns to give a magnetic flux of 550 μWb, and the relative permeability of the iron under this condition.

$$\Phi = 550 \times 10^{-6} \text{ Wb}$$

$$B = \frac{\Phi}{A} = \frac{550 \times 10^{-6}}{500 \times 10^{-6}} = 1.1 \text{ T}$$

From the graph shown

$$B = 1.1, H = 200 \text{ A/m}$$

$$\text{m.m.f.} = NI = Hl$$

$$NI = 200 \times 0.2 = 40 \text{ A}$$

$$I = \frac{40}{400} = 0.1 \text{ A}$$

Absolute permeability

$$\mu = \frac{B}{H} = \frac{1.1}{200} = 5.5 \times 10^{-3} \text{ H/m}$$

Relative permeability

$$\mu_r = \frac{\mu}{\mu_0}$$

$$= \frac{5.5 \times 10^{-3}}{4\pi \times 10^{-7}} = 4.377 \times 10^3$$

Note

$$H = \frac{NI}{l}$$

$$= \frac{400 \times 0.1}{0.2}$$

$$= 200 \text{ A/m (the value from the graph)}$$

```
10 PRINT "PROG 148"
20 PRINT "CURRENT AND RELATIVE PERMEABILIT
30 INPUT "ENTER NUMBER OF TURNS" ; N
40 INPUT "ENTER AREA IN SQR MILLIMETRES" ; A
50 INPUT "ENTER LENGTH IN METRES" ; L
60 INPUT "ENTER TOTAL FLUX IN MICROWEBERS" ; PHI
70 INPUT "ENTER FIELD STRENGTH IN AMPS/METRE" ; H
80 LET B=PHI*10^6/(A*10^6)
90 PRINT "FLUX DENSITY = " B "TESLAS"
100 LET I=H*L/N
110 PRINT "CURRENT = " I "AMPS"
120 U=B/H
130 LET UR=U*10^7/(4*3.141593)
140 PRINT "RELATIVE PERMEABILITY = " UR
```

Example 20 If an air gap 0.5 mm wide is made in the ring of Example 19, find the new value of the current needed to maintain the flux at its original value of 550 μWb.

For the air gap both Φ and A are the same as for the ring, i.e.

$$B = 1.1 \text{ T}$$

But

$$H = \frac{B}{\mu_0} = \frac{1.1}{4\pi \times 10^{-7}} \text{ A/m}$$

$$\text{m.m.f.} = Hl = \frac{1.1 \times 0.5 \times 10^{-3}}{4\pi \times 10^{-7}} = 437.67 \text{ A}$$

Cutting an air gap means that the length of the ring is shortened to 0.1995 m.

For the ring

$NI = 200 \times 0.1995$

$= 39.9$ A

The circuit m.m.f. is the sum of the gap m.m.f. and the ring m.m.f.

$NI = 437.67 + 39.9$

$= 477.57$

$I = \dfrac{477.57}{400} = 1.1939$ A

Note the increase in current required to maintain the original conditions.

```
10 PRINT "PROG 149"
20 PRINT "CURRENT WITH AIR GAP"
30 INPUT "ENTER FLUX DENSITY IN TESLAS" ; B
40 INPUT "ENTER LENGTH OF AIR GAP IN MMS" ; LA
50 INPUT "ENTER RING FIELD STRENGTH IN AMPS/METRE" ; H
60 INPUT "ENTER RING LENGTH IN METRES" ; LR
70 INPUT "ENTER NUMBER OF TURNS" ; N
80 LET HA=B*10^7/(4*3.141593)
90 LET MI=HA*LA/10^3
100 LET M2=H*LR
110 LET I=(MI+M2)/N
120 PRINT "CURRENT = " I "AMPS"
```

Example 21 A magnetic circuit consists of two parts in series. Part A is made of iron 200 mm long, having a B–H characteristic as shown in Figure 1. Part B is an air gap 0.8 mm long. Both parts have the same cross-sectional area of 0.8 cm^2. Find the current required in the coil of 200 turns wound uniformly on part A to produce a flux of 104 μWb in the air gap. Neglect losses due to magnetic leakage and fringing.

Air gap

$$B = \dfrac{\Phi}{A} = \dfrac{104 \times 10^{-6}}{0.8 \times 10^{-4}} = 1.3 \text{ T}$$

$$H = \dfrac{B}{\mu_o} = \dfrac{1.3}{4\pi \times 10^{-7}}$$

$$\text{m.m.f.} = Hl = \dfrac{1.3 \times 0.8 \times 10^{-3}}{4\pi \times 10^{-7}} = 828 \text{ A}$$

Iron

$B = 1.3$ T since Φ and A are the same

$H = 350$ A/m

m.m.f. $= Hl = 350 \times 200 \times 10^{-3} = 70$ A

Total m.m.f. $= 828 + 70 = 898 = NI$

$$I = \dfrac{898}{200} = 4.49 \text{ A}$$

```
10 PRINT "PROG 150"
20 PRINT "CURRENT WITH AIR GAP"
30 INPUT "ENTER FLUX IN AIR GAP IN MICROWEBERS" ; PHI
40 INPUT "ENTER LENGTH OF AIR GAP IN MMS" ; LA
50 INPUT "ENTER RING FIELD STRENGTH IN AMPS/METRE" ; H
60 INPUT "ENTER RING LENGTH IN MILLIMETRES" ; LR
70 INPUT "ENTER NUMBER OF TURNS" ; N
80 INPUT "ENTER AREA IN SQR CENTIMETRES" ; A
90 LET B=PHI*10^4/(A*10^6)
100 LET HA=B*10^7/(4*3.141593)
110 LET M1=HA*LA/10^3
120 LET M2=H*LR/10^3
130 LET I=(M1+M2)/N
140 PRINT "CURRENT = " I "AMPS"
```

10.8 Reluctance, R_m

In Section 10.3 the comparison between the electric circuit and the magnetic circuit was made, and the reluctance is analogous to the resistance. The equation for the reluctance from Section 10.3 is

$$R_m = \frac{m.m.f.}{\Phi} = \frac{l}{\mu_r \mu_o A}$$

= the reluctance of the magnetic circuit in amperes per weber, A/Wb

Example 22 A steel ring has a cross-sectional area of 400 mm² and a mean circumference of 300 mm. It is wound with a coil of 150 turns. Calculate: (i) the reluctance of the ring, (ii) the current required to produce a flux of 480 µWb in the ring. The relative permeability of the ring is shown in Figure 10.2.

$$\text{Flux density } (B) = \frac{480 \times 10^{-6} \text{ Wb}}{400 \times 10^{-6} \text{ m}^2} = 1.2 \text{ T}$$

From the graph 1.2 T = 500 = μ_r

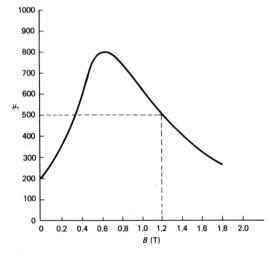

Figure 10.2

(i) $S = \dfrac{0.3}{500 \times 4\pi \times 10^{-7} \times 400 \times 10^{-6}}$

$= 1.194 \times 10^6$ A/Wb

(ii) $F = \Phi R_m = 480 \times 10^{-6} \times 1.194 \times 10^6 = 573.12$ A

$573.12 = NI = 150I$

$I = \dfrac{573.12}{150} = 3.82$ A

An alternative solution is to consider that

$NI = Hl$ and $H = \dfrac{B}{\mu_r \mu_o}$

$NI = \dfrac{1.2 \times 0.3}{500 \times 4\pi \times 10^{-7}}$

$= 572.96$

$I = \dfrac{572.96}{150} = 3.82$ A

```
10 PRINT "PROG 151"
20 PRINT "RELUCTANCE AND PERMEABILITY"
30 INPUT "ENTER AREA IN SQR MILLIMETRES" ; A
40 INPUT "ENTER CIRCUMFERENCE IN MILLIMETRES" ; L
50 INPUT "ENTER NUMBER OF TURNS" ; N
60 INPUT "ENTER RELATIVE PERMEABILITY" ; UR
70 INPUT "ENTER TOTAL FLUX IN MICROWEBERS" ; PHI
80 LET S=L*10^7*10^6/(UR*4*3.141593*A*10^3)
90 PRINT "RELUCTANCE = " S "AMPS/WEBER"
100 LET I=PHI*S/(N*10^6)
110 PRINT "CURRENT I = " I "AMPS"
```

Example 23 Example 20 in Section 10.7 can be solved by the reluctance method.

Total $R_m = R_m(\text{air}) + R_m(\text{iron})$

$= \dfrac{l}{\mu_o A} + \dfrac{l}{\mu_o \mu_r A}$

$= \dfrac{0.5 \times 10^{-3}}{4\pi \times 10^{-7} \times 500 \times 10^{-6}}$

$\quad + \dfrac{0.1995}{4\pi \times 10^{-7} \times 4.377 \times 10^3 \times 500 \times 10^{-6}}$

$= 795775 + 72541.5$

$= 868316.5$

$NI = R_m \Phi$

$= 868316.5 \times 550 \times 10^{-6}$

$= 477.57$

$I = \dfrac{477.57}{400} = 1.19$ A

```
10 PRINT "PROG 152"
20 PRINT "RELUCTANCE METHOD"
30 INPUT "ENTER FLUX DENSITY IN TESLAS" ; B
40 INPUT "ENTER LENGTH OF AIR GAP IN MMS" ; LA
50 INPUT "ENTER RING FIELD STRENGTH IN AMPS/METRE" ; H
60 INPUT "ENTER RING LENGTH IN METRES" ; LR
70 INPUT "ENTER AREA IN SQR MILLIMETRES" ; A
80 INPUT "ENTER TOTAL FLUX IN MICROWEBERS" ; PHI
90 INPUT "ENTER NUMBER OF TURNS" ; N
100 LET UR=B*10^7/(4*3.141593*H)
110 LET S1=LA*10^7*10^6/(4*3.141593*A*10^3)
120 LET S2=LR*10^7*10^6/(4*3.141593*UR*A)
130 LET S=S1+S2
140 LET I=S*PHI/(N*10^6)
150 PRINT "CURRENT = " I "AMPS"
```

Problems

1. A conductor carrying 25 A is at right angles to a magnetic field having a density of 0.75 T. Calculate the force on the conductor in newtons per metre length.
2. The force on a conductor 65 mm long is 4.5 N, and the current in the conductor is 1.2 A. Calculate the flux density.
3. Find the length of a conductor which has a force of 3 N exerted on it when carrying a current of 6 A in a magnetic field of 6 T.
4. The force on a conductor 2 m long is 16 N. If the flux density is 0.35 T, find the current.
5. Calculate the voltage induced in a conductor 100 mm long moving with a velocity of 12 m/s, at right angles to a magnetic field having a flux density of 1.5 T.
6. If 4 V is induced in a conductor 240 mm long, moving at a velocity of 8 m/s, calculate the flux density.
7. Find the length of a conductor if the voltage induced in it is 180 mV when moving at a velocity of 4 m/s in a magnetic field of 1.0 T.
8. Calculate the velocity of a conductor 1.6 m long in a magnetic field of 5.5 T if the voltage induced is 10 V.
9. A magnetic flux of 750 μWb acts through a coil of 2000 turns and the field is reversed in 0.2 s. Calculate the induced voltage.
10. A coil of 650 turns is wound on a wooden ring having a mean length of 500 mm. If the current in the coil is 3 A, calculate the magnetic field strength.
11. A coil 250 mm long has a field strength of 2500 A/m when the current flowing is 10 A. Calculate the number of turns.
12. Find the current required to produce a field strength of 1000 A/m in a coil which has 60 turns and is 800 mm long.
13. Calculate the circumference of a ring having 350 turns if a current of 7.5 A produces a field strength of 5000 A/m.
14. A coil of 160 turns is wound on a non-magnetic former which has a mean circumference of 320 mm and a cross-sectional area of 960 mm^2. If the current in the coil is 16 A, calculate: (a) the field strength, H, (b) the flux density, B, (c) the total flux, Φ.
15. A coil is wound on a wooden bobbin and it has a magnetic field strength of 3000 A/m. If the total flux is 45 μWb, calculate the cross-sectional area of the bobbin.
16. A coil wound on a plastic former has a total flux of 45 μWb. If the cross-sectional area is 4500 mm^2, calculate the field strength.
17. Calculate the flux density in Problems 15 and 16.

11 A.c. circuits

11.1 Introduction

Chapter 6 looked at the nature of the a.c. waveform. This chapter looks at the behaviour of the resistor, the inductor and the capacitor, as individual components, and then in both series and parallel combinations when driven by a.c. voltages.

The aim is to show that an a.c. circuit can involve the calculation of up to 20 parameters. Generally, if four parameters are known, the others can be deduced. The worked examples are chosen to show how the formulas are used and transposed.

11.2 Resistor, R

The treatment of a purely resistive circuit shown in Figure 11.1 is identical to the treatment given in Chapter 3 under d.c. conditions. The formulas are the same for the voltage, current, resistance and power. The units used are the same. The voltage and current are in phase, as shown in Figure 11.2. If a voltage is quoted as 240, say, it is always assumed to be 240 r.m.s. It is quite in order, of course, to carry out the calculations using peak or peak to peak values, but care must be taken not to mix the units.

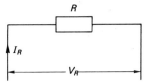

Figure 11.1

Figure 11.2

11.3 Inductor, L

Example 1 In the circuit of Figure 11.3, if $V_L = 100$ V, $f = 50$ Hz, and $L = 200$ mH, calculate the current I_L.

$$I = \frac{V}{X_L}$$

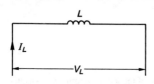

Figure 11.3

where I_L = current in amps
V_L = applied voltage in volts
X_L = inductive reactance in ohms
    = $2\pi fL$
f = frequency in hertz
L = inductance in henrys

Thus

$$X_L = 2 \times \pi \times 50 \times 200 \times 10^{-3} \, \Omega$$
$$= 62.83 \, \Omega$$
$$I_L = \frac{100}{62.83} = 1.59 \, \text{A}$$

The current I_L lags the voltage V_L by 90° as shown in Figure 11.4.

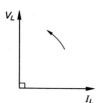

Figure 11.4

```
10 PRINT "PROG 153"
20 PRINT "INDUCTOR CURRENT"
30 INPUT "ENTER VOLTAGE" ; VL
40 INPUT "ENTER INDUCTANCE IN HENRYS" ; L
50 INPUT "ENTER FREQUENCY" ; F
60 LET XL=2*3.141593*F*L
70 LET IL=VL/XL
80 PRINT "CURRENT = " IL "AMPS"
```

Example 2 Let $I_L = 2$ A, and $L = 0.2$ H. Calculate the value of V_L at 100 Hz.

$$V_L = IX_L$$
$$X_L = 2 \times \pi \times 100 \times 0.2 = 125.66 \, \Omega$$
$$V_L = 2 \times 125.66 = 251.33 \, \text{V}$$

```
10 PRINT "PROG 154"
20 PRINT "INDUCTOR VOLTAGE"
30 INPUT "ENTER CURRENT IN AMPS" ; IL
40 INPUT "ENTER INDUCTANCE IN HENRYS" ; L
50 INPUT "ENTER FREQUENCY" ; F
60 LET XL=2*3.141593*F*L
70 LET VL=IL*XL
80 PRINT "VOLTAGE = " VL "VOLTS"
```

Example 3 Let $V_L = 100$ V, 250 Hz, and $I_L = 2$ A. Calculate L.

$$X_L = \frac{V}{I} = \frac{100}{2} = 50 \text{ }\Omega$$

$$2\pi f L = 50 \text{ }\Omega$$

$$L = \frac{50}{2\pi f} = \frac{50}{2 \times \pi \times 250} = 0.0318 \text{ H}$$

$$= 31.8 \text{ mH}$$

```
10 PRINT "PROG 155"
20 PRINT "INDUCTANCE"
30 INPUT "ENTER CURRENT IN AMPS" ; IL
40 INPUT "ENTER VOLTAGE" ; VL
50 INPUT "ENTER FREQUENCY" ; F
60 LET XL=VL/IL
70 LET L=XL/(2*3.141593*F)
80 PRINT "INDUCTANCE = " L "HENRYS"
```

Example 4 Let $V_L = 200$ V, $I_L = 3$ A, $L = 250$ μH $= 250 \times 10^{-6}$ H. Calculate f.

$$X_L = \frac{V_L}{I_L} = \frac{200}{3} = 66.67 \text{ }\Omega$$

$$2\pi f L = 66.67 \text{ }\Omega$$

$$f = \frac{66.67}{2 \times \pi \times 250 \times 10^{-6}}$$

$$= \frac{6.667 \times 10^6}{2 \times \pi \times 250} = 42\text{ }443 \text{ Hz} = 42.443 \text{ kHz}$$

```
10 PRINT "PROG 156"
20 PRINT "INDUCTOR FREQUENCY"
30 INPUT "ENTER CURRENT IN AMPS" ; IL
40 INPUT "ENTER VOLTAGE" ; VL
50 INPUT "ENTER INDUCTANCE IN MICROHENRYS" ; L
60 LET XL=VL/IL
70 LET F=XL*10^6/(2*3.141593*L)
80 PRINT "FREQUENCY = " F "HERTZ"
```

11.4 Capacitor C

Example 5 In the circuit of Figure 11.5, if $V_C = 200$ V, 100 Hz and $C = 10$ μF, calculate the current I_C.

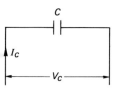

Figure 11.5

$$I_C = \frac{V}{X_C}$$

where I_C = current in amps
V_C = applied voltage in volts
X_C = capacitive reactance in ohms

$$= \frac{1}{2\pi f C}$$

f = frequency in hertz
C = capacitance in farads

$$X_C = \frac{1}{2 \times \pi \times 100 \times 10 \times 10^{-6}}$$

$$= \frac{10^6}{2 \times \pi \times 100 \times 10}$$

$$= 159.15 \, \Omega$$

$$I_C = \frac{200}{159.15} = 1.26 \, \text{A}$$

The current I_C leads the voltage V_C by 90°, as shown in Figure 11.6

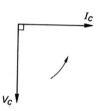

Figure 11.6

```
10 PRINT "PROG 157"
20 PRINT "CAPACITOR CURRENT"
30 INPUT "ENTER VOLTAGE" ; VC
40 INPUT "ENTER CAPACITANCE IN MICROFARADS" ; C
50 INPUT "ENTER FREQUENCY" ; F
60 LET XC=10^6/(2*3.141593*F*C)
70 LET IC=VC/XC
80 PRINT "CURRENT = " IC "AMPS"
```

Example 6 Let $I_C = 1.5$ A and $C = 25$ μF. Calculate V_C at 100 Hz.

$$V_C = IX_C \text{ where } X_C = \frac{1}{2\pi f c}$$

$$X_C = \frac{1}{2 \times \pi \times 100 \times 25 \times 10^{-6}} = \frac{10^6}{2 \times \pi \times 100 \times 25}$$

$$= 63.66 \, \Omega$$

$$V_C = 1.5 \times 63.66 = 95.5 \, \text{V}$$

```
10 PRINT "PROG 158"
20 PRINT "CAPACITOR VOLTAGE"
30 INPUT "ENTER CURRENT IN AMPS" ; IC
40 INPUT "ENTER CAPACITANCE IN MICROFARADS" ; C
50 INPUT "ENTER FREQUENCY" ; F
60 LET XC=10^6/(2*3.141593*F*C)
70 LET VC=IC*XC
80 PRINT "VOLTAGE = " VC "VOLTS"
```

Example 7 Let $V_C = 240$ V, 50 Hz, and $I_C = 0.5$ A. Calculate C.

$$X_C = \frac{V}{I_C} = \frac{240}{0.5} = 480 \ \Omega$$

$$480 = \frac{1}{2\pi fC} = \frac{1}{2 \times \pi \times 50 \times C}$$

$$C = \frac{1}{2 \times \pi \times 50 \times 480} \ \text{F}$$

$$= \frac{10^6}{2 \times \pi \times 50 \times 480} \ \mu\text{F} = 6.63 \ \mu\text{F}$$

```
10 PRINT "PROG 159"
20 PRINT "CAPACITANCE"
30 INPUT "ENTER CURRENT IN AMPS" ; IC
40 INPUT "ENTER VOLTAGE" ; VC
50 INPUT "ENTER FREQUENCY" ; F
60 LET XC=VC/IC
70 LET C=10^6/(2*3.141593*F*XC)
80 PRINT "CAPACITANCE = " C "MICROFARADS"
```

Example 8 Let $V_C = 2000$ V, $I_C = 1$ mA $= 1 \times 10^{-3}$ A, and $C = 100$ pf $= 100 \times 10^{-12}$ F. Calculate f.

$$X_C = \frac{V_C}{I_C} = \frac{2000}{1 \times 10^{-3}} = 2000 \times 10^3 = 2 \times 10^6 \ \Omega$$

$$2 \times 10^6 = \frac{1}{2\pi fC}$$

$$f = \frac{1}{2 \times \pi \times C \times 2 \times 10^6}$$

$$= \frac{1}{2 \times \pi \times 100 \times 10^{-12} \times 2 \times 10^6}$$

$$= \frac{10^{12}}{2 \times \pi \times 100 \times 2 \times 10^6} = 796 \ \text{Hz}$$

```
10 PRINT "PROG 160"
20 PRINT "CAPACITOR FREQUENCY"
30 INPUT "ENTER CURRENT IN AMPS" ; IC
40 INPUT "ENTER VOLTAGE" ; VC
50 INPUT "ENTER CAPACITANCE IN PICOFARADS" ; C
60 LET XC=VC/IC
70 LET F=10^12/(2*3.141593*C*XC)
80 PRINT "FREQUENCY = " F "HERTZ"
```

11.5 Series circuits

(a) R and L

Given the values of R, L, V and f, we can calculate the circuit parameters. These are:

X_L, reactance of the coil $= 2\pi fL \ \Omega$
Z, circuit impedance $= \sqrt{(R^2 + X_L^2)} \ \Omega$

I, current $= \dfrac{V}{Z}$ A

V_R, resistor voltage drop = IR V
V_L, inductor voltage drop = IX_L V
  True power   = $V_R I$ W
  Reactive power = $V_L I$ VA
  Apparent power = VI VA

$$\text{Power factor p.f.} = \frac{\text{True power}}{\text{Apparent power}}$$

Phase angle $\phi°$ or radians

Example 9 In Figure 11.7, let $R = 12\ \Omega$, $L = 0.1$ H, $V = 100$ V, $f = 50$ Hz.

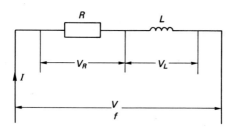

Figure 11.7

Coil reactance X_L

$$X_L = 2 \times \pi \times 50 \times 0.1 = 31.4\ \Omega$$

Circuit impedance Z

$$Z = \sqrt{[(12)^2 + (31.4)^2]} = 33.62\ \Omega$$

Current I

$$I = \frac{100}{33.62} = 2.975\ \text{A}$$

Voltage drop across the resistor V_R

$$V_R = IR = 2.975 \times 12 = 35.7\ \text{V}$$

Voltage drop across the inductor V_L

$$V_L = IX_L = 2.975 \times 31.4 = 93.4\ \text{V}$$

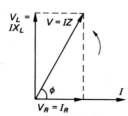

Figure 11.8

Notes:

1   V is the phasor sum of V_R and V_L (figure 11.8).

   $V^2 = V_R^2 + V_L^2$, i.e. $100^2 = 35.7^2 + 93.4^2$

2   Dividing the voltages in Figure 11.8 by I gives us the 'impedance triangle' (Figure 11.9).

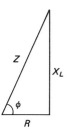

Figure 11.9

$$Z^2 = R^2 + X_L^2, Z = \surd(R^2 + X_L^2)$$

True power. This is sometimes called the active power, and is the power consumed by the resistor.

$P = V_R I = 35.7 \times 2.975 = 106.2$ W

Reactive power. This is sometimes called wattless power.

$P = V_L I = 93.4 \times 2.975 = 277.87$ VA

Apparent power. This equals applied voltage × current expressed in volt/amps.

$P = VI = 100 \times 2.975 = 297.5$ VA

Notes:

1   VI is the phasor sum of $V_R |I|$ and $V_L |I|$ (figure 11.10).

   $(VI)^2 = (V_R I)^2 + (V_L I)^2$ ($|I|$ is the modulus of I)

   i.e. $(297.5)^2 = (106.2)^2 + (277.87)^2$

2   Since this is an inductive circuit the current I lags the voltage by an angle $\phi°$. This angle will always lie between 0° and 90° depending on the values of R and L.

Power factor. There are three ways of expressing the power factor, p.f.

$$\text{p.f.} = \frac{V_R}{V} = \frac{35.7}{100} = 0.357 \text{ Figure 11.8}$$

$$\text{p.f.} = \frac{R}{Z} = \frac{12}{33.62} = 0.357 \text{ Figure 11.9}$$

$$\text{p.f.} = \frac{V_R I}{VI} = \frac{106.2}{297.5} = 0.357 \text{ Figure 11.10}$$

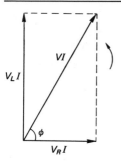

Figure 11.10

Phase angle. Examining the three triangles we notice that the power factor in all cases $= \cos \phi$.

$\cos \phi = $ p.f. $= 0.357$

$\phi = $ arcos $0.357 = 69°$, i.e. $\cos 69° = 0.357$

Notes:

1 An alternative method of determining the value of ϕ would be to draw any one of Figures 11.8, 11.9 or 11.10 to scale and measure the angle.
2 Reverting to Figure 11.10, note that

$$\cos \phi = \frac{V_R I}{VI}$$

$VI \cos \phi = VI = $ true power $= I^2 R$

3 This detailed example has shown that to solve a problem, the values of four of the five parameters R, L, V, I and f must be known. The next examples will cover the other four possibilities

```
10 PRINT "PROG 161"
20 PRINT "SERIES R-L CIRCUIT"
30 INPUT "ENTER RESISTOR VALUE" ; R
40 INPUT "ENTER INDUCTANCE IN HENRYS" ; L
50 INPUT "ENTER VOLTAGE" ; V
60 INPUT "ENTER FREQUENCY" ; F
70 LET XL=2*3.141593*F*L
80 LET Z=SQR((R^2)+(XL^2))
90 LET I=V/Z
100 LET VR=I*R
110 LET VL=I*XL
120 LET P1=VR*I
130 LET P2=VL*I
140 LET P3=V*I
150 LET PF=VR/V
160 LET PA=ATN(VL/VR)*180/3.141593
170 PRINT "XL="XL "Z="Z "I="I "VR="VR "VL="VL
180 PRINT "TRUE POWER="P1 "REACTIVE POWER="P2
190 PRINT "APPARENT POWER="P3 "POWER FACTOR="PF
200 PRINT "PHASE ANGLE="PA"DEGREES"
```

Example 10 A coil having an inductance of 0.2 H takes a current of 1.0 A from a 200-V supply at a frequency of 100 Hz. Calculate the resistance of the coil.

The circuit can be drawn as a series R and L, see Figure 11.11.

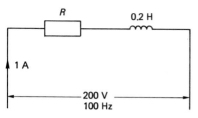

Figure 11.11

$$X_L = 2 \times \pi \times 100 \times 0.2 = 125.66 \ \Omega$$

$$Z = \frac{V}{I} = \frac{200}{1} = 200 \ \Omega$$

$$Z^2 = R^2 + X_L^2$$

$$R^2 = Z^2 - X_L^2$$

$$R^2 = (200)^2 - (125.66)^2$$

$$R = \sqrt{[(200)^2 - (125.66)^2]} = 155.6 \ \Omega$$

Having reached this point the other circuit parameters can be calculated.

$$V_R = 1 \times 155.6 = 155.6 \ \text{V}$$

$$V_L = 1 \times 125.66 = 125.66 \ \text{V}$$

True power $= 155.6 \times 1 = 155.6$ W

Reactive power $= 125.66 \times 1 = 125.66$ VA

Apparent power $= 200 \times 1 = 200$ VA

$$\text{p.f.} = \frac{V_R}{V} = \frac{155.6}{200} = 0.778$$

$$\phi = \arccos 0.778 = 38.9° \ (0.68 \text{ radians})$$

Note also that

True power $= VI \cos \phi = 200 \times 1 \times 0.778 = 155.6$ W

The voltage, impedance and power phasor diagrams could now be drawn if required, as shown in Example 9.

```
10  PRINT "PROG 162"
20  PRINT "SERIES R-L CIRCUIT"
30  INPUT "ENTER CURRENT IN AMPS" ; I
40  INPUT "ENTER INDUCTANCE IN HENRYS" ; L
50  INPUT "ENTER VOLTAGE" ; V
60  INPUT "ENTER FREQUENCY" ; F
70  LET XL=2*3.141593*F*L
80  LET Z=V/I
90  LET R=SQR(Z^2-XL^2)
100 LET VR=I*R
110 LET VL=I*XL
120 LET P1=VR*I
130 LET P2=VL*I
140 LET P3=V*I
150 LET PF=VR/V
160 LET PA=ATN(VL/VR)*180/3.141593
170 PRINT "XL="XL "Z="Z "R="R "VR="VR "VL="VL
180 PRINT "TRUE POWER="P1 "REACTIVE POWER="P2
190 PRINT "APPARENT POWER="P3 "POWER FACTOR="PF
200 PRINT "PHASE ANGLE="PA"DEGREES"
```

Example 11 A 150 Ω resistor in series with an inductor takes a current of 2.5 A from a 500 V supply at a frequency of 50 Hz. Find the inductance of the coil. The circuit is shown in Figure 11.12.

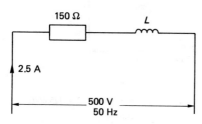

Figure 11.12

$$Z = \frac{V}{I} = \frac{500}{2.5} = 200 \text{ Ω}$$
$$Z^2 = R^2 + X_L^2$$
$$X_L^2 = Z^2 - R^2$$
$$X_L^2 = (200)^2 - (150)^2$$
$$X_L = \sqrt{[(200)^2 - (150)^2]} = 132.3 \text{ Ω}$$
$$132.3 = 2 \times \pi \times 50 \times L$$
$$L = \frac{132.3}{2 \times \pi \times 50} = 0.42 \text{ H} = 420 \text{ mH}$$
$$V_R = 2.5 \times 150 = 375 \text{ V}$$
$$V_L = 2.5 \times 132.3 = 330.75 \text{ V}$$

True power   = 937.5 W $(V_R I)$

Reactive power = 826.88 VA $(V_L I)$

Apparent power = 1250 VA (VI)

$$\text{p.f.} = \frac{V_R}{V} = \frac{375}{500} = 0.75$$
$$\text{or } \frac{R}{Z} = \frac{150}{200} = 0.75$$
$$\text{or } \frac{V_R I}{VI} = \frac{937.5}{1250} = 0.75$$
$$\phi = \text{arcos } 0.75 = 41.4° \text{ (0.722 radians)}$$

Note also that

True power   = $VI \cos \phi$
                 = $500 \times 2.5 \times 0.75 = 937.5$ W

a.c. circuits

```
10 PRINT "PROG 16 3"
20 PRINT "SERIES R-L CIRCUIT"
30 INPUT "ENTER CURRENT IN AMPS" ; I
40 INPUT "ENTER RESISTOR VALUE" ; R
50 INPUT "ENTER VOLTAGE" ; V
60 INPUT "ENTER FREQUENCY" ; F
70 LET Z=V/I
80 LET XL=SQR(Z^2-R^2)
90 LET L=XL/(2*3.141593*F)
100 LET VR=I*R
110 LET VL=I*XL
120 LET P1=VR*I
130 LET P2=VL*I
140 LET P3=V*I
150 LET PF=VR/V
160 LET PA=ATN(VL/VR)*180/3.141593
170 PRINT "XL="XL "Z="Z "L="L "VR="VR "VL="VL
180 PRINT "TRUE POWER="P1 "REACTIVE POWER="P2
190 PRINT "APPARENT POWER="P3 "POWER FACTOR="PF
200 PRINT "PHASE ANGLE="PA"DEGREES"
```

Example 12 A voltage applied to a 6 kΩ resistor in series with a 1 H inductor produces a current of 1 mA at a frequency of 1 kHz. Calculate the magnitude of the voltage. The circuit is shown in Figure 11.13.

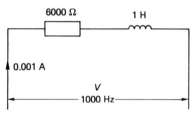

Figure 11.13

$$X_L = 2 \times \pi \times 1000 \times 1 = 6283.2 \ \Omega$$

$$Z = \sqrt{[(6000)^2 + (6283.2)^2]}$$

$$= 8687.83 \ \Omega$$

$$\frac{V}{Z} = I$$

$$V = IZ$$

$$= 0.001 \times 8687.83 = 8.68 \ \text{V}$$

$$V_R = 0.001 \times 6000 = 6 \ \text{V}$$

$$V_L = 0.001 \times 6283 = 6.28 \ \text{V}$$

True power $= 6 \times 0.001 = 0.006 \ \text{W} = 6 \ \text{mW}$

Reactive power $= 6.28 \times 0.001 = 0.00628 \ \text{VA}$

Apparent power $= 8.68 \times 0.001 = 0.00868 \ \text{VA}$

$$\text{p.f.} = \frac{V_R}{V} = \frac{6}{8.68} = 0.69$$

$$= \frac{R}{Z} = \frac{6000}{8687.83} = 0.69$$

$$= \frac{V_R I}{VI} = \frac{0.006}{0.00868} = 0.69$$

$$\phi = \arccos 0.69 = 46.27° \text{ (0.81 radians)}$$

Note also that

$$\text{True power} = VI \cos \phi = 8.68 \times 0.001 \times 0.69$$

$$= 0.006 \text{ W} = 6 \text{ mW}$$

```
10 PRINT "PROG 164"
20 PRINT "SERIES R-L CIRCUIT"
30 INPUT "ENTER CURRENT IN AMPS" ; I
40 INPUT "ENTER RESISTOR VALUE" ; R
50 INPUT "ENTER INDUCTANCE IN HENRYS" ; L
60 INPUT "ENTER FREQUENCY" ; F
70 LET XL=2*3.141593*F*L
80 LET Z=SQR(R^2+XL^2)
90 LET V=I*Z
100 LET VR=I*R
110 LET VL=I*XL
120 LET P1=VR*I
130 LET P2=VL*I
140 LET P3=V*I
150 LET PF=VR/V
160 LET PA=ATN(VL/VR)*180/3.141593
170 PRINT "XL="XL "Z="Z "V="V "VR="VR "VL="VL
180 PRINT "TRUE POWER="P1 "REACTIVE POWER="P2
190 PRINT "APPARENT POWER="P3 "POWER FACTOR="PF
200 PRINT "PHASE ANGLE="PA"DEGREES"
```

Example 13 Given that a voltage of 300 V applied to a 100 Ω resistor in series with a 0.02 H inductor produces a current of 0.2 A, calculate the frequency. See Figure 11.14.

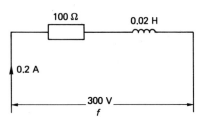

Figure 11.14

$$\frac{V}{I} = Z = \frac{300}{0.2} = 1500 \text{ Ω}$$

$$Z^2 = R^2 + X_L^2$$

$$X_L^2 = Z^2 - R^2 = (1500)^2 - (100)^2$$

$$X_L = \sqrt{[(1500)^2 - (100)^2]} = 1496.67 \text{ Ω}$$

$$1496.67 = 2 \times \pi \times f \times 0.02$$

$$f = \frac{1496.67}{2 \times \pi \times 0.02} = 11\,910 \text{ Hz}$$

$$= 11.91 \text{ kHz}$$

a.c. circuits **185**

$$V_R = 0.2 \times 100 = 20 \text{ V}$$

$$V_L = 0.2 \times 1496.67 = 299.3 \text{ V}$$

True power $= 20 \times 0.2 = 4$ W

Reactive power $= 299.3 \times 0.2 = 59.9$ VA

Apparent power $= 300 \times 0.2 = 60$ VA

$$\text{p.f.} = \frac{V_R}{V} = \frac{20}{300} = 0.067$$

$$= \frac{R}{Z} = \frac{100}{1500} = 0.067$$

$$= \frac{V_R I}{VI} = \frac{4}{60} = 0.067$$

$$\phi = \arccos 0.067 = 86.2° \text{ (1.5 radians)}$$

Note also that

True power $= VI \cos \phi$

$= 300 \times 0.2 \times 0.067 = 4$ W

```
10 PRINT "PROG 165"
20 PRINT "SERIES R-L CIRCUIT"
30 INPUT "ENTER CURRENT IN AMPS" ; I
40 INPUT "ENTER RESISTOR VALUE" ; R
50 INPUT "ENTER INDUCTANCE IN HENRYS" ; L
60 INPUT "ENTER VOLTAGE" ; V
70 LET Z=V/I
80 LET XL=SQR(Z^2-R^2)
90 LET F=XL/(2*3.141593*L)
100 LET VR=I*R
110 LET VL=I*XL
120 LET P1=VR*I
130 LET P2=VL*I
140 LET P3=V*I
150 LET PF=VR/V
160 LET PA=ATN(VL/VR)*180/3.141593
170 PRINT "XL="XL "Z="Z "F="F "VR="VR "VL="VL
180 PRINT "TRUE POWER="P1 "REACTIVE POWER="P2
190 PRINT "APPARENT POWER="P3 "POWER FACTOR="PF
200 PRINT "PHASE ANGLE="PA"DEGREES"
```

Example 14 It is of course possible to take the values of four other parameter combinations to completely solve a problem. Consider the following problem.

An inductor is connected to a 200 V supply, frequency 100 Hz, and takes a current of 5 A at a phase angle of 30°. Calculate the resistance and the inductance of the coil and the power consumed.

$$Z = \frac{V}{I} = \frac{200}{5} = 40 \text{ } \Omega$$

$$\cos 30 = \frac{R}{Z} \text{ See Figure 11.9}$$

$$R = Z \cos 30 = 40 \times 0.866 = 34.64 \text{ } \Omega$$

$$\sin 30 = \frac{X_L}{Z}$$

$$X_L = Z \sin 30 = 40 \times 0.5 = 20 \text{ } \Omega$$

$$20 = 2 \times \pi \times 100 \times L$$

$$L = \frac{20}{2 \times \pi \times 100} = 0.032 \text{ H}$$

Power consumed = power consumed in the resistance = true power

$$= V_R I = IRI = I^2 R = 5^2 \times 34.64 = 866 \text{ W}$$

Having reached this point all the other parameters can be calculated as shown in Examples 9 to 13.

```
10 PRINT "PROG 166"
20 PRINT "SERIES R-L CIRCUIT"
30 INPUT "ENTER CURRENT IN AMPS" ; I
40 INPUT "ENTER PHASE ANGLE IN DEGREES" ; PHI
50 INPUT "ENTER VOLTAGE" ; V
60 INPUT "ENTER FREQUENCY" ; F
70 LET Z=V/I
80 LET R=Z*COS(PHI*3.141593/180)
90 LET XL=Z*SIN(PHI*3.141593/180)
100 LET VR=I*R
110 LET VL=I*XL
120 LET P1=VR*I
130 LET P2=VL*I
140 LET P3=V*I
150 LET PF=VR/V
160 LET L=XL/(2*3.141593*F)
170 PRINT "XL="XL "Z="Z "L="L "VR="VR "VL="VL
180 PRINT "TRUE POWER="P1 "REACTIVE POWER="P2
190 PRINT "APPARENT POWER="P3 "POWER FACTOR="PF
200 PRINT "R="R
```

Example 15 In a series circuit $V_R = 60$ V and $V_L = 80$ V at a frequency of 50 Hz. If the true power is 600 W calculate V, I, R and L.

$$V^2 = V_R^2 + V_L^2 = 60^2 + 80^2 = 10\ 000$$

$$V = \sqrt{10\ 000} = 100 \text{ V}$$

True power = $V_R I$, $600 = 60I$

$$I = \frac{600}{60} = 10 \text{ A}$$

$$\frac{V}{I} = Z, Z = \frac{100}{10} = 10 \text{ }\Omega$$

$$V_R = IR, R = \frac{V_R}{I} = \frac{60}{10} = 6 \text{ }\Omega$$

$$Z^2 = R^2 + X_L^2$$

$$X_L^2 = Z^2 - R^2 = (10)^2 - (6)^2 = 64$$

$$X_L = \sqrt{64} = 8 \text{ }\Omega$$

$$X_L = 2\pi f L, L = \frac{X_L}{2\pi f}$$

$$= \frac{8}{2 \times \pi \times 50} = 0.0255 \text{ H}$$

```
10 PRINT "PROG 167"
20 PRINT "SERIES R-L CIRCUIT"
30 INPUT "ENTER VR" ; VR
40 INPUT "ENTER VL" ; VL
50 INPUT "TRUE POWER" ; P1
60 INPUT "ENTER FREQUENCY" ; F
70 LET V=SQR(VR^2+VL^2)
80 LET I=P1/VR
90 LET Z=V/I
100 LET R=VR/I
110 LET XL=SQR(Z^2-R^2)
120 LET L=XL/(2*3.141593*F)
130 PRINT "V="V "I="I "R="R "L="L
```

(b) R and C
The approach to problem solving is identical to that for the R–L circuit in Section 11.5(a). The only difference is that the reactance of the capacitor

$$C = X_C = \frac{1}{2\pi f C}$$

where C is in farads, and I_C leads V_C by 90° ($\pi/2$ radians).

Example 16 Let $R = 100\ \Omega$, $C = 10\ \mu F$, $V = 50\ V$, $f = 50\ Hz$. Capacitor reactance X_C. See circuit in Figure 11.15.

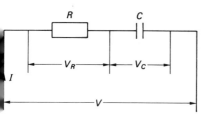

Figure 11.15

$$X_C = \frac{1}{2 \times \pi \times 50 \times 10 \times 10^{-6}} = \frac{10^6}{2 \times \pi \times 50 \times 10}$$
$$= 318.3\ \Omega$$

Circuit impedance Z

$$Z = \sqrt{[(200)^2 + (318.3)^2]} = 376\ \Omega$$

Current I

$$I = \frac{50}{376} = 0.133\ A$$

Voltage drop across the resistor, V_R

$$V_R = IR = 0.133 \times 200 = 26.6\ V$$

Voltage drop across the capacitor, V_C

$$V_C = IX_C = 0.133 \times 318.3 = 42.3\ V$$

Notes:

- V is the phasor sum of V_R and V_C (figure 11.16).

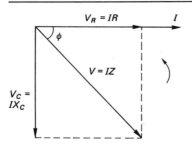

Figure 11.16

$$V^2 = V_R^2 + V_C^2$$
i.e. $(50)^2 = (26.6)^2 + (42.30)^2$

2 Dividing the voltages in Figure 11.16 by I gives us the 'impedance triangle' (Figure 11.17).

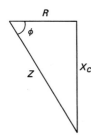

Figure 11.17

$$Z^2 = R^2 + X_C^2$$
$$Z = \sqrt{(R^2 + X_C^2)}$$

True power. This is sometimes called active power and is the power consumed by the resistor.

$$P = V_R I = 26.6 \times 0.133 = 3.54 \text{ W}$$

Reactive power. This is sometimes called wattless power.

$$P = V_C I = 42.3 \times 0.133 = 5.63 \text{ VA}$$

Apparent power. This equals applied voltage × current expressed in volt/amps.

$$P = VI = 50 \times 0.133 = 6.65 \text{ VA}$$

Notes:

1 VI is the phasor sum of $V_R I$ and $V_C I$, Figure 11.18.
$$(VI)^2 = (V_R I)^2 + (V_C I)^2$$
i.e. $(6.65)^2 = (3.54)^2 + (5.63)^2$

2 Since this is a capacitor circuit the current I leads the applied

voltage by the angle ϕ. This angle will always lie between 0 and 90°, depending on the values of R and C.

Power factor. There are three ways of expressing the power factor, p.f.

$$\text{p.f.} = \frac{V_R}{V} = \frac{26.6}{50} = 0.53$$

$$\text{p.f.} = \frac{R}{Z} = \frac{200}{3.76} = 0.53$$

$$\text{p.f.} = \frac{V_R I}{VI} = \frac{3.54}{6.65} = 0.53$$

Phase angle. Examining the three triangles we notice that the power factor in all cases $= \cos \phi$.

$$\cos \phi = \text{p.f.} = 0.53$$
$$\phi = \arccos 0.53 = 58° \text{ (1.01 radians)}$$

Notes:

1. An alternative method of determining the value of ϕ would be to draw any one of Figures 11.16, 11.17 or 11.18 to scale and measure the angle.

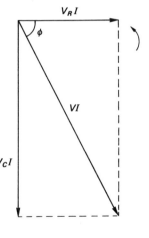

Figure 11.18

2. Reverting to Figure 11.18 note that

$$\cos \phi = \frac{V_R I}{VI}$$

$VI \cos \phi = V_R I$ = true power

3. This detailed example has again shown that to solve a problem, the values of four of the parameters R, C, V, I and f must be known. The next examples will cover the other four possibilities.

```
10 PRINT "PROG 168"
20 PRINT "SERIES R-C CIRCUIT"
30 INPUT "ENTER RESISTOR VALUE" ; R
40 INPUT "ENTER CAPACITANCE IN MICROFARADS" ; C
50 INPUT "ENTER VOLTAGE" ; V
60 INPUT "ENTER FREQUENCY" ; F
70 LET XC=10^6/(2*3.141593*F*C)
80 LET Z=SQR(R^2+XC^2)
90 LET I=V/Z
100 LET VR=I*R
110 LET VC=I*XC
120 LET P1=VR*I
130 LET P2=VC*I
140 LET P3=V*I
150 LET PF=VR/V
160 LET PA=ATN(VC/VR)*180/3.141593
170 PRINT "XC="XC "Z="Z "I="I "VR="VR "VC="VC
180 PRINT "TRUE POWER="P1 "REACTIVE POWER="P2
190 PRINT "APPARENT POWER="P3 "POWER FACTOR="PF
200 PRINT "PHASE ANGLE="PA"DEGREES"
```

Example 17 A resistor connected in series with a 4 µF capacitor draws a current of 0.5 A from a 100 V supply at a frequency of 250 Hz. Find the value of the resistor and all other parameters. See Figure 11.19.

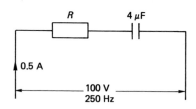

Figure 11.19

$$X_C = \frac{1}{2 \times \pi \times 250 \times 4 \times 10^{-6}}$$

$$= \frac{10^6}{2 \times \pi \times 250 \times 4} = 159.2 \ \Omega$$

$$Z = \frac{V}{I} = \frac{100}{0.5} = 200 \ \Omega$$

$$Z^2 = R^2 + X_C^2$$

$$R^2 = Z^2 - X_C^2 = (200)^2 - (159.2)^2$$

$$R = \sqrt{[(200)^2 - (159.2)^2]} = 121 \ \Omega$$

$$V_R = 0.5 \times 121 = 60.5 \ \text{V}$$

$$V_C = 0.5 \times 159.2 = 79.6 \ \text{V}$$

True power $= 60.5 \times 0.5 = 30.25$ W

Reactive power $= 79.6 \times 0.5 = 39.8$ VA

Apparent power $= 100 \times 0.5 = 50$ VA

$$\text{p.f.} = \frac{V_R}{V} = \frac{60.5}{100} = 0.605$$

$$\phi = \arccos 0.605 = 52.8° \ (0.92 \text{ radians})$$

Note also that

$$\text{True power} = VI \cos \phi$$
$$= 100 \times 0.5 \times 0.605 = 30.25 \text{ W}$$

```
10 PRINT "PROG 169"
20 PRINT "SERIES R-C CIRCUIT"
30 INPUT "ENTER CURRENT IN AMPS" : I
40 INPUT "ENTER CAPACITANCE IN MICROFARADS" ; C
50 INPUT "ENTER VOLTAGE" : V
60 INPUT "ENTER FREQUENCY" : F
70 LET XC=10^6/(2*3.141593*F*C)
80 LET Z=V/I
90 LET R=SQR(Z^2-XC^2)
100 LET VR=I*R
110 LET VC=I*XC
120 LET P1=VR*I
130 LET P2=VC*I
140 LET P3=V*I
150 LET PF=VR/V
160 LET PA=ATN(VC/VR)*180/3.141593
170 PRINT "XC="XC "Z="Z "R="R "VR="VR "VC="VC
180 PRINT "TRUE POWER="P1 "REACTIVE POWER="P2
190 PRINT "APPARENT POWER="P3 "POWER FACTOR="PF
200 PRINT "PHASE ANGLE="PA "DEGREES"
```

Example 18 An 80-Ω resistor in series with a capacitor takes a current of 2 A from a 240-V supply at a frequency of 50 Hz. See Figure 11.20. Calculate the value of C.

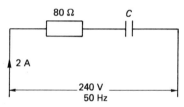

Figure 11.20

$$Z = \frac{V}{I} = \frac{240}{2} = 120 \ \Omega$$

$$Z^2 = R^2 + X_C^2$$

$$X_C^2 = Z^2 - R^2$$

$$= (120)^2 - (80)^2$$

$$X_C = \sqrt{[(120)^2 - (80)^2]} = 89.44 \ \Omega$$

$$89.44 = \frac{1}{2 \times \pi \times 50 \times C \times 10^{-6}} \ \mu\text{F}$$

$$C = \frac{10^6}{2 \times \pi \times 50 \times 89.44} = 35.6 \ \mu\text{F}$$

At this point all the other circuit values could be calculated if required, in the same way as in Examples 16 and 17.

```
10 PRINT "PROG 170"
20 PRINT "SERIES R-C CIRCUIT"
30 INPUT "ENTER CURRENT IN AMPS" ; I
40 INPUT "ENTER RESISTANCE" ; R
50 INPUT "ENTER VOLTAGE" ; V
60 INPUT "ENTER FREQUENCY" ; F
70 LET Z=V/I
80 LET XC=SQR(Z^2-R^2)
90 LET C=10^6/(2*3.141593*F*XC)
100 LET VR=I*R
110 LET VC=I*XC
120 LET P1=VR*I
130 LET P2=VC*I
140 LET P3=V*I
150 LET PF=VR/V
160 LET PA=ATN(VC/VR)*180/3.141593
170 PRINT "XC="XC "Z="Z "C="C "VR="VR "VC="VC
180 PRINT "TRUE POWER="P1 "REACTIVE POWER="P2
190 PRINT "APPARENT POWER="P3 "POWER FACTOR="PF
200 PRINT "PHASE ANGLE="PA"DEGREES"
```

Example 19 A voltage applied to an 8-kΩ resistor in series with a 200-pF capacitor produces a current of 5 mA at a frequency of 100 kHz. Find the value of the voltage. See Figure 11.21.

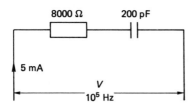

Figure 11.21

$$X_C = \frac{1}{2 \times \pi \times 10^5 \times 200 \times 10^{-12}}$$

$$= \frac{10^{12}}{2 \times \pi \times 10^5 \times 200} = 7958 \; \Omega$$

$$Z = \sqrt{(8000)^2 + (7958)^2} = 11\,284 \; \Omega$$

$$I = \frac{V}{Z}$$

$$V = IZ = 5 \times 10^{-3} \times 11\,284 \; V$$

$$= \frac{5 \times 11\,284}{10^3} = 56.42 \; V$$

All other parameters could be calculated as shown in Examples 16 and 17.

```
10 PRINT "PROG 171"
20 PRINT "SERIES R-C CIRCUIT"
30 INPUT "ENTER CURRENT IN AMPS" ; I
40 INPUT "ENTER RESISTANCE" ; R
50 INPUT "ENTER FREQUENCY" ; F
60 INPUT "ENTER CAPACITANCE IN PICOFARADS" ; C
70 LET XC=10^12/(2*3.141593*F*C)
80 LET Z=SQR(R^2+XC^2)
90 LET V=I*Z
100 LET VR=I*R
110 LET VC=I*XC
120 LET P1=VR*I
130 LET P2=VC*I
140 LET P3=V*I
150 LET PF=VR/V
160 LET PA=ATN(VC/VR)*180/3.141593
170 PRINT "XC="XC "Z="Z "V="V "VR="VR "VC="VC
180 PRINT "TRUE POWER="P1 "REACTIVE POWER="P2
190 PRINT "APPARENT POWER="P3 "POWER FACTOR="PF
200 PRINT "PHASE ANGLE="PA"DEGREES"
```

Example 20 A voltage of 400 V applied to a 2-kΩ resistor in series with a 0.5-μF capacitor produces a current of 0.1 A. Calculate the frequency. See figure 11.22.

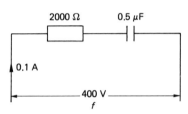

Figure 11.22

$$\frac{V}{I} = Z = \frac{400}{0.1} = 4000 \ \Omega$$

$$Z^2 = R^2 + X_C^2$$

$$X_C^2 = Z^2 - R^2$$

$$= (4000)^2 - (2000)^2$$

$$X_C = \sqrt{[(4000)^2 - (2000)^2]}$$

$$= 3464 \ \Omega$$

$$3464 = \frac{1}{2 \times \pi \times f \times 0.5 \times 10^{-6}}$$

$$3464 = \frac{10^6}{2 \times \pi \times f \times 0.5}$$

$$f = \frac{10^6}{2 \times \pi \times 0.5 \times 3464} = 92 \ \text{Hz}$$

All other parameters could be calculated as shown in Examples 16 and 17.

```
10 PRINT "PROG 172"
20 PRINT "SERIES R-C CIRCUIT"
30 INPUT "ENTER CURRENT IN AMPS" ; I
40 INPUT "ENTER RESISTANCE" ; R
50 INPUT "ENTER VOLTAGE" ; V
60 INPUT "ENTER CAPACITANCE IN MICROFARADS" ; C
70 LET Z=V/I
80 LET XC=SQR(Z^2-R^2)
90 LET F=10^6/(2*3.141593*C*XC)
100 LET VR=I*R
110 LET VC=I*XC
120 LET P1=VR*I
130 LET P2=VC*I
140 LET P3=V*I
150 LET PF=VR/V
160 LET PA=ATN(VC/VR)*180/3.141593
170 PRINT "XC="XC "Z="Z "F="F "VR="VR "VC="VC
180 PRINT "TRUE POWER="P1 "REACTIVE POWER="P2
190 PRINT "APPARENT POWER="P3 "POWER FACTOR="PF
200 PRINT "PHASE ANGLE="PA"DEGREES"
```

Example 21 This example and the next illustrate again that given any four parameters the problems can be solved.

A series R–C circuit draws current from a 50-V 50-Hz supply with a phase angle of 60°. If the reactance of the capacitor is 400 Ω, calculate the values of C, I and R.

$$X_C = 400 = \frac{1}{2 \times \pi \times 50 \times C \times 10^{-6}} \, \mu F$$

$$C = \frac{10^6}{2 \times \pi \times 50 \times 400} = 7.96 \, \mu F$$

$$\text{p.f.} = \arccos 60° = 0.5 = \frac{V_R}{V}$$

$$V_R = 50 \times 0.5 = 25 \, V$$

$$V^2 = V_R^2 + V_C^2$$

$$V_C^2 = V^2 - V_R^2$$

$$= 50^2 - 25^2, \; V_C = \sqrt{(50^2 - 25^2)} = 43.3 \, V$$

$$IX_C = V_C \quad I = \frac{V_C}{X_C}$$

$$= \frac{43.3}{400} = 0.108 \, A$$

$$V_R = IR \quad R = \frac{V_R}{I}$$

$$= \frac{25}{0.108} = 231 \, \Omega$$

```
10 PRINT "PROG 173"
20 PRINT "SERIES R-C CIRCUIT"
30 INPUT "ENTER PHASE ANGLE" ; PHI
40 INPUT "ENTER FREQUENCY" ; F
50 INPUT "ENTER VOLTAGE" ; V
60 INPUT "ENTER REACTANCE" ; XC
70 LET C=10^6/(2*3.141593*F*XC)
80 LET VR=V*COS(PHI*3.141593/180)
90 LET VC=SQR(V^2-VR^2)
100 LET I=VC/XC
110 LET R=VR/I
120 PRINT "C="C  "I="I  "R="R
```

Example 22 A resistor is connected in series with a 2-µF capacitor. The circuit impedance is 100 Ω. Measurements of V_R and V_C show both to be 100 V. Calculate the amplitude of the applied voltage, the frequency, the current, the power factor and the value of R.

$$V^2 = V_R^2 + V_C^2 = (100)^2 + (100)^2$$

$$V = \sqrt{[(100))))^{22} + (100)^2]} = 141.42 \text{ V}$$

$$V = IZ, I = \frac{V}{Z} = \frac{141.42}{100} = 1.414 \text{ A}$$

$$\text{p.f.} = \frac{V_R}{V} = \frac{100}{141.42} = 0.707$$

$$\text{p.f.} = \frac{R}{Z}, R = (\text{p.f.}) \times Z = 0.707 \times 100 = 70.7 \text{ Ω}$$

$$Z^2 = R^2 + X_C^2, X_C^2 = Z^2 - R^2 = 100^2 - 70.7^2$$

$$X_C = \sqrt{(100^2 - 70.7^2)} = 70.7 \text{ Ω}$$

$$70.7 = \frac{1}{2 \times \pi \times f \times 2 \times 10^{-6}} = \frac{10^6}{2 \times \pi \times f \times 2}$$

$$f = \frac{10^6}{2 \times \pi \times 2 \times 70.7} = 1126 \text{ Hz}$$

```
10 PRINT "PROG 174"
20 PRINT "SERIES R-C CIRCUIT"
30 INPUT "ENTER CAPACITANCE IN MICROFARADS" ; C
40 INPUT "ENTER IMPEDANCE" ; Z
50 INPUT "ENTER VOLTAGE VR" ; VR
60 INPUT "ENTER VOLTAGE VC" ; VC
70 LET V=SQR(VR^2+VC^2)
80 LET I=V/Z
90 LET PF=VR/V
100 LET R=PF*Z
110 LET XC=SQR(Z^2-R^2)
120 LET F=10^6/(2*3.141593*C*XC)
130 PRINT "V="V "F="F "I="I "PF="PF
```

(c) *R, L* and *C*

The treatment of a series circuit containing *R, L* and *C* is identical to circuits containing only *R* and *L*, or *R* and *C*, except that the total reactance (*X*) will be the difference between X_L and X_C, i.e.

$$Z = \sqrt{[R^2 + (X_L - X_C)^2]} \text{ when } X_L > X_C$$

$$Z = \sqrt{[R^2 + (X_C - X_L)^2]} \text{ when } X_C > X_L$$

Example 23 The circuit in Figure 11.23 has the following values: $R = 5$ Ω, $L = 0.05$ H, $C = 300$ µF, $V = 100$ V, $f = 50$ Hz. We can therefore carry out the following calculations as before.

$$X_L = 2\pi f L = 2 \times \pi \times 50 \times 0.05 = 15.7 \text{ Ω}$$

$$X_C = \frac{1}{2\pi f C} = \frac{1}{2 \times \pi \times 50 \times 300 \times 10^{-6}}$$

$$= \frac{10^6}{2 \times \pi \times 50 \times 300} = 10.61 \text{ Ω}$$

$$X = X_L - X_C = 15.7 - 10.61 \qquad = 5.09 \text{ Ω}$$

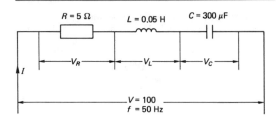

Figure 11.23

$$Z = \sqrt{(R^2 + X^2)} = \sqrt{[(5)^2 + (5.09)^2]} = 7.13 \; \Omega$$
$$I = \frac{V}{Z} = \frac{100}{7.13} = 14.03 \; A$$
$$V_R = IR = 14.03 \times 5 = 70.13 \; V$$
$$V_L = IX_L = 14.03 \times 15.7 = 220.2 \; V$$
$$V_C = IX_C = 14.03 \times 10.61 = 148.8 \; V$$

Notes:

1. In this example $X_L > X_C$.
2. We can now draw the voltage phasor diagram, remembering that the current I is always in phase with V_R, and that V_L leads the current by 90° and V_C lags the current by 90°.
3. V is the phasor sum of V_R and $(V_L - V_C)$, see Figure 11.24.

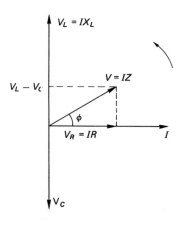

Figure 11.24

$$\text{True power} = V_R I = 70.13 \times 14.03 = 983.92 \; W$$
$$\text{Reactive power} = (V_L - V_C)I = 71.4 \times 14.03$$
$$= 1001.74 \; VA$$
$$\text{Apparent power} = VI = 100 \times 14.03 = 1403 \; VA$$

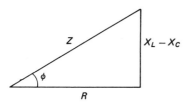

Figure 11.25

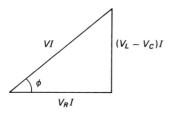

Figure 11.26

From figures 11.24, 11.25 and 11.26 we have

$$\text{p.f.} = \frac{V_R}{V} = \frac{70.13}{100} = 0.7013$$

$$\text{p.f.} = \frac{R}{Z} = \frac{5}{7.13} = 0.7013$$

$$\text{p.f.} = \frac{V_R I}{VI} = \frac{983.72}{1403} = 0.7013$$

Phase angle ϕ = arcos $0.7013 = 45.5°$ (0.793 radians)

Note from Figure 11.26

$$\cos \phi = \frac{V_R I}{VI}$$

$$\sin \phi = \frac{(V_L - V_C)I}{VI}$$

$VI \cos \phi = V_R I$ = true power

$VI \sin \phi = (V_L - V_C)I$ = reactive power

Note: from the phasor diagram, Figure 11.24, we see that the current lags the applied voltage by the angle ϕ.

```
10 PRINT "PROG 175"
20 PRINT "SERIES R-L-C CIRCUIT"
30 INPUT "ENTER RESISTOR VALUE" ; R
40 INPUT "ENTER CAPACITANCE IN MICROFARADS" ; C
45 INPUT "ENTER INDUCTANCE IN HENRYS" ; L
50 INPUT "ENTER VOLTAGE" ; V
60 INPUT "ENTER FREQUENCY" ; F
65 LET XL=2*3.141593*F*L
70 LET XC=10^6/(2*3.141593*F*C)
80 LET Z=SQR(R^2+(XL-XC)^2)
90 LET I=V/Z
100 LET VR=I*R
105 LET VL=I*XL
110 LET VC=I*XC
120 LET P1=VR*I
130 LET P2=(VL-VC)*I
140 LET P3=V*I
150 LET PF=VR/V
160 LET PA=ATN((VL-VC)/VR)*180/3.141593
170 PRINT "XL="XL "XC="XC "XL-XC="XL-XC "Z="Z
175 PRINT "I="I "VR="VR "VL="VL "VC="VC
180 PRINT "TRUE POWER="P1 "REACTIVE POWER="P2
190 PRINT "APPARENT POWER="P3 "POWER FACTOR="PF
200 PRINT "PHASE ANGLE="PA"DEGREES"
```

Example 24 Find the parameters of the circuit in Figure 11.27.

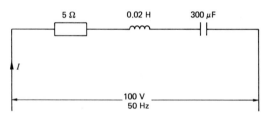

Figure 11.27

$$X_L = 2\pi fL = 2 \times \pi \times 50 \times 0.02 = 6.283 \ \Omega$$

$$X_C = \frac{1}{2\pi fC} = \frac{1}{2 \times \pi \times 50 \times 300 \times 10^{-6}}$$

$$= \frac{10^6}{2 \times \pi \times 50 \times 300} = 10.61 \ \Omega$$

$$X = X_C - X_L = 10.61 - 6.283 = 4.327 \ \Omega$$

$$Z = \sqrt{(R^2 + X^2)} = \sqrt{[(5)^2 + (4.327)^2]} = 6.612 \ \Omega$$

$$I = \frac{V}{Z} = \frac{100}{6.612} = 15.124 \ \text{A}$$

$$V_R = IR = 15.124 \times 5 = 75.62 \ \text{V}$$

$$V_L = IX_L = 15.124 \times 6.283 = 95.02 \ \text{V}$$

$$V_C = IX_C = 15.124 \times 10.61 = 160.46 \ \text{V}$$

True power $= V_R I = 75.62 \times 15.124 = 1144 \ \text{W}$

Reactive power $= (V_C - V_L)I = 65.44 \times 15.124 = 990 \ \text{VA}$

Apparent power $= VI = 100 \times 15.124 = 1512 \ \text{VA}$

a.c. circuits

$$\text{p.f.} = \frac{V_R}{V} = \frac{75.62}{100} = 0.756$$

$$\text{p.f.} = \frac{R}{Z} = \frac{5}{6.612} = 0.756$$

$$\text{p.f.} = \frac{V_R I}{VI} = \frac{1144}{1512} = 0.756$$

phase angle ϕ = arcos 0.756
   = 40.9° (0.713 radians)

$$\cos \phi = \frac{V_R I}{VI}$$
   $= VI \cos \phi = V_R I$ = true power

$$\sin \phi = \frac{(V_C - V_L)I}{VI}$$
   $= VI \sin \phi = (V_C - V_L)I$ = reactive power

Note: the phasor diagrams are shown in Figures 11.28(a), (b) and (c). In this example $X_C > X_L$.

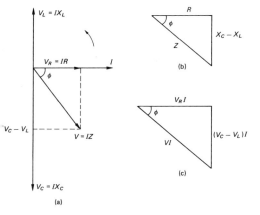

Figure 11.28

```
10 PRINT "PROG 176"
20 PRINT "SERIES R-L-C CIRCUIT"
30 INPUT "ENTER RESISTOR VALUE" ; R
40 INPUT "ENTER CAPACITANCE IN MICROFARADS" ; C
45 INPUT "ENTER INDUCTANCE IN HENRYS" ; L
50 INPUT "ENTER VOLTAGE" : V
60 INPUT "ENTER FREQUENCY" : F
65 LET XL=2*3.141593*F*L
70 LET XC=10^6/(2*3.141593*F*C)
80 LET Z=SQR(R^2+(XC-XL)^2)
90 LET I=V/Z
100 LET VR=I*R
105 LET VL=I*XL
110 LET VC=I*XC
120 LET P1=VR*I
130 LET P2=(VC-VL)*I
```

```
140 LET P3=V*I
150 LET PF=VR/V
160 LET PA=ATN((VC-VL)/VR)*180/3.141593
170 PRINT "XL="XL "XC="XC "XC-XL="XC-XL "Z="Z
175 PRINT "I="I "VR="VR "VL="VL "VC="VC
180 PRINT "TRUE POWER="P1 "REACTIVE POWER="P2
190 PRINT "APPARENT POWER="P3 "POWER FACTOR="PF
200 PRINT "PHASE ANGLE="PA"DEGREES"
```

Example 25 (Resonance) Find the parameters in Figure 11.29.

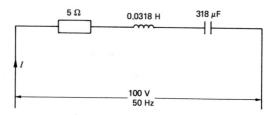

Figure 11.29

$$X_L = 2\pi f L = 2 \times \pi \times 50 \times 0.0318 = 10 \ \Omega$$

$$X_C = \frac{1}{2\pi f C} = \frac{1}{2 \times \pi \times 50 \times 318 \times 10^{-6}}$$

$$= \frac{10^6}{2 \times \pi \times 50 \times 318} = 10 \ \Omega$$

$$X = X_L - X_C = 10 - 10 = 0$$

$$Z = \sqrt{(R^2 + X^2)} = \sqrt{[(5)^2 + 0]} = 5 \ \Omega$$

$$I = \frac{V}{Z} = \frac{100}{5} = 20 \ \text{A}$$

$$V_R = IR = 20 \times 5 = 100 \ \text{V}$$

$$V_L = IX_L = 20 \times 10 = 200 \ \text{V}$$

$$V_C = IX_C = 20 \times 10 = 200 \ \text{V}$$

True power $= V_R I = 100 \times 20 = 2000 \ \text{W}$

Reactive power $= (V_L - V_C)I = 0 \times 20 = 0$

Apparent power $VI = 100 \times 20 = 2000 \ \text{VA}$

$$\text{p.f.} = \frac{V_R}{V} = \frac{100}{100} = 1$$

$$\text{p.f.} = \frac{R}{Z} = \frac{5}{5} = 1$$

$$\text{p.f.} = \frac{V_R I}{VI} = \frac{2000}{2000} = 1$$

phase angle $\phi = \arccos 1 = 0$

Also

True power $= VI \cos \phi = 100 \times 20 \times 1 = 2000 \ \text{W}$

Reactive power $= VI \sin \phi = 100 \times 20 \times 0 = 0$

Notes:

1 In this example the values of L and C were chosen so that
$$X_L = X_C$$

2 When $X_L = X_C$ the phasor diagram (see Figure 11.30) shows that the current I is in phase with the applied voltage V, which equals V_R. This condition is called series resonance.

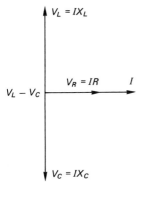

Figure 11.30

3 $V_L = -V_C$.
4 $Z = R$, which is the minimum possible impedance for the circuit.
5 The current I is determined purely by the value of R, i.e. $I = V/R$ and this is the maximum possible current for this circuit.
6 Since $X_L = -X_C$

$$2\pi f L = \frac{1}{2\pi f C}$$

$$ff = \frac{1}{2 \times \pi \times L \times 2 \times \pi \times C}$$

$$f^2 = \frac{1}{4\pi^2 LC}$$

$$\sqrt{f^2} = \sqrt{\left(\frac{1}{4\pi^2 LC}\right)}$$

$$f = \frac{1}{2\pi} \times \frac{1}{\sqrt{(LC)}}$$

where f = resonant frequency.

Using the circuit values

$$f = \frac{1}{2\pi} \times \frac{1}{\sqrt{(0.0318 \times 318 \times 10^{-6})}} = \frac{314.46}{2\pi}$$

$$= 50 \text{ Hz}$$

which confirms the frequency used in the problem.

7 One other property of a resonant circuit is the Q factor. When R

is small compared to X_L and X_C then V_L and V_C are greater than the supply voltage.

Voltage magnification at resonance

$$= \frac{\text{voltage across } L \text{ or } C}{\text{supply voltage}} = Q$$

Hence

$$Q = \frac{V_L}{V} = \frac{200}{100} = 2 = \frac{IX_L}{IR} = \frac{X_L}{R}$$

$$Q = \frac{V_C}{V} = \frac{200}{100} = 2 = \frac{IX_C}{IR} = \frac{X_C}{R}$$

8 The true power = apparent power. This means that all the power is consumed in the resistance, and a power factor of 1 therefore means maximum efficiency.

```
10 PRINT "PROG 177"
20 PRINT "SERIES RESONANCE"
30 INPUT "ENTER INDUCTANCE IN HENRYS" ; L
40 INPUT "ENTER CAPACITANCE IN MICROFARADS" ; C
50 LET X=2*3.141593*SQR(L*C/10^6)
60 LET FR=1/X
70 PRINT "RESONANT FREQUENCY = " FR "HERTZ"
```

11.6 Parallel circuits

(a) R and L

Example 26 Figure 11.31 shows a parallel R–L circuit. Calculate all the parameters.

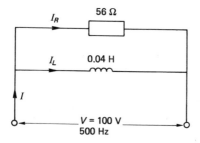

Figure 11.31

(i) Resistor current

$$I_R = \frac{V}{R} = \frac{100}{56} = 1.79 \text{ A}$$

(ii) Inductor current

$$I_L = \frac{V}{X_L}$$

$$X_L = 2\pi fL = 2 \times \pi \times 500 \times 0.04 = 125.66 \ \Omega$$

$$I_L = \frac{100}{125.66} = 0.796 \text{ A}$$

(iii) Total current

$$I = \sqrt{(I_R^2 + I_L^2)}$$

$$I = \sqrt{(1.79^2 + 0.796^2)} = 1.96 \text{ A}$$

See phasor diagram, Figure 11.32.

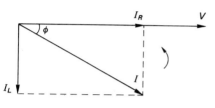

Figure 11.32

(iv) Phase angle

$$\phi = \arctan \frac{I_L}{I_R}$$

$$= \arctan \frac{0.796}{1.79} = 24° \ (0.418 \text{ radians})$$

(v) Circuit impedance

$$Z = \frac{V}{I}$$

$$= \frac{100}{1.96} = 51.02 \ \Omega$$

(vi) True power

$$\text{True power} = V_R I_R$$

$$= 100 \times 1.79 = 179 \text{ W}$$

Alternatively

$$\text{True power} = VI \cos \phi$$

$$= 100 \times 1.96 \times 0.9135 = 179 \text{ W}$$

(vii) Reactive power

$$\text{Reactive power} = V_L I_L$$

$$= 100 \times 0.796 = 79.6 \text{ VA}$$

(viii) Apparent power

$$\text{Apparent power} = VI$$

$$= 100 \times 1.96 = 196 \text{ VA}$$

```
10 PRINT "PROG 178"
20 PRINT "PARALLEL R-L CIRCUIT"
30 INPUT "ENTER RESISTOR VALUE" ; R
40 INPUT "ENTER INDUCTANCE IN HENRYS" ; L
50 INPUT "ENTER VOLTAGE" ; V
60 INPUT "ENTER FREQUENCY" ; F
70 LET IR=V/R
80 LET XL=2*3.141593*F*L
90 LET IL=V/XL
100 I=SQR(IR^2+IL^2)
110 LET PA=ATN(IL/IR)*180/3.141593
120 LET Z=V/I
130 LET P1=IR^2*R
140 LET P2=IL^2*XL
150 LET P3=V*I
160 PRINT "IR="IR "XL="XL "IL="IL "I="I
170 PRINT "PHASE ANGLE = " PA "Z="Z
180 PRINT "TRUE POWER="P1 "REACTIVE POWER="P2
190 PRINT "APPARENT POWER="P3
```

(b) *R and C*

Example 27 Figure 11.33 shows a parallel R–C circuit. Calculate all the parameters.

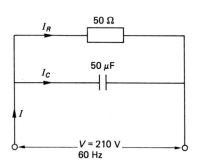

Figure 11.33

(i) Resistor current

$$I_R = \frac{V}{R}$$

$$= \frac{210}{50} = 4.2 \text{ A}$$

(ii) Capacitor current

$$I_C = \frac{V}{X_C}$$

$$X_C = \frac{1}{2\pi fC} = \frac{1}{2 \times \pi \times 60 \times 50 \times 10^{-6}} = 53.05 \ \Omega$$

$$I_C = \frac{210}{53.05} = 3.96 \text{ A}$$

(iii) Total current

$$I = \sqrt{(I_R^2 + I_C^2)}$$
$$I = \sqrt{(4.2^2 + 3.96^2)} = 5.77 \text{ A}$$

See phasor diagram, Figure 11.34.

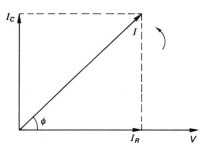

Figure 11.34

(iv) Phase angle

$$\phi = \arctan \frac{I_C}{I_R}$$
$$= \arctan \frac{3.96}{4.2} = 43.32° \text{ (0.76 radians)}$$

(v) Circuit impedance

$$Z = \frac{V}{I}$$
$$= \frac{210}{5.77} = 36.39 \ \Omega$$

(vi) True power

True power $= V_R I_R$
$$= 210 \times 4.2 = 882 \text{ W}$$

Alternatively

True power $= VI \cos \phi$
$$= 210 \times 5.77 \times 0.728 = 882 \text{ W}$$

(vii) Reactive power

Reactive power $= V_C I_C$
$$= 210 \times 3.96 = 831.6 \text{ VA}$$

(viii) Apparent power

Apparent power $= VI$
$$= 210 \times 5.77 = 1211.7 \text{ VA}$$

```
10 PRINT "PROG 179"
20 PRINT "PARALLEL R-C CIRCUIT"
30 INPUT "ENTER RESISTOR VALUE" ; R
40 INPUT "ENTER CAPACITANCE IN MICROFARADS" ; C
50 INPUT "ENTER VOLTAGE" ; V
60 INPUT "ENTER FREQUENCY" ; F
70 LET IR=V/R
80 LET XC=10^6/(2*3.141593*F*C)
90 LET IC=V/XC
100 I=SQR(IR^2+IC^2)
110 LET PA=ATN(IC/IR)*180/3.141593
120 LET Z=V/I
130 LET P1=IR^2*R
140 LET P2=IC^2*XC
150 LET P3=V*I
160 PRINT "IR="IR "XC="XC "IC="IC "I="I
170 PRINT "PHASE ANGLE = " PA "Z="Z
180 PRINT "TRUE POWER="P1 "REACTIVE POWER="P2
190 PRINT "APPARENT POWER="P3
```

(c) *L* and *C*

Example 28 Figure 11.35 shows a parallel *L–C* circuit. Calculate all the parameters.

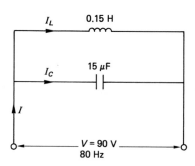

Figure 11.35

(i) Inductor current
$$I_L = \frac{V}{X_L}$$
$$X_L = 2\pi f L = 2 \times \pi \times 80 \times 0.15 = 75.39 \ \Omega$$
$$I_L = \frac{90}{75.39} = 1.194 \text{ A}$$

(ii) Capacitor current
$$I_C = \frac{V}{X_C}$$
$$X_C = \frac{1}{2\pi f C} = \frac{1}{2 \times \pi \times 80 \times 15 \times 10^{-6}} = 132.63 \ \Omega$$
$$I_C = \frac{90}{132.63} = 0.679 \text{ A}$$

(iii) Total current

$$I = I_L - I_C$$
$$= 1.194 - 0.679 = 0.515 \text{ A}$$

See phasor diagram, Figure 11.36.

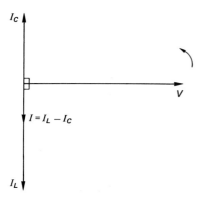

Figure 11.36

(iv) Phase angle. The phase angle between the voltage V and current I is 90° ($\pi/2$ radians).

(v) Circuit impedance

$$Z = \frac{V}{I}$$
$$= \frac{90}{0.515} = 174.76 \text{ }\Omega$$

(vi) The power consumed will be zero since

$$VI \cos \phi = 90 \times 0.515 \times \cos 90° = 0 \text{ W}$$

since neither L nor C can dissipate energy.

```
9 LIST
10 PRINT "PROG 180"
20 PRINT "PARALLEL L-C CIRCUIT"
30 INPUT "ENTER INDUCTANCE IN HENRYS" ; L
40 INPUT "ENTER CAPACITANCE IN MICROFARADS" ; C
50 INPUT "ENTER VOLTAGE" ; V
60 INPUT "ENTER FREQUENCY" ; F
70 LET XL=2*3.141593*F*L
80 LET XC=10^6/(2*3.141593*F*C)
90 LET IL=V/XL
100 LET IC=V/XC
110 LET I=IL-IC
120 LET Z=V/I
130 PRINT "XL="XL "XC="XC "IL="IL "IC="IC "I="I "Z="Z
```

Example 29 Find the resonant frequency of the circuit in Figure 11.35.

$$f_r = \frac{1}{2\pi\sqrt{LC}}$$

$$f_r^2 = \frac{1}{4\pi^2 LC}$$

$$= \frac{1}{4 \times \pi^2 \times 0.15 \times 15 \times 10^{-6}}$$

$$= 11257.9$$

$$f_r = \sqrt{11257.9} = 106 \text{ Hz}$$

(See program 177, page 202)

(d) R–L–C

Example 30 Figure 11.37 shows a coil L with a resistance R connected in parallel with a capacitor C. Calculate the circuit parameters.

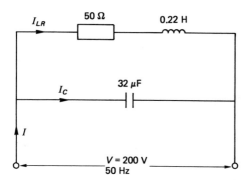

Figure 11.37

The circuit consists of two branches, the inductor and the capacitor branches.

Inductor branch

$$X_L = 2 \times \pi \times 50 \times 0.22 = 69.115 \, \Omega$$

$$Z_L = \sqrt{(50^2 + 69.115^2)} = 85.3 \, \Omega$$

$$I_{LR} = \frac{200}{85.3} = 2.345 \text{ A}$$

$$\phi = \arctan \frac{69.115}{50} = 54.12°$$

Capacitor branch

$$X_C = \frac{1}{2 \times \pi \times 50 \times 32 \times 10^{-6}} = 99.47 \, \Omega$$

$$I_C = \frac{200}{99.47} = 2.01 \text{ A}$$

Complete circuit. The phasor diagram is shown in Figure 11.38 and it shows how the current I can be calculated.

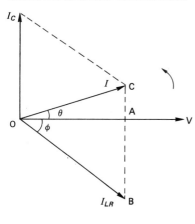

Figure 11.38

$$AB = 2.345 \sin 54.12 = 1.9 \text{ A}$$

$$AC = I_C - AB = 2.01 - 1.9 = 0.11 \text{ A}$$

$$OA = \frac{1.9}{\tan 54.12} = 1.374 \text{ A}$$

$$I = \sqrt{(1.374^2 + 0.11^2)} = 1.378 \text{ A}$$

$$\theta = \arctan \frac{0.11}{1.378} = 4.6°$$

True power $= I_{LR}^2 R = 2.345^2 \times 50 = 274.9$ W

Apparent power $= VI = 200 \times 1.378 = 275.6$ VA

```
10 PRINT "PROG 181"
20 PRINT "PARALLEL R-L-C CIRCUIT"
30 INPUT "ENTER RESISTOR VALUE" ; R
40 INPUT "ENTER CAPACITANCE IN MICROFARADS" ; C
45 INPUT "ENTER INDUCTANCE IN HENRYS" ; L
50 INPUT "ENTER VOLTAGE" ; V
60 INPUT "ENTER FREQUENCY" ; F
65 LET XL=2*3.141593*F*L
70 LET XC=10^6/(2*3.141593*F*C)
80 LET ZLR=SQR(R^2+XL^2)
90 LET ILR=V/ZLR
100 LET PA=ATN(XL/R)*180/3.141593
110 LET IC=V/XC
120 LET A=ILR*COS(PA*3.141593/180)
130 LET B=IC-(ILR*SIN(PA*3.141593/180))
140 LET I=SQR((A^2)+(B^2))
150 LET PHI=ATN(B/A)*180/3.141593
160 LET Z=V/I
170 LET P1=ILR^2*R
180 LET P2=V*I
190 PRINT "XL="XL "XC="XC "ZLR="ZLR "ILR="ILR
200 PRINT "COIL PHASE ANGLE="PA "IC="IC "I="I "Z="Z
210 PRINT "CIRCUIT PHASE ANGLE="PHI
220 PRINT "TRUE POWER="P1 "APPARENT POWER="P2
```

Example 31 Given a parallel circuit similar to the one in Figure 11.37 with a 14-H inductor, a 14 µF capacitor and a coil resistance of 140 Ω, calculate the resonant frequency.

$$f_r = \frac{1}{2\pi} \sqrt{\left(\frac{1}{LC} - \frac{R^2}{L^2}\right)}$$

$$\frac{1}{LC} = \frac{1}{14 \times 14 \times 10^{-6}} = 5102$$

$$\frac{R^2}{L^2} = \frac{140^2}{14^2} = 100$$

$$\sqrt{(5102 - 100)} = 70.725$$

$$f_r = \frac{70.725}{2\pi} = 11.26 \text{ Hz}$$

```
10 PRINT "PROG 182"
20 PRINT "PARALLEL RESONANCE"
30 INPUT "ENTER INDUCTANCE IN HENRYS" ; L
40 INPUT "ENTER CAPACITANCE IN MICROFARADS" ; C
45 INPUT "ENTER RESISTANCE" ; R
50 LET A=10^6/(L*C)
60 LET B=R^2/L^2
70 LET FR=SQR(A-B)/(2*3.141593)
80 PRINT "RESONANT FREQUENCY=" FR "HERTZ"
```

Problems

1. A 0.25-H inductor is connected to a 110 V, 60-Hz supply. Calculate the current flowing.
2. A current of 1.75 A flows in a 4-H inductor. Calculate the applied voltage given that the frequency is 400-Hz.
3. A 50-V, 300-Hz supply is applied to an inductor. If the current flowing is 3 A, calculate the value of the inductance.
4. A 240-V supply produces a current of 1.5 A in a 500-µH coil. Calculate the frequency of the supply voltage.
5. A 12-µF capacitor is connected to a 180-V, 120-Hz supply. Calculate the current flowing.
6. A current of 1.4 A flows in a 30-µF capacitor. Calculate the applied voltage, given that the frequency is 150-Hz.
7. A 140-V, 60-Hz supply produces a current of 2 A in a capacitor. Find the value of the capacitance.
8. A 1500-V supply produces a current of 0.05 A in a 250-pF capacitor. Calculate the supply frequency.
9. A series R–L circuit consists of a 10-Ω resistor and a 0.15-H inductor. The circuit is connected to a 110-V, 60-Hz supply. Calculate: (a) the inductive reactance, (b) the circuit impedance, (c) the current, (d) the voltage drops across both the resistor and inductor, (e) the true, reactive and apparent powers, (f) the power factor, (e) the phase angle.
10. A 0.2-H inductor is connected to a 180-V, 110-Hz supply. The current flowing is 0.3 A. Calculate: (a) the inductive reactance, (b) the impedance, (c) the coil resistance, (d) the voltage drops, (e) the true, reactive and apparent power, (f) the power factor, (g) the phase angle.
11. An R–L circuit is connected to a 300-V, 50-Hz supply. The resistor value is 60 Ω, and the current flowing is 3 A. Calculate: (a) the inductive reactance, (b) the circuit impedance, (c) the

a.c. circuits 211

inductance of the coil, (d) the voltage drops, (e) the true, reactive and apparent powers, (f) the power factor, (e) the phase angle.

12 A series R–L circuit is connected to a 850-Hz supply. The resistor has a value of 4700-Ω and the coil inductance is 2-H. If the current flowing is 0.04 A, calculate: (a) the inductive reactance, (b) the circuit impedance, (c) the applied voltage, (d) the voltage drops across R and L, (e) the true, reactive and apparent powers, (f) the power factor, (g) the phase angle.

13 A series R–L circuit is driven by a 450-V supply. R is 270 Ω and L is 0.04 H. If the current flowing is 0.6 A, calculate: (a) the inductive reactance, (b) the circuit impedance, (c) the frequency, (d) the voltage drops across R and L, (e) the true, reactive and apparent powers, (f) the power factor, (e) the phase angle.

14 A coil is connected to a 240-V, 110-Hz supply, and takes a current of 6 A at a phase angle of 45°. Calculate the resistance and the inductance of the coil.

15 If the voltage drop across a resistor is 45 V, and the voltage drop across the inductor is 60 V in a series circuit, calculate the values of the applied voltage, the current, the resistor and the inductance, given that the true power is 500 W and the frequency 50 Hz.

16 A series R–C circuit is connected to a 60-V, 60-Hz supply. If R is 180 Ω and C is 15 μF, calculate: (a) the capacitive reactance, (b) the circuit impedance, (c) the current, (d) the voltage drops across R and C, (e) the true, reactive and apparent powers, (f) the power factor and the phase angle.

17 A series R–C circuit is connected to a 210-V, 300-Hz supply. The capacitance is 5 μF, and the current flowing is 0.35 A. Calculate: (a) the capacitive reactance, (b) the impedance, (c) the value of the resistor, (d) the voltage drops across R and C, (e) the true, reactive and apparent powers, (f) the power factor and the phase angle.

18 A series R–C circuit consumes 4 A from a 440-V, 60-Hz supply. If R is 68 Ω, find the value of C.

19 A voltage applied to a 10-kΩ resistor in series with a 470-pF capacitor produces a current of 0.1 A at a frequency of 60 kHz. Find the value of the applied voltage.

20 A voltage of 600 V applied to a 1500-Ω resistor in series with a 0.4-μF capacitor produces a current of 0.12 A. Calculate the frequency.

21 A series R–C circuit is connected to a 45-V, 60-Hz supply. If the reactance of the capacitor is 500 Ω, and the phase angle 50°, calculate: (a) the value of C, (b) the value of I, (c) the value of R.

22 In a series R–C circuit C is 4.5 μF. If the circuit impedance is 140 Ω and V_R and V_C are 80 and 100 V respectively, calculate: (a) the applied voltage, (b) the frequency, (c) the current.

23 A series R–L–C has the following parameters: $R = 6$ Ω, $L = 0.08$ H, $C = 280$ μF, $V = 150$ V, $f = 50$ Hz. Calculate: (a) the inductive reactance, (b) the capacitive reactance, (c) the circuit impedance, (d) the current, (e) the voltage drops across R, L and C, (f) the true, reactive and apparent powers, (g) the power factor and phase angle.

24 Repeat Problem 23 with $R = 15$ Ω, $L = 0.025$ H, $C = 160$ μF, $V = 120$ V, $f = 60$ Hz.

25 Calculate the resonant frequency of a circuit where the inductance is 1 H and the capacitance is 1 μF.

26 A parallel R–L circuit has the following parameters:

$R = 200\ \Omega$, $L = 2.0$ H, $V = 1000$ V, $f = 50$ Hz. Calculate: (a) the resistor current, (b) the inductive reactance, (c) the inductor current, (d) the total current, (e) the phase angle, (f) the circuit impedance, (g) the true, reactive and apparent powers.

27 A parallel R–C circuit has the following parameters: $R = 250\ \Omega$, $C = 200\ \mu\text{F}$, $V = 240$ V, $f = 50$ Hz. Calculate: (a) the resistor current, (b) the capacitive reactance, (c) the capacitor current, (d) the total current, (e) the phase angle, (f) the impedance, (g) the true, reactive and apparent powers.

28 A parallel L–C circuit has the following parameters: $L = 0.22$ H, $C = 20\ \mu\text{F}$, $V = 110$ V, $f = 60$ Hz. Calculate: (a) the inductive reactance, (b) the capacitive reactance, (c) the inductor current, (d) the capacitor current, (e) the total current, (f) the circuit impedance.

29 A parallel R–L–C circuit identical to Figure 11.37 has the following parameters: $R = 65\ \Omega$, $L = 1.0$ H, $C = 21\ \mu\text{F}$, $V = 250$ V, $f = 100$ Hz. Calculate: (a) inductive reactance, (b) capacitive reactance, (c) impedance of L–R branch, (d) current in L–R branch, (e) phase angle in L–R branch, (f) capacitor current, (g) total current, (h) circuit impedance, (i) circuit phase angle, (j) true power and apparent power.

30 Given a parallel circuit identical to Figure 11.37 with a 0.02-H inductor, a 0.4-μF capacitor and a 6-Ω resistor, calculate the resonant frequency.

12 Phasors

12.1 Introduction

Chapter 11 showed that problems in a.c. circuits can be solved using phasors. A phasor has both magnitude and direction and it can be resolved into two components at right angles to each other. Figure 12.1 shows a phasor OA, and this can be written as

$$x + jy$$

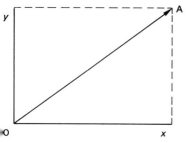

Figure 12.1

x is the horizontal component, and the operator j is introduced to indicate that the other component is along the y axis.

$x + jy$ is called a complex number. x, the horizontal component, is a real number. If the vertical component is an imaginary number, Figure 12.1 shows that the length of the phasor OA is

$$\sqrt{(x^2 + y^2)} \text{ (modulus)}$$

and the phase angle is

$$\tan^{-1} \frac{y}{x} \text{ (argument)}$$

By introducing the symbol j, a phasor such as OA in Figure 12.2 can be written as

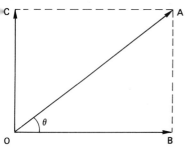

Figure 12.2

$$OA = OB + jOC$$

This is called the rectangular or cartesian notation.

$$\text{since } \cos\theta = \frac{OB}{OA}, \ OB = OA \cos\theta$$

$$\sin\theta = \frac{AB}{OA}, \ AB = OA \sin\theta$$

and

$$\text{since } AB = OC, \ OC = OA \sin\theta$$

$$OA = OA \cos\theta + jOA \sin\theta$$
$$= OA(\cos\theta + j\sin\theta)$$

This is a trigonometric notation. A phasor can also be written as

$$OA \angle\theta$$

where θ is the angle between the phasor and the positive x axis. This is called the polar notation.

The operator j rotates a phasor anticlockwise through an angle of 90°, giving the phasor jx as shown in Figure 12.3.

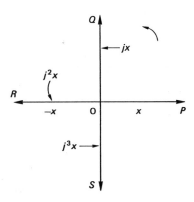

Figure 12.3

Operating on phasor jx by j rotates the phasor through another 90°, giving the phasor $j(jx)$ or j^2x. Phasor j^2x is the negative of the original phasor x, i.e.

$$j^2x = -x$$

i.e. $j^2 = -1$

and $j = \sqrt{-1}$ which is an imaginary number

if $j^2 = -1$

$$j^3 = j^2 j = -1 \times j = -j$$
$$j^4 = j^2 j^2 = -1 \times -1 = 1$$

Summarizing

$$j = \sqrt{-1}$$
$$j^2 = -1$$
$$j^3 = -j,$$
$$j^4 = 1$$
$$j^5 = j$$

Example 1 Convert to polar form $3 + j4$.

$$3 + j4 = \sqrt{(3^2 + 4^2)} \tan^{-1} \frac{4}{3}$$
$$= 5 \underline{/53.13°}$$

```
10 PRINT "PROG 183"
20 PRINT "CONVERSION TO POLAR FORM"
30 INPUT "ENTER X" ; X
40 INPUT "ENTER Y" ; Y
50 LET A=SQR(X^2+Y^2)
60 LET B=ATN(Y/X)*180/3.141593
70 PRINT "MODULUS = " A
80 PRINT "ARGUMENT = " B "DEGREES"
```

Example 2 Convert $5 \underline{/53.13°}$ to rectangular form.

$$5 \underline{/53.13°} = 5(\cos 53.13° + j \sin 53.13°)$$
$$= 5(0.6 + j0.8)$$
$$= 3 + j4$$

```
10 PRINT "PROG 184"
20 PRINT "CONVERSION TO RECTANGULAR FORM"
30 INPUT "ENTER MODULUS" ; A
40 INPUT "ENTER ARGUMENT" ; B
50 LET C=B*3.141593/180
60 LET D=A*COS(C)
70 LET E=A*SIN(C)
80 PRINT "X COMPONENT = " D
90 PRINT "Y COMPONENT = " E
```

Phasors can be in any of the four quadrants, as shown in Figure 12.4. The general forms are

quadrant 1 $x + jy$

quadrant 2 $-x + jy$

quadrant 3 $-x - jy$

quadrant 4 $x - jy$

Figure 12.4 shows a phasor in each quadrant. These can be converted to the polar form, OA $\underline{/\theta}$, OB $\underline{/\theta}$, OC $\underline{/\theta}$, OD $\underline{/\theta}$ where θ in each case is the angle between the phasor and the positive x axis.

In quadrant 1

$$\text{OA} = 4 + j3 = \sqrt{(4^2 + 3^2)} \tan^{-1} \frac{3}{4}$$
$$= 5 \underline{/36.87°}$$

(See program 183, above.)

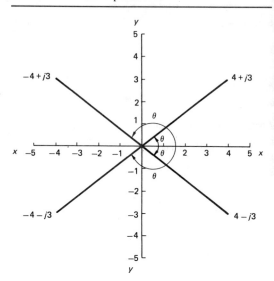

Figure 12.4

In quadrant 2

$$OB = -4 + j3 = \sqrt{(-4)^2 + 3^2} \left(180° + \tan^{-1}\frac{3}{4}\right)$$

$$= 5 \; \underline{/143.13°}$$

```
10 PRINT "PROG 185"
20 PRINT "CONVERSION TO POLAR FORM"
30 INPUT "ENTER X" ; X
40 INPUT "ENTER Y" ; Y
50 LET A=SQR(X^2+Y^2)
60 LET B=180+ATN(Y/X)*180/3.141593
70 PRINT "MODULUS = " A
80 PRINT "ARGUMENT = " B "DEGREES"
```

In quadrant 3

$$OC = -4 - j3 = \sqrt{[(-4)^2 + (-3)^2]} \left(\tan^{-1}\frac{-3}{-4}\right) - 180°$$

$$= 5 \; \underline{/-143.13°}$$

```
10 PRINT "PROG 186"
20 PRINT "CONVERSION TO POLAR FORM"
30 INPUT "ENTER X" ; X
40 INPUT "ENTER Y" ; Y
50 LET A=SQR(X^2+Y^2)
60 LET B=ATN(Y/X)*180/3.141593-180
70 PRINT "MODULUS = " A
80 PRINT "ARGUMENT = " B "DEGREES"
```

In quadrant 4

$$OD = 4 - j3 = \sqrt{[4^2 + (-3)^2]} \tan^{-1} \frac{-3}{4}$$

$$= 5 \angle -36.87°$$

(See program 183, page 215.)

Example 3 Convert the answers in Example 2 to the rectangular form.

(i) $5 \angle 36.87° = 5(\cos 36.87 + j \sin 36.87)$
$= 5(0.8 + j0.6) = 4 + j3$

(See program 184, page 215.)

(ii) $5 \angle 143.13° = 5(\cos 143.13 + j \sin 143.13)$
$= 5(-0.8 + j0.6) = -4 + j3$

(See program 184, page 215.)

(iii) $5 \angle -143.13° = 5(\cos - 143.13 + j \sin - 143.13)$
$= 5[-0.8 + (-j0.6)] = -4 - j3$

(See program 184, page 215.)

(iv) $5 \angle -36.87° = 5(\cos - 36.87 + j \sin - 36.87)$
$= 5[0.8 + (-j0.6)] = 4 - j3$

(See program 184, page 215.)

12.2 Addition

When two complex numbers are to be added together the real and imaginary parts are added separately.

Two numbers given in polar form cannot be directly added together. Both must be converted to the rectangular notation first.

Example 4 Add $2 + j3$ and $4 - j$.

$$2 + j3 + 4 - j = 6 + j2$$

The graphical addition is shown in Figure 12.5.

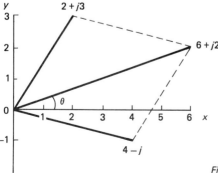

Figure 12.5

$$6 + j2 = \sqrt{(6^2 + 2^2)} \tan^{-1} \frac{2}{6}$$

$$= 6.32 \underline{/18.43°}$$

6.32 is the length of the resultant phasor OA, and 18.43° is the angle θ between the x axis and the phasor.

```
10  PRINT "PROG 187"
20  PRINT "ADDITION OF COMPLEX NUMBERS"
30  INPUT "ENTER X1" ; X1
40  INPUT "ENTER Y1" ; Y1
50  INPUT "ENTER X2" ; X2
60  INPUT "ENTER Y2" ; Y2
70  LET A=X1+X2
80  LET B=Y1+Y2
90  LET C=SQR(A^2+B^2)
100 PRINT "MODULUS = " C
110 IF A>=0 THEN GOTO 190
120 IF B<0 THEN GOTO 160
130 LET D=180+ATN(B/A)*180/3.141593
140 PRINT "ARGUMENT = " D "DEGREES"
150 STOP
160 LET D=ATN(B/A)*180/3.141593-180
170 PRINT "ARGUMENT = " D "DEGREES"
180 STOP
190 LET D=ATN(B/A)*180/3.141593
200 PRINT "ARGUMENT = " D "DEGREES"
```

Example 5 Add $2 + j3$ and $-3 + j2$.

$$2 + j3 - 3 + j2 = -1 + j5$$

$$-1 + j5 = \sqrt{(-1^2 + 5^2)}\ 180° - \tan^{-1} \frac{5}{-1}$$

$$= 5.099 \underline{/101.3°}$$

(See program 187, above.)

The graphical addition is shown in Figure 12.6.

Example 6 Add $-2 + j2$ and $-3 - j3$.

$$-2 + j2 - 3 - j3 = -5 - j$$

$$-5 - j = \sqrt{[-5^2 + (-1)^2]} \tan^{-1} \frac{-1}{-5} - 180°$$

$$= 5.099 \underline{/-168.69°}$$

(See program 187, above.)

The graphical addition is shown in Figure 12.7.

Example 7 Add $3 - j$ and $2 - j3$.

$$3 - j + 2 - j3 = 5 - j4$$

$$5 - j4 = \sqrt{[5^2 + (-4)^2]} \tan^{-1} \frac{-4}{5}$$

$$= 6.403 \underline{/-38.66°}$$

(See program 187, above.)

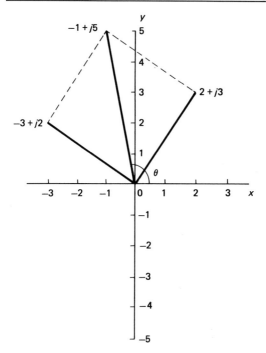

Figure 12.6

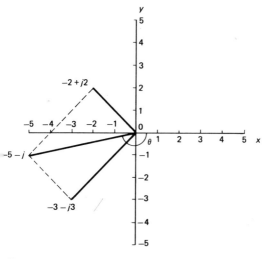

Figure 12.7

The graphical addition is shown in Figure 12.8.

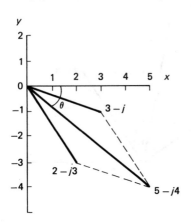

Figure 12.8

Example 8 Add 20 ∠30° and 40 ∠60°.

$$20 \cos 30° + j\, 20 \sin 30° = 17.32 + j10$$

$$40 \cos 60° + j\, 40 \sin 60° = 20 + j34.64$$

$$17.32 + j10 + 20 + j34.64$$

$$= 37.32 + j44.64$$

$$= \sqrt{(37.32^2 + 44.64^2)}\, \tan^{-1} \frac{44.64}{37.32}$$

$$= 58.19 \angle 50.1°$$

```
10 PRINT "PROG 188"
20 PRINT "ADDITION OF COMPLEX NUMBERS"
30 INPUT "ENTER X1" ; X1
40 INPUT "ENTER Y1" ; Y1
50 INPUT "ENTER X2" ; X2
60 INPUT "ENTER Y2" ; Y2
70 LET A=Y1*3.141593/180
80 LET B=X1*COS(A)
90 LET C=X1*SIN(A)
100 LET D=Y2*3.141593/180
110 LET E=X2*COS(D)
120 LET F=X2*SIN(D)
130 LET G=B+E
140 LET H=C+F
150 LET I=SQR(G^2+H^2)
160 PRINT "MODULUS = " I
170 IF G>=0 THEN GOTO 250
180 IF H<0 THEN GOTO 220
190 LET J=180+ATN(H/G)*180/3.141593
200 PRINT "ARGUMENT = " J "DEGREES"
210 STOP
220 LET K=ATN(H/G)*180/3.141593-180
230 PRINT "ARGUMENT = " K "DEGREES"
240 STOP
250 LET L=ATN(H/G)*180/3.141593
260 PRINT "ARGUMENT = " L "DEGREES"
```

Example 9 Add 20 $\underline{/150°}$ and 40 $\underline{/120°}$.

$$20 \cos 150° + j\, 20 \sin 150° = -17.3 + j10$$
$$40 \cos 120° + j\, 40 \sin 120° = -20 + j34.64$$
$$-17.3 + j10 - 20 + j34.64$$
$$= -37.3 + j44.64$$
$$= \sqrt{(-37.3^2 + 44.64^2)}\; 180 + \tan^{-1} \frac{44.64}{-37.3}$$
$$= 58.17\; \underline{/129.89°}$$

(See program 188, page 220.)

Example 10 Add 20 $\underline{/-100°}$ and 40 $\underline{/-140°}$.

$$20 \cos -100° + j\, 20 \sin -100° = -3.47 - j19.7$$
$$40 \cos -140° + j\, 40 \sin -140° = -30.64 - j25.7$$
$$-3.47 - j19.7 - 30.64 - j25.7$$
$$= -34.11 - j45.4$$
$$= \sqrt{[-34.11^2 + (-45.4)^2]}\; \tan^{-1} \frac{-45.4}{-34.11} - 180$$
$$= 56.78\; \underline{/-126.92°}$$

(See program 188, page 220.)

Example 11 Add 10 $\underline{/-45°}$ and 15 $\underline{/-30°}$.

$$10 \cos -45° + j\, 10 \sin -45° = 7.07 - j7.07$$
$$15 \cos -30° + j\, 15 \sin -30° = 12.99 - j7.5$$
$$7.07 - j7.07 + 12.99 - j7.5$$
$$= 20.06 - j14.57$$
$$= \sqrt{[20.06^2 + (-14.57)^2]}\; \tan^{-1} \frac{-14.57}{20.06}$$
$$= 24.79\; \underline{/-35.99°}$$

(See program 188, page 220.)

12.3 Subtraction

The subtraction of two complex numbers follows the same rules as for addition. Complex numbers in polar form must be changed to rectangular form before subtraction.

Example 12 Subtract $3 + j5$ from $4 - j2$.

$$(4 - j2) - (3 + j5)$$
$$= 4 - j2 - 3 - j5 = 1 - j7$$

The graphical subtraction is shown in Figure 12.9.

$$1 - j7 = \sqrt{[1^2 + (-7)^2]}\; \tan^{-1} \frac{-7}{1}$$
$$= 7.07\; \underline{/-81.87°}$$

7.07 is the length of the resultant phasor OA, $\theta = -81.87°$, and is

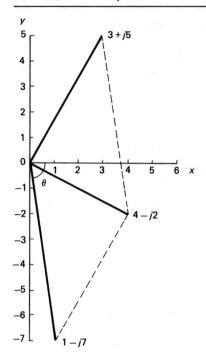

Figure 12.9

the angle between the phasor and the positve x axis. Note that the negative sign indicates that the phasor is in the fourth quadrant.

```
10 PRINT "PROG 189"
20 PRINT "SUBTRACTION OF COMPLEX NUMBERS"
30 INPUT "ENTER X1" ; X1
40 INPUT "ENTER Y1" ; Y1
50 INPUT "ENTER X2" ; X2
60 INPUT "ENTER Y2" ; Y2
70 LET A=X1-X2
80 LET B=Y1-Y2
90 LET C=SQR(A^2+B^2)
100 PRINT "MODULUS = " C
110 IF A>=0 THEN GOTO 190
120 IF B<0 THEN GOTO 160
130 LET D=180+ATN(B/A)*180/3.141593
140 PRINT "ARGUMENT = " D "DEGREES"
150 STOP
160 LET D=ATN(B/A)*180/3.141593-180
170 PRINT "ARGUMENT = " D "DEGREES"
180 STOP
190 LET D=ATN(B/A)*180/3.141593
200 PRINT "ARGUMENT = " D "DEGREES"
```

Example 13 Subtract $4 - j2$ from $6 + j4$.

$$(6 + j4) - (4 - j2)$$
$$= 6 + j4 - 4 + j2 = 2 + j6$$

$$= \sqrt{(2^2 + 6^2)} \tan^{-1}\frac{6}{2}$$

$$= 6.32 \ \underline{/71.56°}$$

(See program 189, page 222.)

Example 14 Subtract $6 + j3$ from $4 + j2$.

$$(4 + j2) - (6 + j3)$$
$$4 + j2 - 6 - j3 = -2 - j$$
$$= \sqrt{[-2^2 + (-1)^2]} \tan^{-1}\frac{-1}{-2} - 180$$
$$= 2.236 \ \underline{/-153.435°}$$

(See program 189, page 222.)

Example 15 Subtract $4 + j$ from $2 + j3$

$$(2 + j3) - (4 + j)$$
$$2 + j3 - 4 - j = -2 + j2$$
$$= \sqrt{(-2^2 + 2^2)} \ 180 + \tan^{-1}\frac{2}{-2}$$
$$= 2.83 \ \underline{/135°}$$

(See program 189, page 222.)

Example 16 Subtract $10 \ \underline{/30°}$ from $10 \ \underline{/60°}$

$$10 \ \underline{/60°} = 10(\cos 60° + j \sin 60°)$$
$$= 10(0.5 + j0.866)$$
$$= 5 + j8.66$$

$$10 \ \underline{/30°} = 10(\cos 30° + j \sin 30°)$$
$$= 10(0.866 + j0.5)$$
$$= 8.66 + j5$$

$$10 \ \underline{/60°} - 10 \ \underline{/30°}$$
$$= (5 + j8.66) - (8.66 + j5)$$
$$= 5 + j8.66 - 8.66 - j5$$
$$= -3.66 + j3.66$$

$$= \sqrt{[(-3.66)^2 + 3.66^2]} \ 180 + \tan^{-1}\frac{3.66}{-3.66}$$

$$= 5.17 \ \underline{/135°}$$

```
10 PRINT "PROG 190"
20 PRINT "SUBTRACTION OF COMPLEX NUMBERS"
30 INPUT "ENTER X1" ; X1
40 INPUT "ENTER Y1" ; Y1
50 INPUT "ENTER X2" : X2
60 INPUT "ENTER Y2" : Y2
70 LET A=Y1*3.141593/180
80 LET B=X1*COS(A)
90 LET C=X1*SIN(A)
100 LET D=Y2*3.141593/180
110 LET E=X2*COS(D)
```

```
120 LET F=X2*SIN(D)
130 LET G=B-E
140 LET H=C-F
150 LET I=SQR(G^2+H^2)
160 PRINT "MODULUS = " I
170 IF G>=0 THEN GOTO 250
180 IF H<0 THEN GOTO 220
190 LET J=180+ATN(H/G)*180/3.141593
200 PRINT "ARGUMENT = " J "DEGREES"
210 STOP
220 LET K=ATN(H/G)*180/3.141593-180
230 PRINT "ARGUMENT = " K "DEGREES"
240 STOP
250 LET L=ATN(H/G)*180/3.141593
260 PRINT "ARGUMENT = " L "DEGREES"
```

12.4 Multiplication

Complex numbers can be multiplied in either the rectangular or the polar form.

In the rectangualr form

$$(a + jb)(c + jd) = ac + jad + jbc + j^2bd$$

in the polar form

$$N \underline{/\theta°} + M \underline{/\phi°} = NM \underline{/\theta° + \phi°}$$

Example 17 Evaluate $(2 + j3)(4 - j)$.

$$\begin{array}{r} 2 + j3 \\ \underline{4 - j} \\ 8 + j12 \\ \underline{-j2 - j^23} \\ 8 + j10 - j^23 \end{array}$$

$8 + j10 - j^23 = 8 + j10 - 3(-1) = 11 + j10$

```
10 PRINT "PROG 191"
20 PRINT "MULTIPLICATION"
30 INPUT "ENTER A" ; A
40 INPUT "ENTER B" ; B
50 INPUT "ENTER C" ; C
60 INPUT "ENTER D" ; D
70 LET E=A*C
80 LET F=A*D
90 LET G=B*C
100 LET H=B*D
110 LET I=F+G
120 LET K=-H
130 LET L=E+K
140 PRINT L "J*" I
```

$11 + j10$ in polar form = $\sqrt{(11^2 + 10^2)} \tan^{-1}\dfrac{10}{11}$

$$= 14.87 \underline{/42.27°}$$

$2 + j3 = \sqrt{(2^2 + 3^2)} \tan^{-1}\dfrac{3}{2}$

$$= 3.61 \underline{/56.31°}$$

$4 - j = \sqrt{[4^2 + (-1)^2]} \tan^{-1}\dfrac{-1}{4} = 4.12 \underline{/-14.04°}$

Note

$$3.61 \times 4.12 \; \underline{/56.31° - 14.04°} = 14.87 \; \underline{/42.27°}$$

Multiplication in polar form is considerably easier than multiplying in rectangular from.

12.5 Division

Division can be carried out using either the rectangular or the polar form.

In the rectangular form the expression is rationalized by multiplying the numerator and the denominator by the conjugate of the denominator thus

$$\frac{a + jb}{c + jd} = \frac{(a + jb)(c - jd)}{(c + jd)(c - jd)}$$

$$= \frac{ac + jbc - jad - j^2bd}{c^2 + jcd - jcd - j^2d^2}$$

$$= \frac{ac + jbc - jad + bd}{c^2 + d^2} \; \text{since} \; j^2 = -1$$

Note $c - jd$ is the conjugate of $c + jd$.

$j^2 = -1$ see Section 12.1

In the polar form

$$\frac{N \; \underline{/\theta°}}{M \; \underline{/\phi°}} = \frac{N}{M} \; \underline{/\theta° - \phi°}$$

Example 18 Evaluate

$$\frac{2 + j9}{5 - j2}$$

$$= \frac{10 + j45 + j4 + j^2 18}{25 + 4} = \frac{10 + j45 + j4 - 18}{25 + 4}$$

$$= \frac{-8 + j49}{29} = -0.276 + j1.69$$

```
10 PRINT "PROG.192"
20 PRINT "DIVISION"
30 INPUT "ENTER A" ; A
40 INPUT "ENTER B" ; B
50 INPUT "ENTER C" ; C
60 INPUT "ENTER D" ; D
70 LET E=A*C
80 LET F=A*D
90 LET G=B*C
100 LET H=B*D
110 LET I=-F
120 LET K=I+G
130 LET L=E+H
140 LET M=C^2+D^2
150 LET N=L/M
160 LET O=K/M
170 PRINT N "J*" O
```

$$-0.276 + j1.69 = \sqrt{[(-0.276)^2 + (1.69)^2]} \; 180$$

$$- \tan^{-1} \frac{1.69}{-0.276} \; \text{(quadrant 2)}$$

$$= 1.712 \, \underline{/99.27°}$$

$$2 + j9 = \sqrt{(2^2 + 9^2)} \tan^{-1} \frac{9}{2} = 9.22 \, \underline{/77.47°}$$

$$5 - j2 = \sqrt{[5^2 + (-2)^2]} \tan^{-1} \frac{-2}{5} = 5.38 \, \underline{/-21.80°}$$

Note

$$\frac{9.22}{5.38} \, \underline{/77.47°} - (-21.80°) = 1.712 \, \underline{/99.27°}$$

12.6 Further addition

Example 19 Derive an expression for the sum of the two waves

$$80 \sin 2\pi ft$$

$$85 \sin \left(2\pi ft - \frac{\pi}{4}\right)$$

There are three methods for solving a problem of this kind.

Method 1. By drawing.
Figure 12.10 shows the phasor diagram where OA and OB represent the maximum values to scale. OB lags OA by $\pi/4$ radians (45°). Completing the parallelogram OABC gives the resultant OC.

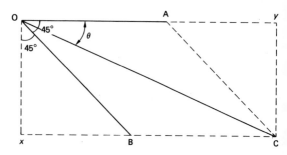

Figure 12.10

By measurement OC = 152

By measurement $\theta = -23.2°$ (-0.405 radians)

The resultant is

$$152 \sin (2\pi ft - 0.405)$$

Method 2. By trigonometry
First complete OYCX in Figure 12.10.

since XOB = 45°

$$\sin 45 = \frac{XB}{85}$$

$$XB = 85 \sin 45 = 60$$

$$\cos 45 = \frac{OX}{85}$$

$$OX = 85 \cos 45 = 60$$

$$OC^2 = XC^2 + OX^2$$

$$= 140^2 + 60^2 \text{ since } BC = OA$$

$$OC = \sqrt{(140^2 + 60^2)} = 152$$

$$\tan \theta = \frac{YC}{OY} = \frac{-60}{140}$$

$$\theta = \arctan \frac{-60}{140} = -23.2° = -0.405 \text{ radians}$$

Method. By j notation
First resolve each phasor into the horizontal and vertical components.

$$OA = 80 + j0 \text{ since it lies along the } x \text{ axis}$$

$$OB = 60 - j60$$

$$OC = OA + OB = 80 + j0 + 60 - j60$$

$$= 140 - j60$$

$$= \sqrt{(140^2 + -60^2)} \tan^{-1} \frac{-60}{140}$$

$$= 152 \angle -23.2° = 152, -0.405 \text{ radians}$$

The answer is expressed in polar notation.

Example 20 A sinusoidal waveform is represented by $e = 283 \sin 1256t$ V. Calculate: (i) The peak value, (ii) The r.m.s. value, (iii) The frequency, (iv) The periodic time.

(i) peak value = 283 V

(ii) r.m.s. value = $\frac{283}{\sqrt{2}} = 200$ V

(iii) $1256 = 2\pi f$

$$f = \frac{1256}{2\pi} = 200 \text{ Hz}$$

$$t = \frac{1}{f} = \frac{1 \times 10^3}{200} = 5 \text{ ms}$$

Problems

1 Convert to polar form:
   (a) $12 + j24$
   (b) $8 - j6$
   (c) $20 + j20$
   (d) $-2 + j4$
   (e) $-25 + j25$
   (f) $-0.5 + j1.5$
   (g) $-6 - j3$
   (h) $-50 - j50$
   (i) $-1.5 - j2.5$

2. Convert to rectangular form:
   (a) 10 ∠45°
   (b) 100 ∠30°
   (c) 12 ∠−45°
   (d) 120 ∠−30°
   (e) 10 ∠80°
3. Add the following and express the answers in polar form:
   (a) $3 + j4$ and $3 - j5$
   (b) $-2 + j5$ and $1 + j4$
   (c) $20 - j24$ and $-30 + j36$
   (d) $-40 + j25$ and $25 - j40$
4. Add the following and express the answers in polar form:
   (a) 4 ∠30° and 5 ∠30°
   (b) 25 ∠125° and 30 ∠150°
   (c) 10 ∠−110° and 30 ∠−110°
   (d) 12 ∠−30° and 14 ∠−45°
5. Subtract the following and express the answers in polar form:
   (a) $8 + j2$ from $6 + j2$
   (b) $5 + j2$ from $-6 + j3$
   (c) $-5 - j$ from $-6 - j2$
   (d) $50 + j10$ from $20 - j5$
6. Subtract the following and express the answsers in polar form:
   (a) 12 ∠60° from 15 ∠30°
   (b) 5 ∠60° from 5 ∠120°
   (c) 4 ∠120° from 2 ∠−120°
   (d) 2 ∠−60° from 3 ∠140°
7. Multiply:
   (a) $1 + j4$ by $5 + j6$
   (b) $6 + j$ by $8 + j2$
   (c) $-6 + j4$ by $5 + j9$
   (d) $20 + j14$ by $25 - j10$
   (e) $-4 - j3$ by $-6 - j12$
8. Divide:
   (a) $4 + j6$ by $4 + j6$
   (b) $20 + j5$ by $15 + j10$
   (c) $6 + j2$ by $-4 - j10$
   (d) $-100 - j150$ by $-200 - j300$

13 Transformers

13.1 Introduction

Transformers consist of two coils which are wound on a common magnetic former, and are insulated from one another. The primary winding (see Figure 13.1) is connected to an a.c. supply and the output is taken from the secondary winding. Transformers can 'step up' or 'step down' a voltage depending on the winding arrangements. They are also used for resistance and impedance matching, coupling one circuit to another, and for isolating one circuit from another since the primary and secondary windings are not interconnected.

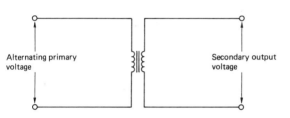

Figure 13.1

13.2 Transformer action

(a) Voltage equations
When an alternating voltage V_p is applied to the primary, a current I_p flows in the winding. I_p produces a magnetic flux Φ, and the changing flux induces an e.m.f. in the primary in opposition to V_p and an e.m.f. V_s in the secondary. It is assumed throughout this chapter that ideal transformers are used. Such transformers have negligible losses, and negligible resistance in the primary and secondary windings, and the magnetic flux produced links both coils. In an ideal transformer, therefore,

$$E_p = V_p \text{ and } E_s = V_s$$

R.m.s. voltages are used in all the calculations.

(b) Induced voltages
The induced voltages are given by

$$E_p = 4.44 n_p f \Phi_{max} \text{ V}$$
$$E_s = 4.44 N_s f \Phi_{max} \text{ V}$$

where N_p = the number of primary turns

f = frequency, Hz

Φ_{max} = maximum flux density, Wb

N_s = the number of secondary turns

Example 1 A transformer is connected to a 120-V, 50-Hz supply and the output voltage is 20 V. The transformer core has a cross-sectional area of 10 cm² and the maximum flux density is 0.5 T. Calculate the number of turns on both windings, assuming an ideal transformer.

$$V_p = 4.44 N_p f \Phi_{max}$$
$$= 4.44 N_p f BA \text{ since } \Phi = BA \text{ See Section 10.2}$$

$$N_p = \frac{V_p}{4.44 f BA}$$

$$= \frac{120}{4.44 \times 50 \times 0.5 \times 10 \times 10^{-4}}$$

$$= 1081 \text{ turns}$$

$$N_s = \frac{20}{4.44 \times 50 \times 0.5 \times 10 \times 10^{-4}}$$

$$= 180 \text{ turns}$$

```
10 PRINT "PROG 193"
20 PRINT "PRIMARY AND SECONDARY TURNS"
30 INPUT "ENTER PRIMARY VOLTAGE" ; VP
40 INPUT "ENTER SECONDARY VOLTAGE" ; VS
50 INPUT "ENTER FREQUENCY" ; F
60 INPUT "ENTER AREA IN SQR CMS" ; A
70 INPUT "ENTER FLUX DENSITY IN TESLAS" ; B
80 LET NP=VP*10^4/(4.44*F*B*A)
90 PRINT "PRIMARY TURNS = " NP
100 LET NS=VS*10^4/(4.44*F*B*A)
110 PRINT "SECONDARY TURNS = " NS
```

Example 2 A transformer has a 150-turn primary winding. If 240 V is applied to the winding, and the maximum flux is 5 mWb, calculate the frequency of the applied voltage.

Rearranging the transformer equation

$$f = \frac{V_p}{4.44 N_p \Phi_{max}}$$

$$= \frac{240}{4.44 \times 150 \times 5 \times 10^{-3}}$$

$$= 72 \text{ Hz}$$

```
10 PRINT "PROG 194"
20 PRINT "FREQUENCY"
30 INPUT "ENTER PRIMARY VOLTAGE" ; VP
40 INPUT "ENTER PRIMARY TURNS" ; NP
50 INPUT "ENTER MAXIMUM FLUX IN MILLIWEBERS" ; PHI
60 LET F=VP*10^3/(4.44*NP*PHI)
70 PRINT "FREQUENCY = " F "HERTZ"
```

Example 3 If a 240-V, 50-Hz mains supply was applied to the transformer in Example 2, what would be the maximum value of the flux in the core?

Rearranging the transformer equation

$$\Phi = \frac{V_p}{4.44 N_p f}$$

$$= \frac{240}{4.44 \times 150 \times 50}$$

$$= 7.2 \times 10^{-3} \text{ Wb} = 7.2 \text{ mWb}$$

```
10 PRINT "PROG 195"
20 PRINT "MAXIMUM FLUX"
30 INPUT "ENTER PRIMARY VOLTAGE" ; VP
40 INPUT "ENTER PRIMARY TURNS" ; NP
50 INPUT "ENTER FREQUENCY" ; F
60 LET PHI=VP*10^3/(4.44*NP*F)
70 PRINT "MAXIMUM FLUX = " PHI "MILLIWEBERS"
```

(c) Input and output voltages

The relationship between these two voltages is given by the equation

$$\frac{V_p}{V_s} = \frac{N_p}{N_s}$$

where V_p = the primary voltage

V_s = the secondary voltage

N_p = the number of primary turns

N_s = the number of secondary turns

Example 4 A transformer has 100 turns on the primary and 250 turns on the secondary. If the input voltage is 6 V, calculate the output voltage.

$$V_s = \frac{V_p N_s}{N_p} = \frac{6 \times 250}{100}$$

$$= 15 \text{ V}$$

```
10 PRINT "PROG 196"
20 PRINT "OUTPUT VOLTAGE"
30 INPUT "ENTER PRIMARY VOLTAGE" ; VP
40 INPUT "ENTER PRIMARY TURNS" ; NP
50 INPUT "ENTER SECONDARY TURNS" ; NS
60 LET VS=VP*NS/NP
70 PRINT "OUTPUT VOLTAGE = " VS "VOLTS"
```

Example 5 The output voltage from a transformer is 750 mV. If the primary winding has 50 turns and the secondary winding has 150 turns, calculate the input voltage.

$$V_p = \frac{V_s N_p}{N_s} = \frac{750 \times 50}{150}$$

$$= 250 \text{ mV}$$

```
10 PRINT "PROG 197"
20 PRINT "INPUT VOLTAGE"
30 INPUT "ENTER SECONDARY VOLTAGE IN MILLIVOLTS" ; VS
40 INPUT "ENTER PRIMARY TURNS" ; NP
50 INPUT "ENTER SECONDARY TURNS" ; NS
60 LET VP=VS*NP/NS
70 PRINT "INPUT VOLTAGE = " VP "MILLIVOLTS"
```

Example 6 If the primary and secondary voltages of a transformer are 2000 V and 500 V, and the secondary winding has 600 turns, calculate the number of primary turns.

$$N_p = \frac{N_s V_p}{V_s} = \frac{600 \times 2000}{500}$$

$$= 2400 \text{ turns}$$

```
10 PRINT "PROG 198"
20 PRINT "PRIMARY TURNS"
30 INPUT "ENTER SECONDARY VOLTAGE" ; VS
40 INPUT "ENTER PRIMARY VOLTAGE" ; VP
50 INPUT "ENTER SECONDARY TURNS" ; NS
60 LET NP=NS*VP/VS
70 PRINT "PRIMARY TURNS = " NP
```

Example 7 A power transformer is driven from the mains, 240 V, and has an output voltage of 12 V. If the primary has 300 turns, calculate the number of secondary turns.

$$N_s = \frac{N_p V_s}{V_p} = \frac{300 \times 12}{240}$$

$$= 15 \text{ turns}$$

```
10 PRINT "PROG 199"
20 PRINT "SECONDARY TURNS"
30 INPUT "ENTER SECONDARY VOLTAGE" ; VS
40 INPUT "ENTER PRIMARY VOLTAGE" ; VP
50 INPUT "ENTER PRIMARY TURNS" ; NP
60 LET NS=NP*VS/VP
70 PRINT "SECONDARY TURNS = " NS
```

Example 8 Find the turns ratio of a transformer that transforms 1200 V into 24 V.

$$\frac{V_p}{V_s} = \frac{N_p}{N_s} = \text{the transformer ratio, } n$$

$$= \frac{1200}{24} = 50 : 1$$

This is a 'step down' transformer.

```
10 PRINT "PROG 200"
20 PRINT "TURNS RATIO"
30 INPUT "ENTER PRIMARY VOLTAGE" ; VP
40 INPUT "ENTER SECONDARY VOLTAGE" ; VS
50 LET N=VP/VS
60 PRINT "TURNS RATIO = " N
```

(d) *Input and output currents*

If a load is connected to the secondary winding as shown in Figure 13.2 an alternating current, I_s, will flow in the secondary circuit. The secondary current is balanced by an additional current, I_p, produced in the primary where

$$I_p N_p = I_s N_s$$

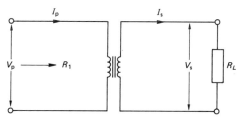

Figure 13.2

and

$$\frac{N_p}{N_s} = \frac{I_s}{I_p}$$

since

$$\frac{V_p}{V_s} = \frac{N_p}{N_s} \text{ from (c)}$$

then

$$\frac{V_p}{V_s} = \frac{N_p}{N_s} = \frac{I_s}{I_p} = n$$

where n = transformer ratio.

(e) Input and output power

The expression in (d) shows that

$$\frac{V_p}{V_s} = \frac{I_s}{I_p}$$

then

$$V_p I_p \text{ VA} = V_s I_s \text{ VA}$$

Since the load is resistive (Figure 13.2), the currents and voltages in both the primary and secondary windings are in phase, hence

$$V_p I_p \text{ W} = V_s I_s \text{ W}$$

i.e. input power = output power.

The efficiency of the transformer is defined as

$$\text{efficiency} = \frac{\text{output power}}{\text{input power}} \times 100\%$$

The efficiency will be 100% for an ideal transformer which has no losses.

When the secondary load is either inductive or capacitive, the primary and secondary currents will be out of phase with the primary and secondary voltages respectively by an angle ϕ (phi). The input and output powers under these conditions will be given by

$$V_p I_p \cos \phi \text{ and } V_s I_s \cos \phi \text{ W}$$

(f) Reflected resistance

If a transformer has a resistive load R_L, as shown in Figure 13.2, then

$$R_L = \frac{V_s}{I_s}$$

and

$$V_p = \frac{V_s N_p}{N_s}$$

and

$$I_p = \frac{I_s N_s}{N_p}$$

Let

$$R_1 = \frac{V_p}{I_p} = \frac{V_s N_p}{N_s} \bigg/ \frac{I_s N_s}{N_p}$$

$$= \frac{V_s N_p}{N_s} \frac{N_p}{I_s N_s}$$

$$= \frac{V_s}{I_s} \frac{N_p^2}{N_s^2}$$

$$= \frac{R_L N_p^2}{N_s^2} = R_L n^2 \text{ where } n \text{ is the turns ratio}$$

R_1 is the value of the effective resistive load presented to the primary source. It is the value of the resistance reflected from the secondary into the primary due to the load R_L. Note that since

$$R_1 = R_L \frac{N_p^2}{N_s^2}$$

$$R_L = R_1 \frac{N_s^2}{N_p^2}$$

13.3 Maximum power transfer

It was shown in Section 13.2(f) that the reflected resistance from the secondary to the primary is

$$\frac{R_L N_p^2}{N_s^2}$$

and if the turns ratio of the transformer is chosen so that this resistance is equal to the internal resistance of the primary source, the secondary load is matched to the primary source. This is the condition for the maximum power transfer to the secondary load.

Example 9 The input voltage to a transformer is 200 V, 50 Hz, and the output voltage is 50 V. The primary has 1200 turns. If a resistive load of 150 Ω is connected to the secondary calculate: (i) the secondary turns, (ii) reflected resistance to the primary, (iii) currents in the primary and secondary.

(i) $N_s = \dfrac{N_p V_s}{V_p}$

$$= \frac{1200 \times 50}{200}$$

$$= 300 \text{ turns}$$

(ii) $R_1 = \frac{R_L N_p^2}{N_s^2}$

$$= \frac{150 \times 1200^2}{300^2}$$

$$= 2400 \ \Omega$$

(iii) $I_p = \frac{V_p}{R_1} = \frac{200}{2400}$

$$= 0.083 \text{ A}$$

$I_s = \frac{V_s}{R_L} = \frac{50}{150}$

$$= 0.333 \text{ A}$$

```
10 PRINT "PROG 201"
20 PRINT "PROG FOR EXAMPLE 9"
30 INPUT "ENTER PRIMARY VOLTAGE" ; VP
40 INPUT "ENTER SECONDARY VOLTAGE" ; VS
50 INPUT "ENTER PRIMARY TURNS" ; NP
60 INPUT "ENTER SECONDARY LOAD" ; RL
70 LET NS=NP*VS/VP
80 PRINT "SECONDARY TURNS = " NS
90 LET R1=RL*(NP^2/NS^2)
100 PRINT "REFLECTED RESISTANCE = " R1 "OHMS"
110 LET IP=VP/R1
120 PRINT "PRIMARY CURRENT = " IP "AMPS"
130 LET IS=VS/RL
140 PRINT "SECONDARY CURRENT = " IS "AMPS"
```

Example 10 A transformer has a load, R_L, of 48 Ω. If the source has a resistance R_1 of 3072 Ω, find the turns ratio of the transformer required to match the load to the source.

$$R_1 = \frac{R_L N_p^2}{N_s^2} = R_L n^2 \text{ where } n = \text{turns ratio}$$

$$n^2 = \frac{R_1}{R_L} = \frac{3072}{48} = 64$$

$$n = \sqrt{64} = 8$$

A turns ratio of 8 means that

1 The input to the primary 'sees' a resistance of

$$R_L(n)^2 = 48 \times 8^2 = 3072 \ \Omega$$

2 Maximum power would be transferred from source to load.

```
10 PRINT "PROG 202"
20 PRINT "LOAD TO SOURCE MATCHING"
30 INPUT "ENTER LOAD RESISTANCE" ; RL
40 INPUT "ENTER SOURCE RESISTANCE" ; R1
50 LET N=SQR(R1/RL)
60 PRINT "TURNS RATIO = " N
```

Example 11 If the number of turns on the secondary in Example 10 equals 65, find the number of turns on the primary.

$$R_1 = \frac{R_L N_p^2}{N_s^2}$$

$$N_p^2 = \frac{R_1 N_s^2}{R_L}$$

$$= \frac{3072 \times 65^2}{48} = 270\,400$$

$$N_p = \sqrt{270\,400} = 520$$

Note

$$65 \times 8 \text{ (turns ratio)} = 520$$

```
10 PRINT "PROG 203"
20 PRINT "PRIMARY TURNS"
30 INPUT "ENTER LOAD RESISTANCE" ; RL
40 INPUT "ENTER SOURCE RESISTANCE" ; R1
50 INPUT "ENTER SECONDARY TURNS" ; NS
60 LET NP=SQR(R1*(NS^2/RL))
70 PRINT "PRIMARY TURNS = " NP
```

Example 12 A transformer has 150 turns on the primary and 750 turns on the secondary. If the source resistance R_1 is 20 Ω, calculate the value of the load resistance R_L for maximum power transfer.

$$R_1 = \frac{R_L N_p^2}{N_s^2}$$

$$R_L = \frac{R_1 N_s^2}{N_p^2} = 20 \times \left(\frac{750^2}{150^2}\right) = 500 \; \Omega$$

```
10 PRINT "PROG 204"
20 PRINT "LOAD RESISTANCE"
30 INPUT "ENTER PRIMARY TURNS" ; NP
40 INPUT "ENTER SOURCE RESISTANCE" ; R1
50 INPUT "ENTER SECONDARY TURNS" ; NS
60 LET RL=R1*(NS^2/NP^2)
70 PRINT "LOAD RESISTANCE = " RL "OHMS"
```

Example 13 A transformer with 1200 primary turns is used to match a 50-Ω load to a voltage source with internal resistance 200 Ω. Calculate the number of secondary turns.

$$R_1 = \frac{R_L N_p^2}{N_s^2}$$

$$N_s^2 = \frac{R_L N_p^2}{R_1}$$

$$N_s = \sqrt{\frac{R_L N_p^2}{R_1}} = \sqrt{\frac{50 \times 1200^2}{200}} = 600 \text{ turns}$$

```
10 PRINT "PROG 205"
20 PRINT "SECONDARY TURNS"
30 INPUT "ENTER PRIMARY TURNS" ; NP
40 INPUT "ENTER SOURCE RESISTANCE" ; R1
50 INPUT "ENTER LOAD RESISTANCE" ; RL
60 LET NS=SQR(RL*(NP^2/R1))
70 PRINT "SECONDARY TURNS = " NS
```

Transformers 237

Example 14 Figure 13.3 shows a 250/1250-V ideal transformer supplied from a 250-V source with a source resistance of 20 Ω. Calculate: (i) the primary current I_p, (ii) the secondary current I_s, (iii) the power dissipated in R and R_L.

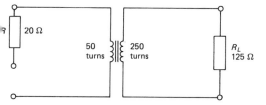

Figure 13.3

(i) Reflected resistance into the primary $R_1 = \dfrac{N_p^2 R_L}{N_s^2}$

$= \dfrac{50^2 \times 125}{250^2} = 5 \text{ Ω}$

Total primary resistance $= R + R_1$

$= 20 + 5 = 25 \text{ Ω}$

$I_p = \dfrac{250}{25} = 10 \text{ A}$

(ii) $I_s = \dfrac{I_p V_p}{V_s} = \dfrac{10 \times 250}{1250} = 2 \text{ A}$

Alternatively note that

Reflected resistance into the secondary $= 20 \times \dfrac{250^2}{50^2}$

$= 500 \text{ Ω}$

Total secondary resistance $= 500 + 125 = 625 \text{ Ω}$

$I_s = \dfrac{1250}{625} = 2 \text{ A}$

(iii) Power dissipated in $R = I_p^2 R$

$= 10^2 \times 20 = 2000 \text{ W}$

Power dissipated in $R_L = I_s^2 R_L$

$= (2)^2 \times 125 = 500 \text{ W}$

Note that since this is an ideal transformer the input power

$V_p I_p = V_s I_s$

i.e. $250 \times 10 = 1250 \times 2 \text{ W}$

The powers dissipated in the two resistances are not the same since the turns ratio does not match the load to the source resistance.

```
10 PRINT "PROG 206"
20 PRINT "SOLUTION TO EXAMPLES 14 AND 15"
30 INPUT "ENTER PRIMARY TURNS" ; NP
40 INPUT "ENTER SOURCE RESISTANCE" ; R
50 INPUT "ENTER LOAD RESISTANCE" ; RL
60 INPUT "ENTER SECONDARY TURNS" ; NS
70 INPUT "ENTER PRIMARY VOLTAGE" ; VP
80 INPUT "ENTER SECONDARY VOLTAGE" ; VS
90 LET R1=RL*(NP^2/NS^2)
100 LET RT=R+R1
110 LET IP=VP/RT
120 PRINT "PRIMARY CURRENT = " IP "AMPS"
130 LET IS=IP*VP/VS
140 PRINT "SECONDARY CURRENT = " IS "AMPS"
150 LET PR=IP^2*R
160 PRINT "POWER DISSIPATED IN R = " PR "WATTS"
170 LET PRL=IS^2*RL
180 PRINT "POWER DISSIPATED IN RL = " PRL "WATTS"
```

Example 15 Repeat Example 14 with a secondary load, R_L, of 500 Ω.

(i) Reflected resistance into the primary $R_1 = \dfrac{N_p^2 R_L}{N_s^2}$

$= \dfrac{50^2 \times 500}{250^2} = 20$ Ω

Total primary resistance $= R + R_1$

$= 20 + 20 = 40$ Ω

$I_p = \dfrac{250}{40} = 6.25$ A

(ii) $I_s = \dfrac{I_p V_p}{V_s} = \dfrac{6.25 \times 250}{1250} = 1.25$ A

Alternatively note that

Reflected resistance into the secondary $= 20 \times \dfrac{250^2}{50^2}$

$= 500$ Ω

Total secondary resistance $= 500 + 500 = 1000$ Ω

$I_s = \dfrac{1250}{1000} = 1.25$ A

(iii) Power dissipated in $R = 6.25^2 \times 20 = 781.25$ W

Power dissipated in $R_L = 1.25^2 \times 500 = 781.25$ W

Note

$V_p I_p = V_s I_s$

i.e. $250 \times 6.25 = 1250 \times 1.25 = 1562.5$ W

The transformer matches the load and source resistances, and hence the power dissipated in both is the same, and the power is the maximum that could be transferred to the load.

(See program 206, above.)

Example 16 A 400-kVa, 8000-V/200-V, 50-Hz transformer has 100 turns on the secondary winding. Assuming an ideal transformer calculate: (i) the maximum permissible primary and secondary currents, (ii) the number of primary turns.

(i) Maximum primary current $= \dfrac{400 \times 10^3 \text{ VA}}{8000 \text{ V}} = 50$ A

Maximum secondary current $= \dfrac{400 \times 10^3 \text{ VA}}{200 \text{ V}}$

$= 2000$ A

(ii) $N_p = \dfrac{N_s V_p}{V_s}$

$= \dfrac{100 \times 8000}{200} = 4000$ turns

400 kVa is the rating of the transformer. This means that the product of the voltage and current must not exceed 400×10^3.

```
10 PRINT "PROG 207"
20 PRINT "SOLUTION TO EXAMPLE 16"
30 INPUT "ENTER TRANSFORMER RATING" ; KVA
40 INPUT "ENTER PRIMARY VOLTAGE" ; VP
50 INPUT "ENTER SECONDARY VOLTAGE" ; VS
60 INPUT "ENTER SECONDARY TURNS" ; NS
70 LET IP=KVA/VP
80 PRINT "PRIMARY CURRENT = " ; IP "AMPS"
90 LET IS=KVA/VS
100 PRINT "SECONDARY CURRENT = " ; IS "AMPS"
110 LET NP=NS*VP/VS
120 PRINT "PRIMARY TURNS = " NP
```

Problems

1. A transformer is connected to a 240-V a.c. 50-Hz supply. The ouput voltage is 120 V. The transformer core has a cross-sectional area of 15 cm^2, and the maximum flux density is 4.5 T. Calculate the number of turns on both windings.
2. The input voltage to the primary of a transformer is 20 V, 100 Hz. The output voltage from the secondary is 200 V. The core has a cross-sectional area of 20 cm^2 and the maximum flux density is 5 T. Calculate the number of turns on both windings.
3. A transformer has 270 V applied to the primary winding. The winding has 90 turns and the maximum flux is 18 mWb. Calculate the frequency of the applied voltage.
4. A transformer has 300 turns on the primary winding. If the applied voltage is 2000 V and this produces a flux of 24 mWb, calculate the frequency of the input voltage.
5. If 220 V, 60 Hz is applied to a transformer having 160 turns on the primary winding, calculate the maximum value of the flux in the core. Express the answer in milliwebers.
6. A transformer has 230 turns on the primary winding. The input voltage is 60 V, 50 Hz. Find the maximum value of the flux in milliwebers.
7. A transformer has 45 primary turns and 180 secondary turns. If the input voltage is 240 V, calculate the output voltage.
8. A 'step down' transformer has 180 turns on the primary winding and 18 turns on the secondary. If 240 V is applied to the primary, calculate the secondary voltage.
9. The output voltage from a transformer is 60 mV. If the primary

winding has 85 turns and the secondary winding has 255 turns, calculate the input voltage.

10 A transformer has 900 turns on the primary and 30 turns on the secondary. If the output voltage is 90 mV, find the value of the input voltage.

11 If the primary and secondary voltages of a transformer are 440 and 110 V, and the secondary winding has 65 turns, calculate the number of primary turns.

12 A voltage of 40 V is applied to a transformer. The output is found to be 240 V. If there are 120 turns on the secondary, calculate the number of primary turns.

13 A transformer driven from a 240-V mains supply has 24 V on the secondary. If the primary has 150 turns, calculate the number of secondary turns.

14 A 'step up' transformer has 210 V on the primary and 400 V on the secondary. If there are 100 turns on the primary, how many turns are there on the secondary?

15 Find the turns ratio of a transformer that 'steps down' 360 V to 12 V.

16 The mains voltage of 240 V is applied to a transformer, and the output voltage is 120 V. If the primary has 1000 turns and the secondary load is 100 Ω, calculate: (a) the number of secondary turns, (b) the reflected resistance to the primary, (c) the currents in the primary and secondary windings.

17 Repeat Problem 16 with primary and secondary voltages of 50 and 200 V respectively, 500 primary turns, and a secondary load of 180 Ω.

18 A transformer is loaded with a resistance of 52 Ω. If it is connected to a source resistance of 4212 Ω, calculate the required turns ratio to match load and source.

19 It is a requirement to match a 10 Ω source to a 250 Ω load. Calculate the required turns ratio.

20 A transformer is used to match a 5 Ω load to an 80 Ω source. If the secondary has 100 turns, calculate the number of primary turns required.

21 Given a load resistance of 5 Ω and a source resistance of 7.2 Ω, and a transformer with 100 turns on the secondary, calculate the number of primary turns for correct matching.

22 A transformer has 45 turns on the primary and 450 turns on the secondary. If the source resistance is 1.5 Ω calculate the value of the load resistance required for maximum power transfer.

23 A transformer with 30 primary turns is used to match a 50-Ω load to a voltage source of 2 Ω. Calculate the number of secondary turns.

24 A transformer has 150 turns on the primary winding. It is used to match a 50-Ω source to a resistive load of 2 Ω. Calculate the number of secondary turns.

25 The circuit shown in Figure 13.3 has the following parameters: primary turns 1500; secondary turns 500; source resistance 5 Ω; load resistance 30 Ω. The input voltage is 240 V. Calculate: (a) the primary current, (b) the secondary current, (c) the power dissipated in the source and in the load resistances.

26 The circuit shown in Figure 13.3 has the following parameters: both source and load resistances equal 1000 Ω; primary turns 60; secondary turns 240. If the secondary voltage is 48, calculate: (a) the primary current, (b) the secondary current, (c) the power dissipated in the source and load resistances.

27 A 200-kVa, 5000/200-V, 50-Hz transformer has 1000 turns on the secondary winding. Assuming no losses calculate: (a) the

maximum permissible primary and secondary currents, (b) the number of primary turns.

28 Repeat Problem 27 for a 7.5-kVA, 150/250-V, 50-Hz transformer with 120 turns on the secondary.

14 D.c. supplies

14.1 Introduction

Chapter 1 referred to the sources of d.c. supplies.

The primary battery, of which the Leclanché cell is an example, is used extensively in portable electronic equipment. It is often called a 'dry battery', and is discarded when exhausted.

The secondary battery, of which the nickel–cadmium cell is an example, can be recharged when exhausted, and these cells can last up to 15 years if handled with care.

Another source of supply is the solar cell which converts light directly into electrical energy. The selenium cell is the best known example.

Batteries can be connected in series and in parallel. Batteries have internal resistances, and the terminal voltage of a battery depends on the internal resistance and the current drawn.

Power supply units convert a.c. voltages into d.c. voltages by means of the rectification process. Stabilizing circuits are added to the output of such devices, to maintain constant outputs, when load currents and/or input a.c. voltages are varying.

Example 1 Figure 14.1 shows a cell with an e.m.f. of 6 V. If the internal resistance r is 2 Ω, and the load resistance R is 22 Ω, find the terminal voltage V between A and B.

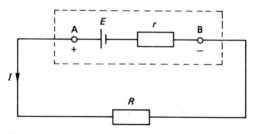

Figure 14.1

$$V = E - Ir$$
$$I = \frac{E}{r + R} = \frac{6}{24} = 0.25 \text{ A}$$
$$Ir = 0.25 \times 2 = 0.5$$
$$V = 6 - 0.5 = 5.5 \text{ V}$$

Note that

$$IR = 0.25 \times 22 = 5.5 \text{ V}$$

which is the expected answer, since the terminal voltage appears across R.

```
10 PRINT "PROG 208"
20 PRINT "TERMINAL VOLTAGE"
30 INPUT "ENTER E.M.F" ; E
40 INPUT "ENTER INTERNAL RESISTANCE" ; R
50 INPUT "ENTER LOAD RESISTANCE" ; RL
60 LET I=E/(R+RL)
70 LET V=E-(I*R)
80 PRINT "TERMINAL VOLTAGE = " V "VOLTS"
```

Example 2 Find the internal resistance r of a battery if the e.m.f. falls from 12 V to 11.5 V when delivering a current of 10 A.

$$V = E - Ir$$
$$Ir = E - V$$
$$r = \frac{E - V}{I}$$
$$= \frac{12 - 11.5}{10} = 0.05 \, \Omega$$

```
10 PRINT "PROG 209"
20 PRINT "INTERNAL RESISTANCE"
30 INPUT "ENTER E.M.F" ; E
40 INPUT "ENTER TERMINAL VOLTAGE" ; V
50 INPUT "ENTER CURRENT IN AMPS" ; I
60 LET R=(E-V)/I
70 PRINT "INTERNAL RESISTANCE = " R "OHMS"
```

Example 3 A battery has a terminal voltage of 23 V when delivering a current of 5 A. If the internal resistance is $0.2 \, \Omega$, what is the e.m.f.?

$$V = E - Ir$$
$$E = V + Ir$$
$$= 23 + (5 \times 0.2) = 24 \text{ V}$$

```
10 PRINT "PROG 210"
20 PRINT "BATTERY E.M.F"
30 INPUT "ENTER INTERNAL RESISTANCE" ; R
40 INPUT "ENTER TERMINAL VOLTAGE" ; V
50 INPUT "ENTER CURRENT IN AMPS" ; I
60 LET E=V+(I*R)
70 PRINT "BATTERY E.M.F = " E "VOLTS"
```

Example 4 Measurements of the terminal voltages and the currents drawn during an experiment on a car battery gave the following data:

V	I (A)
12	10
10	20

Find the internal resistance and the e.m.f.

$$12 = E - 10r$$
$$10 = E - 20r$$

Subtract $\quad\underline{2 = \quad\quad 10r}$

$$r = \frac{2}{10} = 0.2 \ \Omega$$

$$E = V + Ir$$
$$= 12 + (10 \times 0.2) = 14 \ V$$

```
10 PRINT "PROG 211"
20 PRINT "INTERNAL RESISTANCE AND E.M.F"
30 INPUT "ENTER VOLTAGE READING 1" ; V1
40 INPUT "ENTER CURRENT READING 1" ; I1
50 INPUT "ENTER VOLTAGE READING 2" ; V2
60 INPUT "ENTER CURRENT READING 2" ; I2
70 LET R=(V1-V2)/(I2-I1)
80 PRINT "INTERNAL RESISTANCE = " R "OHMS"
90 LET E=V1+(I1*R)
100 PRINT "BATTERY E.M.F = " E "VOLTS"
```

14.2 Battery connections

(a) Series

This is the arrangement encountered most frequently. Figure 14.2 shows three cells, each with an internal resistance r connected in series. The battery voltage is the sum of the individual cell voltages. Since the cell internal resistances are in series, the battery resistance is the sum of the individual resistances. For the battery in Figure 14.2

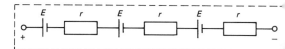

Figure 14.2

Battery voltage $= 3E$

Battery resistance $= 3r$

The battery voltage is the open circuit voltage, and the terminal voltage will depend on the current supplied to the load.

Example 5 If the e.m.f. of each cell is 1.5 V in Figure 14.2, and the internal resistance of each cell is 0.1 Ω, calculate the battery terminal voltage when connected to a load of 4.7 Ω.

Terminal voltage $V = E - Ir$

where $E = 4.5$ and $r = 0.3 \ \Omega$

$$I = \frac{4.5}{4.7 + 0.3} = 0.9 \ A$$

$$V = 4.5 - (0.9 \times 0.3)$$
$$= 4.23 \ V$$

d.c. supplies 245

```
10 PRINT "PROG 212"
20 PRINT "SERIES CONNECTIONS"
30 INPUT "ENTER CELL VOLTAGE" ; E
40 INPUT "ENTER CELL RESISTANCE" ; R
50 INPUT "ENTER LOAD RESISTANCE" ; RL
60 INPUT "ENTER NUMBER OF CELLS" ; N
70 LET E=N*E
80 LET R=N*R
90 LET I=E/(R+RL)
100 LET V=E-(I*R)
110 PRINT "TERMINAL VOLTAGE = " V "VOLTS"
```

(b) Parallel

The arrangement in Figure 14.3 shows three cells connected in parallel. Here the battery and cell voltages are the same, but the capacity of the battery is three times the capacity of a single cell. Since the three cells are in parallel the total internal resistance of the battery is reduced to $r/3$.

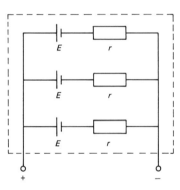

Figure 14.3

Example 6 If each cell in Figure 14.3 has an e.m.f. of 12 V and an internal resistance of 0.6 Ω, calculate the battery terminal voltage when the load resistance is 3 Ω.

Terminal voltage $V = E - Ir$

where $E = 12$ and $r = \dfrac{0.6}{3} = 0.2 \ \Omega$

$$I = \frac{12}{3 + 0.2} = 3.75 \text{ A}$$

$$V = 12 - (3.75 \times 0.2)$$

$$= 11.25 \text{ V}$$

```
10 PRINT "PROG 213"
20 PRINT "PARALLEL CONNECTIONS"
30 INPUT "ENTER CELL VOLTAGE" ; E
40 INPUT "ENTER CELL RESISTANCE" ; R
50 INPUT "ENTER LOAD RESISTANCE" ; RL
60 INPUT "ENTER NUMBER OF CELLS" : N
70 LET R=R/N
80 LET I=E/(R+RL)
90 LET V=E-(I*R)
100 PRINT "TERMINAL VOLTAGE = " V "VOLTS"
```

(c) Series–parallel

Example 7 Figure 14.4 shows a series–parallel arrangement, and Figure 14.5 shows the equivalent circuit. Calculate the terminal voltage of each battery.

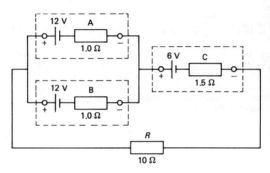

Figure 14.4

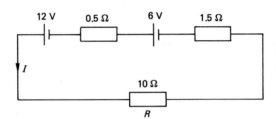

Figure 14.5

From Figure 14.5

$$I = \frac{18}{10 + 2} = 1.5 \text{ A}$$

This means that the distribution of current is

 0.75 A through battery A
 0.75 A through battery B
 1.5 A through battery C

The terminal voltage of both batteries A and B is

$$V = 12 - (0.75 \times 1)$$
$$= 11.25 \text{ V}$$

The terminal voltage of battery C is

$$V = 6 - (1.5 \times 1.5)$$
$$= 3.75 \text{ V}$$

Note that

Total voltage across $R = 11.25 + 3.75 = 15$ V

Current through $R = \dfrac{15}{10} = 1.5$ A

This confirms the earlier current calculation.

```
10 PRINT "PROG 214"
20 PRINT "SERIES PARALLEL"
30 INPUT "ENTER RESISTANCE OF BATTERY A" ; RA
40 INPUT "ENTER RESISTANCE OF BATTERY B" ; RB
50 INPUT "ENTER RESISTANCE OF BATTERY C" ; RC
60 INPUT "ENTER EMF OF BATTERIES A AND B" ; EAB
70 INPUT "ENTER EMF OF BATTERY C" ; EC
80 INPUT "ENTER LOAD RESISTANCE" ; RL
90 LET R1=RA*RB/(RA+RB)
100 LET RT=R1+RC+RL
110 LET I=(EAB+EC)/RT
120 PRINT "TOTAL CURRENT = " I "AMPS"
130 LET VAB=EAB-(I/2*RA)
140 PRINT "VOLTAGES OF BATTS A AND B = " VAB "VOLTS"
150 LET VC=EC-(I*RC)
160 PRINT "VOLTAGE OF BATT C = " VC "VOLTS"
```

Example 8 Each battery in Figure 14.6 has a different internal resistance. Calculate the terminal voltage of each one.

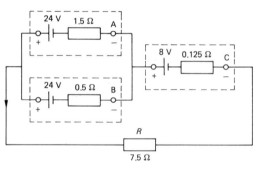

Figure 14.6

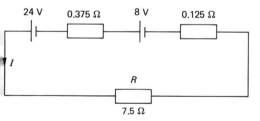

Figure 14.7

Figure 14.7 shows the equivalent circuit. The combined resistance of batteries A and B is given by

$$\frac{1.5 \times 0.5}{1.5 + 0.5} = 0.375 \ \Omega$$

From Figure 14.7

$$I = \frac{32}{7.5 + 0.375 + 0.125} = 4 \text{ A}$$

The terminal voltage of both batteries A and B is

$$V = 24 - (4 \times 0.375)$$
$$= 22.5 \text{ V}$$

The terminal voltage of battery C is

$$V = 8 - (4 \times 0.125)$$
$$= 7.5 \text{ V}$$

Note that

Total voltage across $R = 22.5 + 7.5 = 30$

Current through $R = \dfrac{30}{7.5} = 4$ A

which confirms the earlier current calculation.

The voltage drop in batteries A and B is

$$4 \times 0.375 = 1.5 \text{ V}$$

The current through battery A is

$$\frac{V}{r} = \frac{1.5}{1.5} = 1 \text{ A}$$

and through battery B is

$$\frac{V}{r} = \frac{1.5}{0.5} = 3 \text{ A}$$

```
10  PRINT "PROG 215"
20  PRINT "SERIES PARALLEL"
30  INPUT "ENTER RESISTANCE OF BATTERY A" ; RA
40  INPUT "ENTER RESISTANCE OF BATTERY B" ; RB
50  INPUT "ENTER RESISTANCE OF BATTERY C" ; RC
60  INPUT "ENTER EMF OF BATTERIES A AND B" ; EAB
70  INPUT "ENTER EMF OF BATTERY C" ; EC
80  INPUT "ENTER LOAD RESISTANCE" ; RL
90  LET R1=RA*RB/(RA+RB)
100 LET RT=R1+RC+RL
110 LET I=(EAB+EC)/RT
120 PRINT "TOTAL CURRENT = " I "AMPS"
130 LET VAB=(I*RL)- ((EC-(I*RC)))
140 PRINT "VOLTAGES OF BATTS A AND B = " VAB "VOLTS"
150 LET VC=EC-(I*RC)
160 PRINT "VOLTAGE OF BATT C = " VC "VOLTS"
```

Example 9 What is the total power consumed in the circuit of Figure 14.6?

Power consumed in battery A $= I_1^2 r = (1)^2 \times 1.5 = 1.5$ W

Power consumed in battery B $= I_2^2 r = (3)^2 \times 0.5 = 4.5$ W

Power consumed in battery C $= I^2 r = (4)^2 \times 0.125 = 2$ W

Power consumed in the load $R = I^2 R = (4)^2 \times 7.5 = 120$ W

d.c. supplies 249

Total power $= 1.5 + 4.5 + 2 \times 120$
                $= 128$ W

Note

Total e.m.f. $(24 + 8$ V$) \times$ current $(4$ A$) = 128$ W

```
10 PRINT "PROG 216"
20 PRINT "POWER CONSUMED"
30 INPUT "ENTER RESISTANCE OF BATTERY A" ; RA
40 INPUT "ENTER RESISTANCE OF BATTERY B" ; RB
50 INPUT "ENTER RESISTANCE OF BATTERY C" ; RC
60 INPUT "ENTER CURRENT IN BATTERY A" ; I1
70 INPUT "ENTER CURRENT IN BATTERY B" ; I2
80 INPUT "ENTER CURRENT IN BATTERY C" ; I3
90 LET P1=I1^2*RA
100 LET P2=I2^2*RB
110 LET P3=I3^2*RC
120 INPUT "ENTER LOAD RESISTANCE" ; RL
130 INPUT "ENTER LOAD CURRENT" ; I
140 LET PRL=I^2*RL
150 LET P=P1+P2+P3+PRL
160 PRINT "TOTAL POWER CONSUMED = " P "WATTS"
```

14.3 Battery charger

(a) Constant voltage charging

A secondary cell can be recharged by passing a current through it in the opposite direction to that of the discharge current. Constant voltage charging means that the output voltage of the charger is held constant throughout the charging cycle. The charging current therefore will gradually decrease as the battery voltage rises towards the charger voltage.

Example 10 Figure 14.8 shows a battery charger with an internal resistance of 0.2 Ω charging a car battery. Calculate: (i) the initial charging current, (b) the charging current when the battery e.m.f. has risen to 13.5 V.

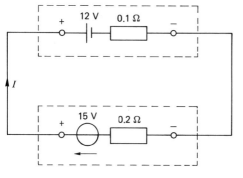

Figure 14.8

(i) Since the charger voltage and the battery voltage are in opposition, initially the voltage producing the current is

$$15 - 12 = 3 \text{ V}$$

Charging current $I = \dfrac{3}{0.2 + 0.1}$

$= 10 \text{ A}$

(ii) Voltage producing the current $= 15 - 13.5 = 1.5$ V

$$I = \dfrac{1.5}{0.2 + 0.1}$$

$= 5 \text{ A}$

```
10 PRINT "PROG 217"
20 PRINT "BATTERY CHARGER"
30 INPUT "ENTER CHARGER VOLTAGE" ; V
40 INPUT "ENTER CHARGER RESISTANCE" ; RC
50 INPUT "ENTER BATTERY VOLTAGE" ; E
60 INPUT "ENTER BATTERY RESISTANCE" ; RB
70 LET I1=(V-E)/(RC+RB)
80 PRINT "INITIAL CURRENT = " ; I1 "AMPS"
90 INPUT "ENTER BATT VOLTAGE AFTER CHARGE" ; EC
100 LET I2=(V-EC)/(RC+RB)
110 PRINT "CURRENT = " I2 "AMPS"
```

Example 11 It is a requirement to limit the initial charging current in Example 10 to 4 A by including a series resistance R in the battery charger.

(i) Find the value of this resistance
(ii) Using this series resistance find the value of the charging current when the e.m.f. has risen to 14.5 V.

(i) $V = 15 - 12 = 3$ V and $I = 4$ A

$$4 = \dfrac{3}{(0.2 + 0.1 + R)} = \dfrac{3}{0.3 + R}$$

$$4(0.3 + R) = 3$$

$$1.2 + 4R = 3$$

$$4R = 3 - 1.2 = 1.8$$

$$R = \dfrac{1.8}{4} = 0.45 \text{ }\Omega$$

(ii) $V = 15 - 14.5 = 0.5$ V

$$I = \dfrac{0.5}{0.2 + 0.1 + 0.45} = 0.667 \text{ A}$$

```
10 PRINT "PROG 218"
20 PRINT "BATTERY CHARGER"
30 INPUT "ENTER CHARGER VOLTAGE" ; V
40 INPUT "ENTER CHARGER RESISTANCE" ; RC
50 INPUT "ENTER BATTERY VOLTAGE" ; E
60 INPUT "ENTER BATTERY RESISTANCE" ; RB
70 INPUT "ENTER INITIAL CHARGING CURRENT" ; I1
80 LET R=(V-E)/I1-RC-RB
90 PRINT "SERIES RESISTANCE = " R "OHMS"
100 INPUT "ENTER INCREASED BATT E.M.F" ; EI
110 LET I2=(V-EI)/(RC+RB+R)
120 PRINT "CURRENT = " I2 "AMPS"
```

Example 12 Find the power wasted as heat in both the battery and the charger in Example 11 when the charging current is 4 A.

$$\text{Power wasted in the battery} = I^2 r = (4)^2 \times 0.1$$
$$= 1.6 \text{ W}$$
$$\text{Power wasted in the charger} = (4)^2 \times (0.2 + 0.45)$$
$$= 10.4 \text{ W}$$

```
10 PRINT "PROG 219"
20 PRINT "POWER CONSUMPTION"
30 INPUT "ENTER CHARGING CURRENT" ; I
40 INPUT "ENTER BATTERY RESISTANCE" ; RB
50 INPUT "ENTER CHARGER RESISTANCE" ; RC
60 INPUT "ENTER SERIES RESISTANCE" ; R
70 LET P1=I^2*RB
80 PRINT "POWER CONSUMED IN BATTERY = " P1
90 LET P2=I^2*(RC+R)
100 PRINT "POWER CONSUMED IN CHARGER" ; P2
```

Example 13 A battery consists of 40 cells in series each with an e.m.f. of 2 V when fully charged. The internal resistance of each cell is 0.02 Ω. If the e.m.f. of each cell has fallen to 1.8 V calculate: (i) the charger voltage needed to fully charge the battery; (ii) the initial charging current given that the internal resistance of the charger is 0.8 Ω; (iii) the charging current when the e.m.f. of each cell is 1.95 V.

(i) Charger voltage = 40 × 2 = 80 V

= battery voltage fully charged

(ii) Discharged voltage of the battery

= 40 × 1.8 = 72 V

Total circuit resistance = battery resistance + charger resistance = (40 × 0.02) + 0.8 = 1.6 Ω

$$\text{Initial current } I = \frac{80 - 72}{1.6} = 5 \text{ A}$$

(iii) Battery voltage = 40 × 1.95 = 78 V

$$I = \frac{80 - 78}{1.6} = 1.25 \text{ A}$$

```
10 PRINT "PROG 220"
20 PRINT "BATTERY CHARGER"
30 INPUT "ENTER NUMBER OF CELLS" ; N
40 INPUT "ENTER BATTERY RESISTANCE" ; RB
50 INPUT "ENTER MAXIMUM BATTERY VOLTAGE" ; E1
60 INPUT "ENTER MINIMUM BATTERY VOLTAGE" ; E2
70 INPUT "ENTER CHARGER RESISTANCE" ; RC
80 LET V1=N*E1
90 PRINT "CHARGER VOLTAGE = " V1 "VOLTS"
100 LET I=(V1-(N*E2))/(N*RB+RC)
110 PRINT "INITIAL CHARGING CURRENT = " I "AMPS"
120 INPUT "ENTER INCREASED BATTERY VOLTAGE" ; E3
130 LET I=(V1-(N*E3))/(N*RB+RC)
140 PRINT "CURRENT = " I "AMPS"
```

b) Constant current charging

Some cells, notably nickel–cadmium, much used in portable electronic equipment, need to be charged at a constant rate of current flow. This current is specified by the manufacturer. The charger voltage must be increased, therefore, in order to maintain constant charging current as the opposing battery e.m.f. increases i.e. the applied voltage must be increased by an amount equal to the increase in the battery e.m.f. throughout the cycle.

Example 14 A battery consists of 50 1.5-V cells each with resistance of 0.01 Ω. The voltage of each cell has fallen to 1.25 V. It is a requirement to charge the battery at a constant current of 2 A from a charger with an internal resistance of 1.5 Ω. Calculate: (i) the applied voltage at the commencement of charge; (ii) the applied voltage at the end of charge.

(i) Initial battery voltage $= 50 \times 1.25 = 62.5$ V

Total resistance R $= (50 \times 0.01) + 1.5 = 2$ Ω

Charging voltage required $= IR = 2 \times 2 = 4$ V

Charger voltage $= 62.5 + 4 = 66.5$ V

(ii) Final battery voltage $= 50 \times 1.5 = 75$ V

Charging voltage $= 4$ V as in (i)

Charger voltage $= 75 + 4 = 79$ V

This means that the change in the supply voltage is 66.5–79. The supply voltage must be greater than the battery by 4 V during the complete charging cycle.

```
10 PRINT "PROG 221"
20 PRINT "BATTERY CHARGER"
30 INPUT "ENTER NUMBER OF CELLS" ; N
40 INPUT "ENTER BATTERY RESISTANCE" ; RB
50 INPUT "ENTER MAXIMUM BATTERY VOLTAGE" ; E1
60 INPUT "ENTER MINIMUM BATTERY VOLTAGE" ; E2
70 INPUT "ENTER CHARGER RESISTANCE" ; RC
80 INPUT "ENTER CHARGING CURRENT" ; I
90 LET VC1=(N*E2)+(I*(N*RB+RC))
100 PRINT "INITIAL CHARGER VOLTAGE = " VC1 "VOLTS"
110 LET VC2=(N*E1)+(I*(N*RB+RC))
120 PRINT "FINAL CHARGER VOLTAGE = " VC2 "VOLTS"
```

(c) Ampere-hour efficiency

If the capacity of a battery is described as 100 ampere-hours at the 10-hour rate, then this means that the battery will, in theory, deliver a current of 10 amperes for 10 hours before it needs recharging. The efficiency of a battery can be expressed in terms of the ampere-hour input and output where

$$\text{Efficiency} = \frac{\text{ampere-hours output}}{\text{ampere-hours input}}$$

Example 15 A battery is charged with a constant current of 24 A for 8 hours. It is discharged at the rate of 16 A for 10 hours.

$$\text{Efficiency} = \frac{16 \times 10}{24 \times 8} = 0.833 \text{ or } 83.3\%$$

```
10 PRINT "PROG 222"
20 PRINT "AMPERE HOUR EFFICIENCY"
30 INPUT "ENTER CHARGING CURRENT" ; I1
40 INPUT "ENTER CHARGING TIME" ; T1
50 INPUT "ENTER DISCHARGE CURRENT" ; I2
60 INPUT "ENTER DISCHARGE TIME" ;T2
70 LET E=I2*T2/(I1*T1)
80 PRINT "EFFICIENCY = " E*100 "%"
```

d.c. supplies 253

Example 16 What would the efficiency of the battery in Example 15 be if the discharge rate had been 30 A for 5 hours?

$$\text{Efficiency} = \frac{30 \times 5}{24 \times 8} = 0.78 \text{ or } 78\%$$

(See program 222, page 252.)

(d) Watt-hour efficiency

$$\text{Efficiency} = \frac{\text{Average output power} \times \text{time}}{\text{Average input power} \times \text{time}}$$

For the case of constant current charging

Watt-hours = current in amperes × average voltage × time in hours

Example 17 A battery is fully charged by a constant current of 15 A flowing for 8 hours. The average charging voltage is 1.8 V. It is discharged at a constant current of 10 A flowing for 9 hours. If the average terminal voltage is 1.3 V, calculate the watt-hour efficiency.

$$\text{Efficiency} = \frac{\text{Energy output}}{\text{Energy input}}$$

$$= \frac{10 \times 1.3 \times 9}{15 \times 1.8 \times 8} = 0.542 \text{ or } 54.2\%$$

```
10 PRINT "PROG 223"
20 PRINT "WATT-HOUR EFFICIENCY"
30 INPUT "ENTER CHARGING CURRENT" ; I1
40 INPUT "ENTER CHARGING TIME" ; T1
50 INPUT "ENTER DISCHARGE CURRENT" ; I2
60 INPUT "ENTER DISCHARGE TIME" ;T2
70 INPUT "ENTER AVERAGE CHARGE VOLTAGE" ; V1
80 INPUT "ENTER AVERAGE DISCHARGE VOLTAGE" ; V2
90 LET E=I2*T2*V2/(I1*T1*V1)
100 PRINT "EFFICIENCY = " E*100 "%"
```

14.4 Rectifier circuits

(a) Half-wave

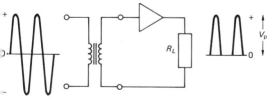

Figure 14.9

Figure 14.9 shows a half-wave circuit together with the input and output waveforms, connected to a resistive load. The output, although still varying, is now unidirectional, and the following parameters can be measured:

(i) The peak voltage V_p

(ii) The mean d.c. voltage where $V_{d.c.} = \dfrac{V_p}{\pi}$

(iii) The r.m.s. voltage where $V_{r.m.s.} = \dfrac{V_p}{2}$

(iv) The a.c. current where $I_{r.m.s.} = \dfrac{i_p}{2}$

(v) The d.c. current where $I_{d.c.} = \dfrac{i_p}{\pi}$

i_p is the peak current and $i_p = V_p/R_L$.

Example 18 Given that the peak output voltage of the circuit in Figure 14.9 is 50 V and that the load resistance is 100 Ω, calculate the values of the voltages across, and the currents flowing through the load resistance.

(i) $i_p = \dfrac{50}{100} = 0.5$ A

(ii) $V_{d.c.} = \dfrac{50}{\pi} = 15.92$ V

(iii) $V_{r.m.s.} = \dfrac{50}{2} = 25$ V

(iv) $I_{r.m.s.} = \dfrac{0.5}{2} = 0.25$ A

(v) $I_{d.c.} = \dfrac{0.5}{\pi} = 0.159$ A

```
10 PRINT "PROG 224"
20 PRINT "HALF WAVE RECTIFIER WITH RESISTIVE LOAD"
30 INPUT "ENTER PEAK VOLTAGE" ; VP
40 INPUT "ENTER LOAD RESISTANCE" ; RL
50 LET IP=VP/RL
60 PRINT "IP = " IP "AMPS"
70 LET IDC=IP/3.141593
80 PRINT "IDC = " IDC "AMPS"
90 LET IRMS=IP/2
100 PRINT "IRMS = " IRMS "AMPS"
110 LET VLP=IP*RL
120 PRINT "VLP = " VLP "VOLTS"
130 LET VLDC=VP/3.141593
140 PRINT "VLDC = " VLDC "VOLTS"
150 LET VLRMS=VP/2
160 PRINT "VLRMS = " VLRMS "VOLTS"
```

(b) Full-wave

Figure 14.10 shows a full-wave circuit together with the input and output waveforms connected to a resistive load. For this circuit

(i) $i_p = \dfrac{V_p}{R_L}$

(ii) $V_{d.c.} = \dfrac{2V_p}{\pi}$

(iii) $V_{r.m.s.} = \dfrac{V_p}{\sqrt{2}}$

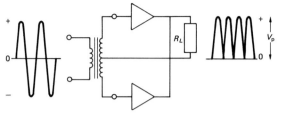

Figure 14.10

(iv) $I_{r.m.s.} = \dfrac{i_p}{\sqrt{2}}$

(v) $I_{d.c.} = \dfrac{2i_p}{\pi}$

Example 19 Given the same values for the peak voltage and the resistive load as those in Example 18, calculate the circuit parameters:

(i) $i_p = \dfrac{50}{100} = 0.5$ A

(ii) $V_{d.c.} = \dfrac{2 \times 50}{\pi} = 31.84$ V

(iii) $V_{r.m.s.} = \dfrac{50}{\sqrt{2}} = 35.36$ V

(iv) $I_{r.m.s.} = \dfrac{0.5}{\sqrt{2}} = 0.354$ A

(v) $I_{d.c.} = \dfrac{2 \times 0.5}{\pi} = 0.318$ A

```
10 PRINT "PROG 225"
20 PRINT "FULL WAVE RECTIFIER WITH RESISTIVE LOAD"
30 INPUT "ENTER PEAK VOLTAGE" ; VP
40 INPUT "ENTER LOAD RESISTANCE" ; RL
50 LET IP=VP/RL
60 PRINT "IP = " IP "AMPS"
70 LET IDC=2*IP/3.141593
80 PRINT "IDC = " IDC "AMPS"
90 LET IRMS=IP/SQR(2)
100 PRINT "IRMS = " IRMS "AMPS"
110 LET VLP=IP*RL
120 PRINT "VLP = " VLP "VOLTS"
130 LET VLDC=2*VP/3.141593
140 PRINT "VLDC = " VLDC "VOLTS"
150 LET VLRMS=VP/SQR(2)
160 PRINT "VLRMS = " VLRMS "VOLTS"
```

14.5 Power supply unit

Figure 14.11 shows a typical full-wave power supply circuit using a diode bridge rectifying arrangement. The rectifier output is fed into a filter unit, consisting of C_1, R_1 and C_2, which removes the alternating component and produces a d.c. voltage across the load R_L.

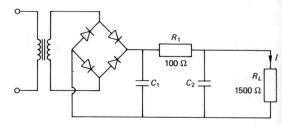

Figure 14.11

Example 20 If the circuit in Figure 14.11 is driven from the mains supply, 240 V r.m.s., and the transformer has a 'step down' ratio of 12 : 1, calculate: (i) the peak voltage output from the bridge rectifier, V_p, (ii) the d.c. load current, I, (iii) the output voltage across the load, V_L. Assume that the capacitor C_1 charges to the peak voltage and that there are no other losses.

(i) $V_p = \dfrac{240}{12} \times \sqrt{2}$ since $V_p = V_{r.m.s.} \times \sqrt{2}$

$\quad\quad = 28.28$ V

(ii) $I = \dfrac{V_p}{R_1 + R_L}$

$\quad\quad = \dfrac{28.28}{100 + 1500} = 0.01767$ A

(iii) $V_L = IR_L$

$\quad\quad = 0.01767 \times 1500$

$\quad\quad = 26.51$ V

Note that

Voltage drop across $R_1 = 0.01767 \times 100 = 1.77$ V

and that

$28.28 - 1.77 = 26.51$ V

```
10 PRINT "PROG 226"
20 PRINT "POWER SUPPLY UNIT"
30 INPUT "ENTER INPUT VOLTAGE" ; VRMS
40 INPUT "ENTER TRANSFORMER RATIO" ; N
50 INPUT "ENTER SERIES RESISTOR" ; R1
60 INPUT "ENTER LOAD RESISTOR" ; RL
70 LET VP=VRMS*SQR(2)/N
80 PRINT "VP = " VP "VOLTS"
90 LET I=VP/(R1+RL)
100 PRINT "I = " I "AMPS"
110 LET VR1=I*R1
120 PRINT "VR1 = " VR1 "VOLTS"
130 LET VRL=I*RL
140 PRINT "OUTPUT VOLTAGE = " VRL "VOLTS"
```

Example 21 If the supply voltage in Example 20 is increased to 960 V, calculate V_p, I and V_L.

(i) $V_p = \dfrac{960 \times \sqrt{2}}{12} = 113.12$ V

(ii) $I = \dfrac{113.12}{100 + 1500} = 0.0707$ A

(iii) $V_L = 0.0707 \times 1500 = 106.05$ V

(See program 226, page 256.)

14.6 Voltage regulators

Voltage regulator circuits are connected in power supplies between the filter output and the load resistance. Figure 14.12 shows one of the most common regulator circuits consisting of a series resistor R_S and a zener diode. The circuit works on the principle that the current flowing through the diode, I_D, can vary considerably with very little change in the voltage across the diode, and hence across the load R_L. If the load current increases or decreases, the current through the diode will decrease or increase by the same amount, thus maintaining a constant voltage drop across R_S and hence a constant voltage drop across R_L. If the input voltage increases or decreases, the diode current will increase or decrease, thus maintaining a fairly constant voltage across the load R_L.

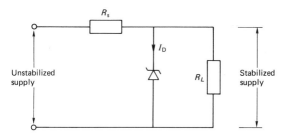

Figure 14.12

Example 22 Figure 14.13 shows a voltage regulator circuit. Calculate the value of the series resistance R_S and the power dissipated in it.

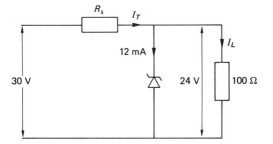

Figure 14.13

$$I_L = \frac{24}{100} = 0.24 \text{ A} = 240 \text{ mA}$$

Total current $I_T = 12 + 240 = 252$ mA $= 0.252$ A

Voltage drop across $R_S = 30 - 24 = 6$ V

$$R_S = \frac{6}{0.252} = 23.8 \text{ }\Omega$$

Power dissipated $= (0.252)^2 \times 23.8 = 1.51$ W

```
10 PRINT "PROG 227"
20 PRINT "VOLTAGE REGULATOR"
30 INPUT "ENTER INPUT VOLTAGE" ; VIN
40 INPUT "ENTER OUTPUT VOLTAGE" ; VOUT
50 INPUT "ENTER LOAD RESISTANCE" ; RL
60 INPUT "ENTER ZENER DIODE CURRENT IN MA" ; IZ
70 LET IRL=VOUT*10^3/RL
80 LET IT=IRL+IZ
90 LET VR1=VIN-VOUT
100 LET R1=VR1*10^3/IT
110 PRINT "R1 = " R1 "OHMS"
120 LET PR1=IT^2*R1/10^6
130 PRINT "POWER DISSIPATED IN R1 = " PR1 "WATTS"
```

Example 23 If the load resistance in Figure 14.13 is increased to 200 Ω, calculate the new current through the diode.

$$I_L = \frac{24}{200} = 0.12 \text{ A}$$

$I_T = 0.252$ A as before

$I_D = 0.252 - 0.12 = 0.132$ A $= 132$ mA

```
10 PRINT "PROG 228"
20 PRINT "VOLTAGE REGULATOR"
30 INPUT "ENTER INPUT VOLTAGE" ; VIN
40 INPUT "ENTER OUTPUT VOLTAGE" ; VOUT
50 INPUT "ENTER LOAD RESISTANCE" ; RL
60 INPUT "ENTER R1" ; R1
70 LET IRL=VOUT*10^3/RL
80 LET IT=(VIN-VOUT)*10^3/R1
90 LET IZ=IT-IRL
100 PRINT "ZENER DIODE CURRENT = " IZ "MILLIAMPS"
```

Example 24 Find the diode current in the circuit shown in Figure 14.14.

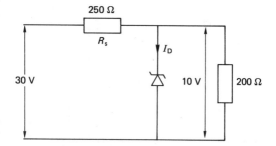

Figure 14.14

d.c. supplies 259

$$\text{Load current} = \frac{10}{200} = 0.05 \text{ A}$$

$$\text{Total current} = \frac{V_{RS}}{R_S} = \frac{20}{250} = 0.08 \text{ A}$$

$$\text{Diode current} = 0.08 - 0.05 = 0.03 \text{ A} = 30 \text{ mA}$$

```
10 PRINT "PROG 229"
20 PRINT "VOLTAGE REGULATOR"
30 INPUT "ENTER INPUT VOLTAGE" ; VIN
40 INPUT "ENTER OUTPUT VOLTAGE" ; VOUT
50 INPUT "ENTER LOAD RESISTANCE" ; RL
60 INPUT "ENTER R1" ; R1
70 LET IRL=VOUT*10^3/RL
80 LET IT=(VIN-VOUT)*10^3/R1
90 LET IZ=IT-IRL
100 PRINT "ZENER DIODE CURRENT = " IZ "MILLIAMPS"
```

Example 25 A zener diode stabilizing circuit is operated with an input voltage of 24 V and a diode current of 10 mA, to provide 12 V across a load of 1800 Ω. Calculate the value of the series resistor.

If the input voltage rises to 25 V calculate the new diode current.

$$\text{Load current} = \frac{12}{1800} = 0.00667 = 6.67 \text{ mA}$$

$$\text{Total current} = 10 + 6.67 = 16.67 \text{ mA} = 0.01667 \text{ A}$$

$$\text{Series resistor} \quad R_S = \frac{24 - 12}{0.01667} = 720 \text{ Ω}$$

With the input voltage at 25 V

$$\text{Voltage drop across } R_S = 25 - 12 = 13 \text{ V}$$

$$\text{Total current} = \frac{13}{720} = 0.01806 = 18.06 \text{ mA}$$

$$\text{New diode current} = 18.06 - 6.67 = 11.39 \text{ mA}$$

```
10 PRINT "PROG 230"
20 PRINT "VOLTAGE REGULATOR"
30 INPUT "ENTER INPUT VOLTAGE" ; VIN
40 INPUT "ENTER OUTPUT VOLTAGE" ; VOUT
50 INPUT "ENTER LOAD RESISTANCE" ; RL
60 INPUT "ENTER ZENER DIODE CURRENT" ; IZ
70 LET IRL=VOUT*10^3/RL
80 LET IT=IZ+IRL
90 LET R1=(VIN-VOUT)*10^3/IT
100 INPUT "ENTER INCREASED INPUT VOLTAGE" ; VIN
110 LET VR1=VIN-VOUT
120 LET IT=VR1*10^3/R1
130 LET IZ=IT-IRL
140 PRINT "R1 = " R1 "OHMS"
150 PRINT "DIODE CURRENT = " IZ "MILLIAMPS"
```

Problems

1 A cell has an e.m.f. of 24 V and an internal resistance of 0.5 Ω. Calculate the terminal voltage when the cell is connected to a 120 Ω load.
2 Find the internal resistance of a battery if the e.m.f. falls from 24 V to 22 V when delivering 7 A.

3 A battery has an internal resistance of 0.05 Ω. The terminal voltage is 1.1 V when delivering a current of 0.1 A. Calculate the battery e.m.f.

4 Measurements of the terminal voltages and the currents drawn during a test on a battery gave the following data:

V	I(A)
50	5
45	10

Calculate the internal resistance and the e.m.f. of the battery.

5 A battery consists of four cells in series, each with an e.m.f. of 6 V and an internal resistance of 0.05 Ω. Calculate the battery terminal voltage when connected to a 12-Ω resistance.

6 A battery consists of four cells in parallel. Each cell has an e.m.f. of 4.5 V and an internal resistance of 0.4 Ω. Calculate the battery terminal voltage when connected to a 6-Ω load.

7 The series–parallel arrangement of Figure 4 consists of the following:

Battery A. e.m.f 6 V. Internal resistance 2 Ω
Battery B. e.m.f 6 V. Internal resistance 2 Ω
Battery C. e.m.f 12 V. Internal resistance 2 Ω

If the combination is connected to a 50-Ω load, calculate: (a) load current, (b) terminal voltage of batteries A and B, (c) terminal voltage of battery C.

8 The battery arrangement in Figure 6 has the following values:

Battery A. 24 V, 0.5 Ω
Battery B. 24 V, 1.0 Ω
Battery C. 12 V, 1.5 Ω

If the load R is 15 Ω calculate: (a) load current, (b) terminal voltage of batteries A and B, (c) terminal voltage of battery C.

9 A circuit identical to Figure 6 has the following parameters:

Resistance of battery A 0.2 Ω
Resistance of battery B 0.4 Ω
Resistance of battery C 0.6 Ω
Current in battery A 2 A
Current in battery B 3 A
Current in battery C 5 A
Load resistance 25 Ω

Calculate the total power consumed.

10 A 36-V battery charger with an internal resistance of 0.35 Ω is used to charge a battery which has an internal resistance of 1.2 Ω. Calculate: (a) the initial charging current if the battery voltage has dropped to 30 V, (b) the charging current when the battery voltage has risen to 34.5 V.

11 It is a requirement to limit the initial charging current in Problem 10 to 2 A, by including a series resistance R in the battery charger. Find the value of this resistance. Using this series resistance find the value of the charging current when the e.m.f. has risen to 34.5 V.

12 A battery is being charged with a current of 4.8 A. The internal resistance of the battery is 0.5 Ω. The charger has an internal resistance of 1.5 Ω, and a series resistance of 12 Ω is included in the charger circuit. Calculate the power consumed in the internal resistance of the battery and in the charger.

13 A battery consists of 50 cells in series each with an e.m.f. of 2 V when fully charged. The internal resistance of each cell is

0.025 Ω. If the e.m.f. of each cell has fallen to 1.75 V, calculate: (a) the charger voltage needed to fully charge the battery, (b) the initial charging current if the internal resistance of the charger is 1.0 Ω, (c) the charging current when the e.m.f. of each cell is 1.96 V.

14 A battery consists of 20 3-V cells each with an internal resistance of 0.015 Ω. The voltage of each cell has fallen to 2.7 V. It is a requirement to charge the battery at a constant current of 1.0 A, from a charger with an internal resistance of 2 Ω. Calculate: (a) the charger voltage at the commencement of charge, (b) the charger voltage at the end of charge.

15 A battery is charged at a constant current of 6.5 A for 10 hours. It is discharged at the rate of 5 A for 8 hours. Calculate the ampere-hour efficiency.

16 What would the efficiency of the battery in Problem 15 be if the discharge rate had been 10 A for 5 hours?

17 A battery is fully charged by a constant current of 10 A over a period of 10 hours, and the average charging voltage is 2.1 V. It is discharged at a constant current of 6 A over 10 hours, with an average terminal voltage of 1.9 V. Calculate the watt-hour efficiency.

18 A half-wave rectifier circuit similar to the one in Figure 9 is connected to a resistive load of 54 Ω. If the input peak voltage is 18 V, calculate: (a) the peak current, (b) the d.c. voltage, (c) the r.m.s. voltage, (d) the r.m.s. current, (e) the d.c. current.

19 A full-wave rectifier circuit similar to the one in Figure 10 is connected to a resistive load of 40Ω. If the input peak voltage is 120 V, calculate: (a) the peak current, (b) the d.c. voltage, (c) the r.m.s. voltage, (d) the r.m.s. current, (e) the d.c. current.

20 A power supply unit similar to the one in Figure 11 is driven from a 110-V r.m.s. supply. The transformer has a 11 : 1 step down ratio. If the series resistor is 50 Ω, and the load resistor 3500 Ω, calculate: (a) the peak output voltage from the bridge rectifier, (b) the d.c. load current I, (c) the output voltage across R_L.

21 Repeat Problem 20 with an input voltage of 550 V r.m.s.

22 A voltage regulator circuit similar to Figure 13 has an unstabilized input of 50 V and a stabilized output of 45 V. If the load resistance is 1000 Ω and the diode current is 20 mA calculate the value of the series resistance R_S and the power dissipated in it.

23 If the load resistance in Problem 22 is increased to 2000 Ω calculate the diode current.

24 A voltage regulator circuit similar to Figure 14 has an input voltage of 20 V and an output voltage of 12 V. If the load resistance is 100 Ω, and the series resistor is 10 Ω, find the value of the diode current.

25 A zener diode stabilizing circuit is operated with an input voltage of 240 V and a diode current of 100 mA, to provide 200 V across a load of 50 Ω. Calculate the value of the series resistor. If the input voltage rises to 300 V, calculate the new diode current.

26 Repeat Problem 25 with the following parameters: input voltage of 1.0 V; output voltage of 0.9 V; load resistance of 1.0 MΩ; diode current of 1.0 mA. If the input voltage rises to 1.2 V, calculate the increased diode current.

15 Transistor amplifiers

15.1 Introduction

Three-element transistors can be connected in a circuit in three different modes. These three modes are referred to as:

1. Common emitter, see Figure 15.1(a)
2. Common base, see Figure 15.1(b)
3. Common collector, see Figure 15.1(c)

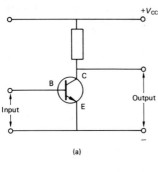

(a)

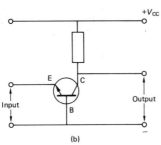

(b)

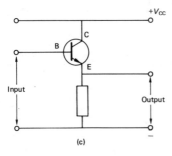

(c)

Figure 15.1

Transistor amplifiers 263

In (1) the signal is applied between the base and emitter and the output taken from the collector.

In (2) the signal is applied between the emitter and base and the output taken from the collector.

In (3) the signal is applied between base and emitter and the output taken from the emitter.

In order to understand how the transistor works, it is necessary to consider the properties of all three, and in particular to calculate the static (d.c.) values and the dynamic (a.c.) values of the various parameters associated with the three methods of connection.

All circuits considered in this chapter use n-p-n transistors. p-n-p transistors may be similarly connected, but the battery polarities and hence the currents will be reversed. Figure 15.2 shows n-p-n polarities.

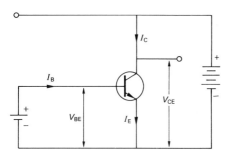

Figure 15.2

15.2 Common emitter – static conditions

Figure 15.3 shows the simplest practical circuit using an n-p-n transistor where

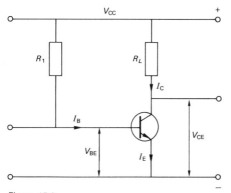

Figure 15.3

R_L = the collector load

R_1 = the bias resistor

V_{CC} = the applied d.c. voltage

V_{CE} = the collector–emitter voltage

V_{BE} = the base–emitter voltage

I_C = the collector current

I_B = the base current

I_E = the emitter current

The following parameters can be calculated:

1 The d.c. input resistance, h_{IE}, where

$$h_{IE} = \frac{V_{BE}}{I_B} \text{ (V_{CE} constant)}$$

2 The d.c. output resistance, h_{OE}, where

$$h_{OE} = \frac{V_{CE}}{I_C} \text{ (I_B constant)}$$

3 The d.c. current gain, h_{FE}, where

$$h_{FE} = \frac{I_C}{I_B} \text{ (V_{CE} constant)}$$

4 Current relationships where

$$I_E = I_C + I_B$$

Example 1 If in Figure 15.3 $R_L = 500\ \Omega$, $I_C = 10$ mA, $V_{CC} = 12$ V, $I_B = 50\ \mu A$, $V_{BE} = 0.5$ V, calculate the values of: (i) I_E, (ii) V_{CE}, (iii) R_1, (iv) h_{IE}, (v) h_{OE}, (vi) h_{FE}.

(i) $I_E = I_C + I_B = 10 + \dfrac{50}{10^3} = 10.05$ mA

(ii) $V_{CE} = V_{CC} - V_{RL}$ where $V_{RL} = I_C R_L$

$$= 12 - \frac{(10 \times 500)}{10^3} = 7 \text{ V}$$

(iii) $V_{R1} = V_{CC} - V_{BE} = 12 - 0.5 = 11.5$ V

$V_{R1} = I_B R_1 = 11.5$ V

$$R_1 = \frac{11.5}{50 \times 10^{-6}} = 230\ \text{k}\Omega$$

(iv) $h_{IE} = \dfrac{V_{BE}}{I_B} = \dfrac{0.5}{50 \times 10^{-6}} = 10\ \text{k}\Omega$

(v) $h_{OE} = \dfrac{V_{CE}}{I_C} = \dfrac{7}{10 \times 10^{-3}} = 700\ \Omega$

(vi) $h_{FE} = \dfrac{I_C}{I_B} = \dfrac{10 \times 10^{-3}}{50 \times 10^{-6}} = 200$

Transistor amplifiers

```
10 PRINT "PROG 231"
20 PRINT "COMMON EMITTER D.C. CONDITIONS"
30 INPUT "ENTER APPLIED VOLTAGE IN VOLTS" ; VCC
35 INPUT "ENTER LOAD RESISTANCE" ; RL
40 INPUT "ENTER BASE-EMITTER VOLTAGE" ; VBE
50 INPUT "ENTER BASE CURRENT IN MICROAMPS" ; IB
60 INPUT "ENTER COLLECTOR CURRENT IN MILLIAMPS" ; IC
65 LET VCE=VCC-(IC*RL/10^3)
66 LET R1=(VCC-VBE)*10^6/IB
70 LET HIE=VBE*10^6/IB
80 LET HOE=VCE*10^3/IC
90 LET HFE=IC*10^3/IB
100 LET IE=IC+(IB/10^3)
101 PRINT "COLLECTOR-EMITTER VOLTAGE = " VCE "VOLTS"
102 PRINT "BIAS RESISTOR = " R1 "OHMS"
110 PRINT "INPUT RESISTANCE = " HIE "OHMS"
120 PRINT "OUTPUT RESISTANCE = " HOE "OHMS"
130 PRINT "CURRENT GAIN = " HFE
140 PRINT "EMITTER CURRENT = " IE "MILLIAMPS"
```

Example 2 If the bias arrangement for the circuit in Figure 15.3 is modified to that shown in Figure 15.4 all other parameters remaining the same, calculate: (i) V_{CE}, (ii) R_1, (iii) h_{OE}.

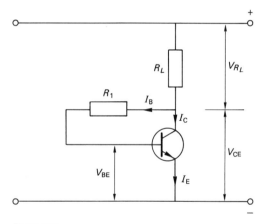

Figure 15.4

(i) $V_{R_L} = (I_C + I_B) R_L$

$$= \frac{10.05 \times 500}{10^3} = 5.025 \text{ V}$$

$$V_{CE} = 12 - 5.025 = 6.975 \text{ V}$$

(ii) $R_1 = \dfrac{6.975 - 0.5}{50 \times 10^{-6}} = 129.5 \text{ k}\Omega$

(iii) $h_{OE} = \dfrac{V_{CE}}{I_C} = \dfrac{6.975}{10 \times 10^{-3}} = 697.5 \text{ }\Omega$

```
10 PRINT "PROG 232"
20 PRINT "COMMON EMITTER D.C. CONDITIONS"
30 INPUT "ENTER APPLIED VOLTAGE IN VOLTS" ; VCC
35 INPUT "ENTER LOAD RESISTANCE" ; RL
40 INPUT "ENTER BASE-EMITTER VOLTAGE" ; VBE
50 INPUT "ENTER BASE CURRENT IN MICROAMPS" ; IB
60 INPUT "ENTER COLLECTOR CURRENT IN MILLIAMPS" ; IC
65 LET VCE=VCC-((IC+(IB/10^3))*RL/10^3)
66 LET R1=(VCE-VBE)*10^6/IB
70 LET HIE=VBE*10^6/IB
80 LET HOE=VCE*10^3/IC
90 LET HFE=IC*10^3/IB
100 LET IE=IC+(IB/10^3)
101 PRINT "COLLECTOR-EMITTER VOLTAGE = " VCE "VOLTS"
102 PRINT "BIAS RESISTOR = " R1 "OHMS"
110 PRINT "INPUT RESISTANCE = " HIE "OHMS"
120 PRINT "OUTPUT RESISTANCE = " HOE "OHMS"
130 PRINT "CURRENT GAIN = " HFE
140 PRINT "EMITTER CURRENT = " IE "MILLIAMPS"
```

Example 3 Figure 15.5 shows a third method of biasing an amplifier circuit. The parameters for this circuit are $I_C = 10$ mA, $I_B = 50$ μA, $V_{BE} = 0.5$ V, $I_1 = 5I_B$, $I_2 = 4I_B$, $V_{CC} = 12$ V. Calculate the values of (i) R_1 and (ii) R_2.

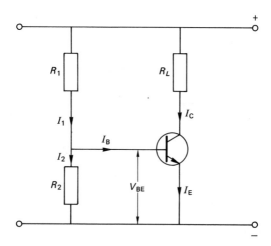

Figure 15.5

(i) $V_{R1} = 12 - 0.5 = 11.5$ V

$I_1 = 5 \times 50 \times 10^{-6} = 250 \times 10^{-6}$ A

$R_1 = \dfrac{11.5}{250 \times 10^{-6}} = 46$ kΩ

(ii) $V_{R2} = V_{BE} = 0.5$ V

$I_2 = I_1 - I_B = 250 - 50$ μA $= 200 \times 10^{-6}$ A $= 4I_B$

$R_2 = \dfrac{0.5}{200 \times 10^{-6}} = 2.5$ kΩ

Transistor amplifiers 267

```
10 PRINT "PROG 233"
20 PRINT "COMMON EMITTER D.C. CONDITIONS"
30 INPUT "ENTER APPLIED VOLTAGE IN VOLTS" ; VCC
40 INPUT "ENTER BASE-EMITTER VOLTAGE" ; VBE
50 INPUT "ENTER BASE CURRENT IN MICROAMPS" ; IB
60 INPUT "ENTER COLLECTOR CURRENT IN MILLIAMPS" ; IC
70 LET R1=(VCC-VBE)*10^6/(5*IB)
80 LET R2=(VBE*10^6)/(4*IB)
90 PRINT "R1 = " R1 "OHMS"
100 PRINT "R2 = " R2 "OHMS"
```

Example 4 Figure 15.6 shows an amplifier where an emitter resistor, R_E, is included to provide negative feedback and to prevent thermal runaway. If $R_E = 150\,\Omega$, $V_{BE} = 0.5$ V, $V_{CC} = 12$ V, $I_B = 50\,\mu$A, $I_C = 10$ mA, $I_1 = 5I_B$, $I_2 = 4I_B$, calculate the values of (i) R_2 and (ii) R_1.

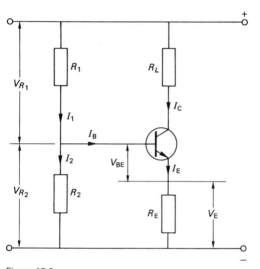

Figure 15.6

(i) $I_E = I_C + I_B = \dfrac{10}{10^3} + \dfrac{50}{10^6}$

$\quad = 0.01005$ A

$V_E = I_E R_E = 0.01005 \times 150 = 1.5075$ V

$V_{R2} = V_{BE} + V_E = 0.5 + 1.5075$

$\quad = 2.0075$ V

$R_2 = \dfrac{V_{R_2}}{I_2} = \dfrac{2.0075}{4 \times 50 \times 10^{-6}} = 10.037$ kΩ

(ii) $V_{R_1} = 12 - 2.0075 = 9.9925$

$I_1 = 5 \times 50 \times 10^{-6}$

$R_1 = \dfrac{V_{R_1}}{I_1} = \dfrac{9.9925}{250 \times 10^{-6}} = 39.97$ kΩ

```
10 PRINT "PROG 234"
20 PRINT "COMMON EMITTER D.C. CONDITIONS"
30 INPUT "ENTER APPLIED VOLTAGE IN VOLTS" ; VCC
40 INPUT "ENTER BASE-EMITTER VOLTAGE" ; VBE
50 INPUT "ENTER BASE CURRENT IN MICROAMPS" ; IB
60 INPUT "ENTER COLLECTOR CURRENT IN MILLIAMPS" ; IC
70 INPUT "ENTER EMITTER RESISTOR" ; RE
80 LET IE=(IC/10^3)+(IB/10^6)
90 LET VE=IE*RE
100 LET VR2=VE+VBE
110 LET R2=VR2*10^6/(4*IB)
120 LET VR1=VCC-VR2
130 LET R1=VR1*10^6/(5*IB)
140 PRINT "R1= " R1 "OHMS"
150 PRINT "R2= " R2 "OHMS"
```

Example 5 Figures 15.7 and 15.8 respectively show the input and output characteristics for the transistor used in the amplifier circuit of Figure 15.3. A load line for R_L has been drawn on the output characteristics and the operating point has been chosen at P. At P, $V_{BE} = 0.6$ V, $I_B = 30$ μA. If $V_{CC} = 10$ V, calculate: (i) the value of R_L, (ii) V_{R_L} and V_{CE}, (iii) d.c. input resistance h_{IE}, (iv) d.c. output resistance h_{OE}, (v) d.c. current gain h_{FE}, (vi) emitter current I_E.

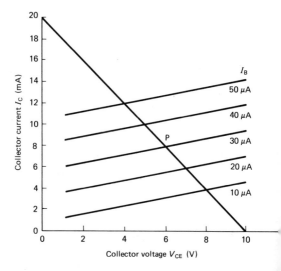

Figure 15.7

(i) $V_{CC} = V_{CE} + V_{R_L}$

$\phantom{V_{CC}} = V_{CE} + I_C R_L$ when $V_{CE} = 0$

$V_{CC} = I_C R_L$ where $I_C = 20$ mA

$10 = \dfrac{20}{10^3} R_L$

$R_L = \dfrac{10 \times 10^3}{20} = 500$ Ω

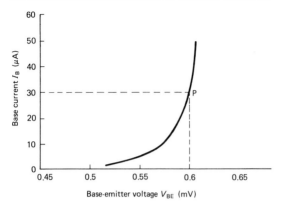

Figure 15.8

(ii) $I_C = 8 \times 10^{-3}$ A at P

$V_{RL} = 8 \times 10^{-3} \times 500 = 4$ V

$V_{CE} = 10 - 4 = 6$ V

(iii) $h_{IE} = \dfrac{V_{BE}}{I_B} = \dfrac{0.6}{30 \times 10^{-6}} = 20 \text{ k}\Omega$

(iv) $h_{OE} = \dfrac{V_{CE}}{I_C} = \dfrac{6}{8 \times 10^{-3}} = 750 \text{ }\Omega$

(v) $h_{FE} = \dfrac{I_C}{I_B} = \dfrac{8 \times 10^{-3}}{30 \times 10^{-6}} = 267$

(vi) $I_E = I_C + I_B$

$= 8 + \dfrac{30}{10^3} = 8.03$ mA

(See program 231, page 265.)

15.3 Common emitter – dynamic conditions

Section 15.2 has looked at the parameters associated with the d.c. conditions of an amplifier circuit. Applying an alternating voltage to an amplifier, in an arrangement such as the one shown in Figure 15.9, means that voltages V_{CE} and V_{BE} and currents I_C, I_B and I_E will vary. This means that the values of the circuit parameters will depend on the amplitude of the changes in the voltages and currents.

The 'change in' will be denoted by δ, the small Greek letter delta. D.c. parameters have been specified using upper case subscripts, A.c. parameters are indicated by lower case subscripts.

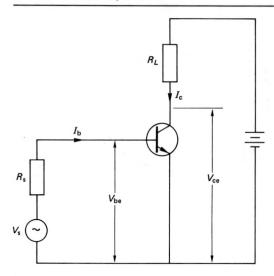

Figure 15.9

The following parameters for the circuit in Figure 15.9 can be calculated:

1. A.c. input resistance, h_{ie}

$$h_{ie} = \frac{\delta V_{BE}}{\delta I_B} = \frac{V_{be}}{I_b}$$

2. A.c. output resistance, h_{oe}

$$h_{oe} = \frac{\delta V_{CE}}{\delta I_C} = \frac{V_{ce}}{I_c}$$

3. A.c. current gain, h_{fe}

$$h_{fe} = \frac{\delta I_C}{\delta I_B} = \frac{I_c}{I_b}$$

Current gain is also denoted by A_i.

4. Output current I_c

$$\text{since } h_{fe} = \frac{I_c}{I_b}$$

$$I_c = h_{fe} I_b$$

$$\text{since } I_b = \frac{V_s}{R_s + R_{in}} = \frac{V_{in}}{R_{in}}$$

where V_s = source voltage

R_s = source resistance

R_{in} = transistor input resistance

V_{in} = transistor input voltage

Transistor amplifiers 271

then $I_c = \dfrac{h_{fe}V_{in}}{R_{in}}$

5 A.c. output voltage, V_{out}. The output voltage is developed across R_L where

$$V_{RL} = I_c R_L$$
$$= \dfrac{h_{fe}V_{in}R_L}{R_{in}}$$

The alternating voltage V_{RL} is also equal to V_{ce}.

6 A.c. voltage gain, A_v

$$A_v = \dfrac{V_{out}}{V_{in}} = \dfrac{V_{ce}}{V_{be}}$$
$$= \dfrac{h_{fe}V_{in}R_L}{V_{in}R_{in}}$$
$$= \dfrac{h_{fe}R_L}{R_{in}}$$

7 A.c. power gain, A_p

$$A_p = \dfrac{P_{out}}{P_{in}} = \dfrac{V_{ce}I_c}{V_{be}I_b}$$

since $P_{in} = I_b^2 R_{in}$

and $P_{out} = I_c^2 R_L = (h_{fe}I_b)^2 R_L$

$$A_p = \dfrac{h_{fe}^2 I_b^2 R_L}{I_b^2 R_{in}}$$
$$= \dfrac{h_{fe}^2 R_L}{R_{in}} = A_v A_i$$

Example 6 An alternating voltage V_{be}, of 36 mV is applied to a circuit similar to Figure 15.9. The change in base current I_b is 60 μA, and the changes in the collector current I_c and the collector to emitter voltage V_{ce} are 10 mA and 4.5 V respectively. Calculate: (i) input resistance, h_{ie}, (ii) output resistance, h_{oe}, (iii) current gain, h_{fe} (A_i), (iv) voltage gain, A_v, (v) power gain, A_p.

(i) $h_{ie} = \dfrac{V_{be}}{I_b} = \dfrac{36 \times 10^{-3}}{60 \times 10^{-6}} = 600\ \Omega$

(ii) $h_{oe} = \dfrac{V_{cc}}{I_c} = \dfrac{4.5}{10 \times 10^{-3}} = 450\ \Omega$

(iii) $A_i = \dfrac{I_c}{I_b} = \dfrac{10 \times 10^{-3}}{60 \times 10^{-6}} = 167$

(iv) $A_v = \dfrac{V_{ce}}{V_{be}} = \dfrac{4.5}{36 \times 10^{-3}} = 125$

(v) $A_p = A_i A_v = 167 \times 125 = 2.09 \times 10^4$

Note also

$$A_p = \dfrac{V_{ce}I_c}{V_{be}I_b}$$

$$= \frac{4.5 \times 10 \times 10^{-6}}{36 \times 10^{-3} \times 60 \times 10^{-6}}$$

$$= 2.09 \times 10^4$$

```
10 PRINT "PROG 235"
20 PRINT "COMMON EMITTER A.C. CONDITIONS"
30 INPUT "ENTER VBE IN MILLIVOLTS" ; VBE
40 INPUT "ENTER IB IN MICROAMPS" ; IB
50 INPUT "ENTER IC IN MILLIAMPS" ; IC
60 INPUT "ENTER VCE IN VOLTS" : VCE
70 LET HIE=VBE*10^3/IB
80 LET HOE=VCE*10^3/IC
90 LET AI=IC*10^3/IB
100 LET AV=VCE*10^3/VBE
110 LET AP=AV*AI
120 PRINT "INPUT RESISTANCE = " HIE "OHMS"
130 PRINT "OUTPUT RESISTANCE = " HOE "OHMS"
140 PRINT "CURRENT GAIN = " AI
150 PRINT "VOLTAGE GAIN = " AV
160 PRINT "POWER GAIN = " AP
```

Example 7 Measurements of the common emitter characteristics of a transistor gave the data in Tables 15.1 and 15.2.

Plot the output characteristic, I_C/V_{CE}, the transfer characteristic, I_C/I_B, and the input characteristic, I_B/V_{BE}.

This transistor is to be used in an amplifier circuit. A load line has been selected which intersects the x axis at $V_{CE} = 10$ V and the y axis at $I_C = 10$ mA. See Figure 15.10(a).

If the operating point of the amplifier P is chosen at the intersection of the load line and the 30 µA line, calculate the d.c. parameters at P.

If a sine wave having an amplitude of 50 mV peak to peak is applied to the input of the amplifier, calculate all the a.c. parameters.

D.c. parameters

(i) D.c. input resistance h_{IE} from Figure 15.10(c)

$$h_{IE} = \frac{V_{BE}}{I_B} = \frac{0.66}{30 \times 10^{-6}} = 22 \text{ k}\Omega$$

(ii) D.c. output resistance h_{OE} from Figure 15.10(a)

$$h_{OE} = \frac{V_{CE}}{I_C} = \frac{4.5}{5.4 \times 10^{-3}} = 833 \ \Omega$$

(iii) D.c. current gain h_{FE} from Figure 15.10(b)

$$h_{FE} = \frac{I_C}{I_B} = \frac{5.4 \times 10^{-3}}{30 \times 10^{-6}} = 180$$

(iv) Emitter current I_E

$$I_E = I_C + I_B = 5.4 + (30 \times 10^{-3}) = 5.43 \text{ mA}$$

A.c. parameters

(i) A.c. input resistance, h_{ie}. Draw a tangent to the point P, Figure 15.10(c), and complete the triangle.

Transistor amplifiers 273

Table 15.1 Output characteristics

V_{CE} (Volts)	Collector current I_C (mA)					
	$I_B = 0$ μA	$I_B = 10$ μA	$I_B = 20$ μA	$I_B = 30$ μA	$I_B = 40$ μA	$I_B = 50$ μA
0	0	0	0	0	0	0
1	0	1.1	3.1	5.1	7.1	9.1
2	0	1.2	3.15	5.18	7.2	9.2
3	0	1.3	3.3	5.25	7.25	9.3
4	0	1.4	3.41	5.4	7.38	9.4
5	0	1.5	3.5	5.5	7.5	9.5
6	0	1.6	3.61	5.62	7.61	9.6
7	0	1.7	3.7	5.7	7.72	9.7
8	0	1.8	3.78	5.8	7.8	9.8
9	0	1.9	3.9	5.9	7.91	9.9
10	0	2.0	4.0	6.0	8.0	10.0

Table 15.2 Input and transfer characteristics

I_B (μA)	0	5	10	20	30	40	50
V_{BE} (volts)	0	0.5	0.56	0.62	0.66	0.67	0.68
I_C (mA)	0	0.9	1.8	3.6	5.4	7.2	9.0

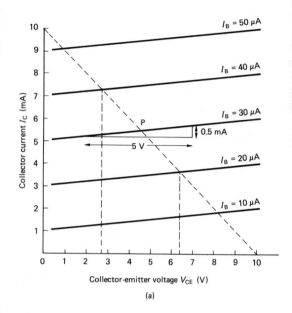

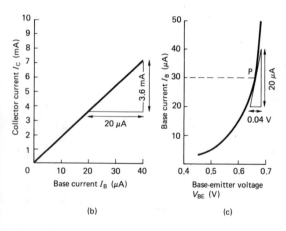

Figure 15.10

$$h_{ie} = \frac{\delta V_{BE}}{\delta V_{IB}} = \frac{V_{be}}{I_b} = \frac{0.68 - 0.64}{(40 - 20) \times 10^{-6}}$$

$$= \frac{0.04}{20 \times 10^{-6}} = 2 \text{ k}\Omega$$

(ii) A.c. output resistance, h_{oe}. Draw a triangle on the 30-μA line in Figure 15.10(a).

$$h_{oe} = \frac{\delta V_{CE}}{\delta I_C} = \frac{V_{ce}}{I_c} = \frac{7 - 2}{(5.7 - 5.2) \times 10^{-3}}$$

$$= \frac{5}{0.5 \times 10^{-3}} = 10 \text{ k}\Omega$$

(iii) A.c. current gain, h_{fe}. Draw a triangle on Figure 15.10(b).

$$h_{fe} = \frac{\delta I_C}{\delta I_B} = \frac{I_c}{I_b} = \frac{(7.2 - 3.6) \times 10^{-3}}{(40 - 20) \times 10^{-6}}$$

$$= \frac{3.6}{20 \times 10^{-3}} = 180$$

(iv) A.c. voltage gain, A_v

$$A_v = \frac{V_{OUT}}{V_{IN}} = \frac{\delta V_{CE}}{\delta V_{BE}} = \frac{V_{ce}}{V_{be}}$$

$$= \frac{6.3 - 2.7}{0.68 - 0.64} \quad \text{See Figure 15.10(a)}$$
$$\phantom{=\frac{6.3 - 2.7}{0.68 - 0.64}} \quad \text{See Figure 15.10(c)}$$

$$= \frac{3.6}{0.04} = 90$$

(v) A.c. power gain A_p

$$A_p = A_v A_i \text{ where } A_i = h_{fe}$$
$$= 90 \times 180 = 16.2 \times 10^3$$

Note also that

$$A_p = \frac{\text{power out}}{\text{power in}} = \frac{I_c V_{ce}}{I_b V_{be}}$$

$$= \frac{3.6 \times 10^{-3} \times 3.6}{20 \times 10^{-6} \times 0.04} = 16.2 \times 10^3$$

Note that the remainder of the equations given earlier in this section confirm the results obtained for the various parameters, including

(i) $\quad I_c = \frac{h_{fe} V_{IN}}{R_{IN}} = \frac{h_{fe} V_{be}}{h_{ie}} = \frac{180 \times 0.04 \times 10^3}{2000} = 3.6$ mA

(ii) when $V_{CE} = 0$ then $V_{CC} = I_C R_L$, Figure 15.10(a)

$$R_L = \frac{V_{CC}}{I_C} = \frac{10}{10 \times 10^{-3}} = 1 \text{ k}\Omega$$

$$V_{RL} = V_{CE} = I_C R_L = 3.6 \times 10^{-3} \times 1000 = 3.6 \text{ V}$$

(iii) $\quad V_{RL} = \frac{h_{fe} V_{IN} R_L}{R_{IN}} = \frac{h_{fe} V_{be} R_L}{h_{ie}}$

$$= \frac{180 \times 0.04 \times 1000}{2000} = 3.6 \text{ V}$$

(iv) $$A_v = \frac{h_{fe}R_L}{R_{IN}} = \frac{h_{fe}R_L}{h_{ie}} = \frac{180 \times 1000}{2000} = 90$$

(v) $$A_p = \frac{h_{fe}^2 R_L}{R_{IN}} = \frac{h_{fe}^2 R_L}{h_{ie}} = \frac{180^2 \times 1000}{2000} = 16.2 \times 10^3$$

```
10 PRINT "PROG 236"
20 PRINT "A.C. PARAMETERS"
30 INPUT "ENTER VBE IN VOLTS" ; VBE
40 INPUT "ENTER IB IN MICROAMPS" ; IB
50 LET HIE=VBE*10^6/IB
60 PRINT "INPUT RESISTANCE = " HIE "OHMS"
70 INPUT "ENTER VCE IN VOLTS" ; VCE
80 INPUT "ENTER IC IN MILLIAMPS" ; IC
90 LET HOE=VCE*10^3/IC
100 PRINT "OUTPUT RESISTANCE = " HOE "OHMS"
110 INPUT "ENTER IC IN MILLIAMPS" ; IC
120 INPUT "ENTER IB IN MICROAMPS" ; IB
130 LET AI=IC*10^3/IB
140 PRINT "CURRENT GAIN = " AI
150 INPUT "ENTER VCE IN VOLTS" ; VCE
160 INPUT "ENTER VBE IN VOLTS" ; VBE
170 LET AV=VCE/VBE
180 PRINT "VOLTAGE GAIN = " AV
190 LET AP=AV*AI
200 PRINT "POWER GAIN = " AP
```

15.4 Common base – static conditions

The circuit of a typical common base amplifier is shown in Figure 15.11. Some typical characteristics are shown in Figure 15.12. Figure 15.12(a) shows the output characteristics, Figure 15.12(b) the input characteristics, and Figure 15.12(d) the transfer characteristics. Since the output characteristics are nearly horizontal, the $I_E = 4$ mA line has been reproduced enlarged for calculation purposes in figure 15.12(c).

The method of calculating the various parameters is identical to the method in Section 15.2 for the common emitter mode. The symbols, however, are different and the calculations are concerned with the collector to base voltage, V_{CB} and the emitter to base voltage, V_{EB}, and the input current is now the emitter current I_E.

Example 8 Assuming that the circuit in Figure 15.11 uses a transistor with the characteristics shown in Figure 15.12, calculate the d.c. parameters. The d.c. load line has been drawn and the operating point chosen at P. The d.c. parameters at P are:

(i) D.c. input resistance

$$h_{IB} = \frac{V_{EB}}{I_E} = \frac{0.6}{4 \times 10^{-3}} = 150 \ \Omega \text{ See Figure 15.12(b)}$$

(ii) D.c. output resistance

$$h_{OB} = \frac{V_{CB}}{I_C} = \frac{6.1}{3.961 \times 10^{-3}} = 1540 \ \Omega$$

See Figure 15.12(c)

(iii) D.c. current gain

$$h_{FB} = \frac{I_C}{I_E} = \frac{3.95}{4.0} = 0.99 \text{ See Figure 15.12(d)}$$

Transistor amplifiers 277

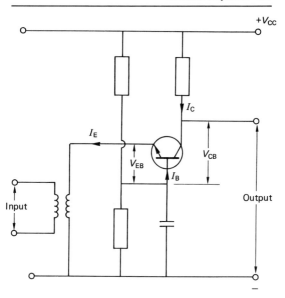

Figure 15.11

(iv) Base current

$$I_B = I_E - I_C = (4 \times 10^{-3}) - (3.961 \times 10^{-3})$$
$$= 3.9 \times 10^{-5} \text{ A}$$

```
10 PRINT "PROG 237"
20 PRINT "COMMON BASE D.C. CONDITIONS"
30 INPUT "ENTER VEB IN VOLTS" ; VEB
40 INPUT "ENTER IE IN MILLIAMPS" ; IE
50 INPUT "ENTER VCB IN VOLTS" ; VCB
60 INPUT "ENTER IC IN MILLIAMPS" ; IC
70 LET HIB=VEB*10^3/IE
80 LET HOB=VCB*10^3/IC
90 LET HFB=IC/IE
100 LET IB=(IE/10^3)-(IC/10^3)
110 PRINT "INPUT RESISTANCE = " HIB "OHMS"
120 PRINT "OUTPUT RESISTANCE = " HOB "OHMS"
130 PRINT "CURRENT GAIN = " HFB
140 PRINT "BASE CURRENT = " IB "AMPS"
```

15.5 Common base – dynamic conditions

The method of calculating the various parameters is identical to the method in Section 15.3 and in Example 7 for the common emitter mode. Lower case subscripts are used to denote varying quantities.

Example 9 If a signal with a peak to peak amplitude of 0.08 V is applied to the circuit in Example 8, calculate the a.c. parameters.

(i) A.c. input resistance

$$h_{ib} = \frac{V_{eb}}{I_e} = \frac{0.64 - 0.56}{(6 - 2) \times 10^{-3}} = \frac{0.08}{4 \times 10^{-3}}$$

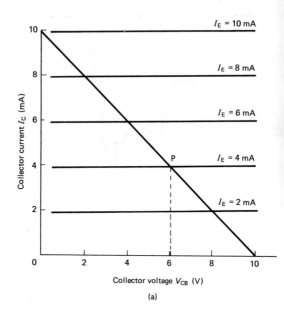

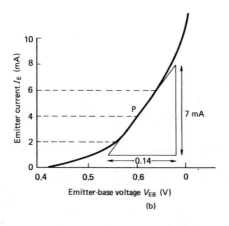

Figure 15.12

Transistor amplifiers 279

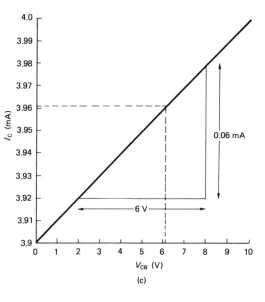

(c)

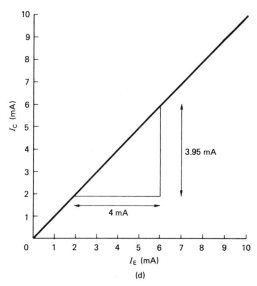

(d)

= 20 Ω See Figure 15.12(b)

(ii) A.c. output resistance

$$h_{ob} = \frac{V_{cb}}{I_c} = \frac{8-2}{(3.98-3.92)\times 10^{-3}} = 100 \text{ k}\Omega$$

See Figure 15.12(c)

(iii) A.c. current gain

$$h_{fb} = \frac{I_c}{I_e} = \frac{(5.9-1.95)\times 10^{-3}}{(6-2)\times 10^{-3}} = \frac{3.95}{4}$$

$$= 0.9875 \text{ See Figure 15.12(d)}$$

(iv) A.c. output current

$$I_c = h_{fb}\frac{V_{in}}{R_{in}} = \frac{0.9875 \times 0.08}{20} = 3.95 \times 10^{-3} \text{ A}$$

(v) A.c. output voltage

When

$$V_{CE} = 0$$

then

$$V_{CC} = I_C R_L$$

$$R_L = \frac{V_{CC}}{I_C} = \frac{10 \times 10^3}{10} = 1 \text{ k}\Omega$$

$$V_{RL} = V_{out} = I_c R_L$$

$$= 3.95 \times 10^{-3} \times 1000 = 3.95 \text{ V}$$

Note also that

$$V_{RL} = \frac{h_{fb} V_{in} R_L}{R_{in}}$$

$$= \frac{0.9875 \times 0.08 \times 1000}{20} = 3.95 \text{ V}$$

(vi) A.c. voltage gain

$$A_v = \frac{V_{cb}}{V_{eb}} = \frac{3.95}{0.08} = 49.375$$

Note also that

$$A_v = \frac{h_{fb} R_L}{R_{in}} = \frac{0.9875 \times 1000}{20}$$

$$= 49.375$$

(vii) A.c. power gain

$$A_p = \frac{P_{out}}{P_{in}} = \frac{V_{cb} I_c}{V_{eb} I_e}$$

$$= \frac{3.95 \times 3.95 \times 10^{-3}}{0.08 \times 4 \times 10^{-3}} = 48.75$$

Note that

$$A_p = \frac{(h_{fb})^2 R_L}{R_{in}}$$

$$= \frac{(0.9875)^2 \times 1000}{20} = 48.75$$

Note also that:

(i) $A_p = A_v A_i$ $(A_i = h_{fb})$

$= 49.375 \times 0.9875 = 48.75$

(ii) $I_c = \dfrac{h_{fb} V_{in}}{R_{in}} = \dfrac{h_{fb} V_{eb}}{h_{ib}} = \dfrac{0.9875 \times 0.08}{20} = 3.95 \times 10^{-3}$ A

$= 3.9$ mA

(iii) $V_{RL} = I_c R_L = 3.95 \times 10^{-3} \times 10^3 = 3.95$ V

$V_{RL} = \dfrac{h_{fb} V_{in} R_L}{R_{in}} = \dfrac{h_{fb} V_{eb} R_L}{h_{ib}}$

$= \dfrac{0.9875 \times 0.08 \times 1000}{20} = 3.95$ V

(iv) $A_v = \dfrac{h_{fb} R_L}{R_{in}} = \dfrac{h_{fb} R_L}{h_{ib}} = \dfrac{0.9875 \times 1000}{20} = 49.375$

(v) $A_p = \dfrac{h_{fb}^2 R_L}{R_{in}} = \dfrac{h_{fb}^2 R_L}{h_{ib}} = \dfrac{0.9875^2 \times 1000}{20} = 48.75$

```
10 PRINT "PROG 238"
20 PRINT "COMMON BASE A.C. CONDITIONS"
30 INPUT "ENTER VEB IN VOLTS" ; VEB
40 INPUT "ENTER IE IN MILLIAMPS" ; IE
45 LET HIB=VEB*10^3/IE
46 PRINT "INPUT RESISTANCE = " HIB "OHMS"
50 INPUT "ENTER VCB IN VOLTS" ; VCB
60 INPUT "ENTER IC IN MILLIAMPS" ; IC
65 LET HOB=VCB*10^3/IC
70 PRINT "OUTPUT RESISTANCE = " HOB "OHMS"
80 INPUT "ENTER IC IN MILLIAMPS" ; IC
90 INPUT "ENTER IE IN MILLIAMPS" ; IE
100 LET HFB=IC/IE
110 PRINT "CURRENT GAIN = " HFB
120 INPUT "ENTER VCB IN VOLTS" ; VCB
130 INPUT "ENTER VEB IN VOLTS" ; VEB
140 LET AV=VCB/VEB
150 PRINT "VOLTAGE GAIN = " AV
160 LET AP=AV*HFB
170 PRINT "POWER GAIN = " AP
```

15.6 Common collector

A common collector amplifier is shown in Figure 15.13. This circuit is also known as the emitter follower, and the main features are:

1. A current gain which is approximately the same as the current gain of a common emitter stage
2. A voltage gain which is approximately unity
3. A power gain which is approximately the same as the current gain
4. High input impedance
5. Low output impedance

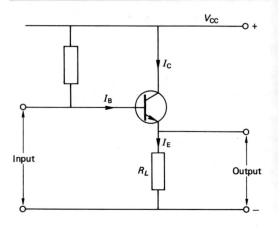

Figure 15.13

6 Ideal for impedance matching
7 Input and output signals in phase

(a) *A.c. current gain,* h_{fc} *or* A_i

$$h_{fc} = \frac{\delta I_E}{\delta I_B} = \frac{I_e}{I_b}$$

$$= \frac{I_e}{I_e - I_c} \quad \text{since } I_b = I_e - I_c$$

$$= \frac{I_e}{I_e[1 - (I_c/I_e)]}$$

$$= \frac{I_e}{I_e(1 - h_{fb})}$$

$$= \frac{1}{1 - h_{fb}}$$

Note also that

$$I_e = I_b + I_c$$

$$\frac{I_e}{I_b} = \frac{I_b}{I_b} + \frac{I_c}{I_b}$$

and therefore

$$h_{fc} = 1 + h_{fe}$$

Note that the current gain is approximately the same as the current gain of the common emitter. The current gain, however, is reduced by an amount which depends on the emitter load.

(b) *A.c. voltage gain,* A_v

$$A_v = \frac{V_{out}}{V_{in}} = \frac{I_e R_L}{I_b R_{in}}$$

$$= h_{fc} \frac{R_L}{R_{in}}$$

(c) A.c. input resistance, h_{ic}

$$h_{ic} = R_{in} = \frac{V_{in}}{I_b}$$

Since the output voltage follows the input voltage, the change in the base emitter voltage V_{be} is small, and hence the base current I_b is small. This means that R_{in} is very high.

(d) A.c. power gain, A_p

$$A_p = \frac{\text{Power out}}{\text{Power in}} = \frac{I_c^2 R_L}{I_b^2 R_{in}}$$

$$= \frac{h_{fc}^2 R_L}{R_{in}}$$

Note that transistor data show that in practice the output characteristics tend to fan out, and that the transfer characteristics are slightly non-linear. Taking these two facts into account means that the answers to the worked examples would be marginally different to those given.

The object of this chapter has been to shown how the various transistor parameters are calculated.

Problems

1. Figure 15.14 shows a small signal amplifier. Calculate the following d.c. parameters: (a) the emitter current, I_E, (b) the collector–emitter voltage, V_{CE}, (c) the value of the bias resistor, R_1, (d) the input resistance, h_{IE}, (e) the output resistance, h_{OE}, (f) the current gain, h_{FE}.

2. If the bias resistor in the circuit of Figure 15.14 is connected to the collector, in an arrangement identical to Figure 15.4, all other parameters remaining the same, calculate: (a) the collector–emitter voltage, V_{CE}, (b) the value of the bias resistor, R_1, (c) the output resistance, h_{OE}.

3. The circuit in Figure 15.5 has the following parameters: $I_C = 5$ mA, $I_B = 30$ µA, $V_{BE} = 0.6$ V, $I_1 = 5I_B$, $I_2 = 4I_B$, $V_{CC} = 12$ V. Calculate the values of the bias resistors R_1 and R_2.

4. Repeat Problem 3 with an emitter resistor of 470 Ω, in an arrangement identical to Figure 15.6.

5. If the operating point P had been chosen at the intersection of the load line and the 20-µA characteristic in Figure 15.7, calculate: (a) the voltage drop across the load resistor, V_{RL}, (b) the collector–emitter voltage, V_{CE}, (c) the input resistance, h_{IE}, (d) the output resistance, h_{OE}, (e) the current gain, h_{FE}, (f) the emitter current, I_E.

6. An alternating voltage, V_{be}, of 40 mV is applied to a common emitter amplifier, identical to Figure 15.9. The change in base current, I_b, is 40 µA, and the changes in the collector current I_c and the collector to emitter voltage V_{ce} are 12 mA and 3.8 V respectively. Calculate: (a) the input resistance, h_{ie}, (b) the

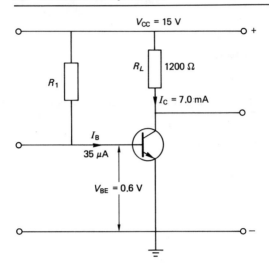

Figure 15.14

output resistance, h_{oe}, (c) the current gain, h_{fe}, (d) the voltage gain, A_v, (e) the power gain, A_p.

7 Measurements of the common emitter characteristics of a transistor produced the data on page 285.

Plot: (i) the output characteristic, I_C/V_{CE}, (ii) the transfer characteristic, I_C/I_B, (iii) the input characteristic, I_B/V_{BE}.

This transistor is to be used in an amplifier circuit. Draw a load line which intersects the x axis at $V_{CE} = 12$ V and the y axis at $I_C = 6$ mA. Select the operating point at the intersection of the load line and the 30-μA characteristic. A sine wave having an amplitude of 100 mV peak to peak is applied to the input of the amplifier. Calculate: (a) the d.c. input resistance, h_{IE}, (b) the d.c. output resistance, h_{OE}, (c) the d.c. current gain, h_{FE}, (d) the emitter current, I_E, (e) the a.c. input resistance, h_{ie}, (f) the a.c. output resistance, h_{oe}, (g) the a.c. current gain, h_{fe}, (h) the a.c. voltage gain, A_v, (i) the a.c. power gain, A_p, (j) the value of the load resistor.

V_{CE} (Volts)	Collector current I_C (mA)					
	$I_B = 0$ µA	$I_B = 10$ µA	$I_B = 20$ µA	$I_B = 30$ µA	$I_B = 40$ µA	$I_B = 50$ µA
0	0	0	0	0	0	0
1	0	0.6	1.6	2.6	3.6	4.6
2	0	0.65	1.65	2.68	3.68	4.68
3	0	0.72	1.7	2.76	3.66	4.67
4	0	0.82	1.8	2.84	3.83	4.84
5	0	0.9	1.9	2.92	3.92	4.92
6	0	1.0	2.0	3.0	4.0	5.0
7	0	1.1	2.1	3.08	4.09	5.08
8	0	1.2	2.21	3.16	4.16	5.15
9	0	1.25	2.26	3.24	4.25	5.24
10	0	1.35	2.35	3.32	4.32	5.32
11	0	1.42	2.41	3.4	4.4	5.4
12	0	1.5	2.5	3.5	4.5	5.5

I_B (µA)	0	10	20	30	40	50	60
V_{BE} (volts)	0	0.47	0.55	0.6	0.65	0.67	0.69
I_C (mA)	0	0.9	1.8	2.7	3.6	4.6	5.5

16 Operational amplifiers

16.1 Open loop amplifiers

The operational amplifier is a high-gain directly coupled voltage amplifier, having a single output and two inputs, one inverting and the other non-inverting. The output signal is proportional to the differential signal voltage between the two input terminals, i.e.

$$V_{OUT} = A_o(V_1 - V_2)$$

where A_o is the low-frequency open loop gain of the operational amplifier, typically 10^5. V_1 and V_2 are the two input voltages.

By grounding the positive (+) terminal (see Figure 16.1(a)) the device can be used as a high-gain inverting amplifier. By grounding the negative (−) terminal (see figure 16.1(b)) the device can be used as a high-gain non-inverting amplifier. It can also be used as a differential amplifier, by feeding signals into both terminals (see Figure 16.1(c)). If both input signals were identical, the output signal ideally would be zero.

Other important characteristics of the operational amplifiers include a high input impedance and a low output impedance. Their many applications include both a.c. and d.c. amplifiers, filters, switches, comparators and oscillators.

16.2 Closed loop amplifiers

In the closed loop mode negative feedback is applied from output to input. Figure 16.2(a) shows the inverting amplifier, and Figure 16.2(b) the non-inverting amplifier. The overall gain is now determined solely by the values of the external components R_1 and R_2. For the inverting amplifier

$$V_{OUT} = -V_{IN}\frac{R_2}{R_1}$$

and voltage gain

$$A_v = \frac{R_2}{R_1}$$

For the non-inverting amplifier

$$V_{OUT} = V_{IN}\frac{R_1 + R_2}{R_2}$$

and the voltage gain

$$A_v = \frac{R_1 + R_2}{R_1} = 1 + \frac{R_2}{R_1}$$

Example 1 In an inverting amplifier R_2 is 15 kΩ and R_1 is 1.5 kΩ. Calculate the gain A_v.

$$A_v = \frac{R_2}{R_1} = \frac{15 \times 10^3}{1.5 \times 10^3} = -10$$

(The minus sign indicates that the output is inverted.)

Operational amplifiers 287

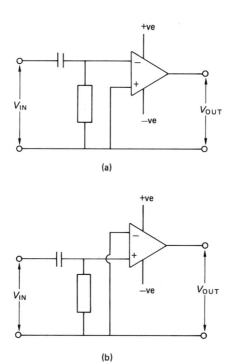

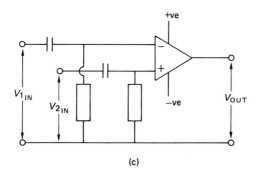

Figure 16.1

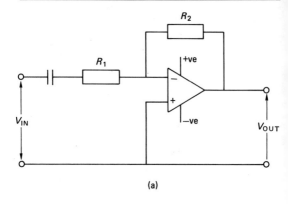

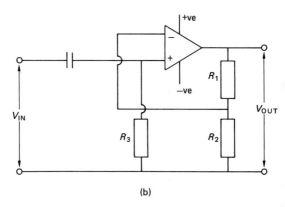

Figure 16.2

```
10 PRINT " PROG 239"
20 PRINT "CLOSED LOOP AMPLIFIER GAIN"
30 INPUT "ENTER R1" ; R1
40 INPUT "ENTER R2" ; R2
50 LET AV=-R2/R1
60 PRINT "VOLTAGE GAIN = " AV
```

Example 2 If both resistances in Example 1 have a value of 10 kΩ, what is the gain?

$$A_v = \frac{R_2}{R_1} = \frac{10 \times 10^3}{10 \times 10^3} = -1$$

This means that the gain is unity, and that the output is 180° out of phase with the input, if the input is an angular function. (See program 239, above.)

Operational amplifiers 289

Example 3 A non-inverting amplifier has the same component values as those in Example 1. Calculate the gain

$$A_v = 1 + \frac{R_2}{R_1} = 1 + \frac{15 \times 10^3}{1.5 \times 10^3} = 11$$

```
10 PRINT " PROG 240"
20 PRINT "CLOSED LOOP AMPLIFIER"
30 INPUT "ENTER R1" ; R1
40 INPUT "ENTER R2" ; R2
50 LET AV=1+(R2/R1)
60 PRINT "VOLTAGE GAIN = " AV
```

Example 4 A non-inverting amplifier has the same component values as those in Example 2. Calculate the gain.

$$A_v = 1 + \frac{R_2}{R_1} = 1 + \frac{10 \times 10^3}{10 \times 10^3} = 2$$

In this example the output is twice the input. Both input and output voltages are in phase. (See program 240, above.)

Example 5 A non-inverting amplifier identical to Figure 16.2(b) has a gain of 21. If R_1 has a value of 2700 Ω, calculate the value of the feedback resistor R_2.

$$A_v = 1 + \frac{R_2}{R_1}$$

$$21 = 1 + \frac{R_2}{2700}, (21 - 1) \times 2700 = R_2$$

$$R_2 = 54 \text{ k}\Omega$$

```
10 PRINT " PROG 241"
20 PRINT "CLOSED LOOP AMPLIFIER"
30 INPUT "ENTER R1" ; R1
40 INPUT "ENTER AV" ; AV
50 LET R2=(AV-1)*R1
60 PRINT "R2=" ; R2
```

Example 6 In a non-inverting amplifier, the feedback resistor R_2 has half the value of R_1. If R_1 is 1200 Ω, calculate the gain.

$$A_v = 1 + \frac{R_2}{R_1} = 1 + \frac{600}{1200} = 1.5$$

This result shows that even when the feedback resistor R_2 is smaller than R_1 the gain is still greater than unity, and if $R_2 = 0$ the gain is 1. (See program 240, above.)

16.3 Summing amplifiers

Figure 16.3(a) shows a two-input summing amplifier, with an inverted output which is equal to the sum of the two input voltages V_1 and V_2. Given the values of the three resistors and the two input voltages, the output voltage can be calculated by the equation

$$V_{OUT} = -\left(\frac{R_3 V_1}{R_1} + \frac{R_3 V_2}{R_2}\right)$$

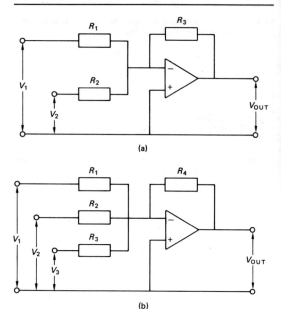

Figure 16.3

Figure 16.3(b) shows a three-input summing amplifier. The output voltage is given by

$$V_{OUT} = -\left(\frac{R_4 V_1}{R_1} + \frac{R_4 V_2}{R_2} + \frac{R_4 V_3}{R_3}\right)$$

Example 7 If R_1 and R_2 and R_3 in Figure 16.3(a) are 10 kΩ, 20 kΩ and 60 kΩ respectively and V_1 is 2 V and V_2 4 V, calculate the value of the output voltage.

$$V_{OUT} = -\left(\frac{R_3 V_1}{R_1} + \frac{R_3 V_2}{R_2}\right)$$
$$= -\left(\frac{60 \times 2}{10} + \frac{60 \times 4}{20}\right) = -24 \text{ V}$$

```
10 PRINT " PROG 242"
20 PRINT "SUMMING AMPLIFIER"
30 INPUT "ENTER R1" ; R1
40 INPUT "ENTER R2" ; R2
50 INPUT "ENTER R3" ; R3
60 INPUT "ENTER V1" ; V1
70 INPUT "ENTER V2" ; V2
80 LET VOUT=(R3*V1/R1)+(R3*V2/R2)
90 PRINT "OUTPUT VOLTAGE = - " VOUT
```

Example 8 If $R_1 = R_2 = R_3 = 5$ kΩ in Figure 16.3(a), calculate the output voltage if $V_1 = 10$ V and $V_2 = 30$ V.

Operational amplifiers

$$V_{OUT} = -\left(\frac{5 \times 10}{5} + \frac{5 \times 30}{5}\right) = -40 \text{ V}$$

Note that when the three resistances have the same value

$$V_{OUT} = -(V_1 + V_2)$$

(See program 242, page 290.)

Example 9 A two-input summing amplifier identical to Figure 16.3(a) has the following parameters: $R_1 = 32$ kΩ, $R_2 = 48$ kΩ, $V_1 = 8$ V, $V_2 = 12$ V, $V_{OUT} = -32$ V. Calculate the value of R_3.

$$V_{OUT} = -\left(\frac{R_3 V_1}{R_1} + \frac{R_3 V_2}{R_2}\right)$$

$$V_{OUT} = -R_3 \left(\frac{V_1}{R_1} + \frac{V_2}{R_2}\right)$$

$$-R_3 = \frac{V_{OUT}}{(V_1/R_1) + (V_2/R_2)}$$

$$-R_3 = \frac{-32}{(8/32) + (12/48)} \text{ (all resistances in k}\Omega\text{)}$$

$$-R_3 = -64$$

$$R_3 = 64 \text{ k}\Omega$$

```
10 PRINT " PROG 243"
20 PRINT "SUMMING AMPLIFIER"
30 INPUT "ENTER R1" ; R1
40 INPUT "ENTER R2" ; R2
50 INPUT "ENTER V1" ; V1
60 INPUT "ENTER V2" ; V2
70 INPUT "ENTER VOUT" ; VOUT
80 LET R3=-VOUT/((V1/R1)+(V2/R2))
90 PRINT "R3=" ; R3
```

Example 10 A two-input summing amplifier identical to Figure 16.3(a) has the following parameters: $R_1 = 8$ kΩ, $R_3 = 60$ kΩ, $V_1 = 2$ V, $V_2 = 10$ v, $V_{OUT} = -30$ V. Calculate the value of R_2.

$$V_{OUT} = -\left(\frac{R_3 V_1}{R_1} + \frac{R_3 V_2}{R_2}\right)$$

$$V_{OUT} + \frac{R_3 V_1}{R_1} = -\frac{R_3 V_2}{R_2}$$

$$R_2 = \frac{-R_3 V_2}{V_{OUT} + (R_3 V_1/R_1)}$$

$$= \frac{-60 \times 10}{-30 + (60 \times 2/8)} = \frac{-600}{-15} = 40 \text{ k}\Omega$$

```
10 PRINT "PROG 244"
20 PRINT "SUMMING AMPLIFIER"
30 INPUT "ENTER R1" ; R1
40 INPUT "ENTER R3" ; R3
50 INPUT "ENTER V1" ; V1
60 INPUT "ENTER V2" ; V2
70 INPUT "ENTER VOUT" ; VOUT
80 LET R2=-(R3*V2)/(VOUT+(R3*V1/R1))
90 PRINT "R2=" ;R2
```

Example 11 A three-input summing amplifier identical to Figure 16.3(b) has the following parameters: $R_1 = 20$ kΩ, $R_2 = 16$ kΩ, $R_3 = 15$ kΩ, $R_4 = 24$ kΩ, $V_1 = 6$ V, $V_2 = 4$ V, $V_3 = 10$ V. Calculate the output voltage.

$$V_{OUT} = -\left(\frac{R_4 V_1}{R_1} + \frac{R_4 V_2}{R_2} + \frac{R_4 V_3}{R_3}\right)$$

$$= -\left(\frac{24 \times 6}{20} + \frac{24 \times 4}{16} + \frac{24 \times 10}{15}\right)$$

$$= -(7.2 + 6 + 16)$$

$$= -29.2 \text{ V}$$

```
10 PRINT "PROG 245"
20 PRINT "SUMMING AMPLIFIER"
30 INPUT "ENTER R1" ; R1
40 INPUT "ENTER R2" ; R2
50 INPUT "ENTER R3" ; R3
60 INPUT "ENTER R4" ; R4
70 INPUT "ENTER V1" ; V1
80 INPUT "ENTER V2" ; V2
90 INPUT "ENTER V3" ; V3
100 LET VOUT=-((R4*V1/R1)+(R4*V2/R2)+(R4*V3/R3))
110 PRINT "VOUT=" ; VOUT
```

Example 12 If the circuit in Figure 16.3(b) has the parameters $R_1 = 10$ kΩ, $R_3 = 15$ kΩ, $R_4 = 20$ kΩ, $V_1 = 2$ V, $V_2 = 1$ V, $V_3 = 3$ V, $V_{OUT} = -12$ V, find the value of R_2

$$V_{OUT} = -\left(\frac{R_4 V_1}{R_1} + \frac{R_4 V_2}{R_2} + \frac{R_4 V_3}{R_3}\right)$$

$$V_{OUT} + \frac{R_4 V_1}{R_1} + \frac{R_4 V_3}{R_3} = -\frac{R_4 V_2}{R_2}$$

$$R_2 = \frac{-R_4 V_2}{V_{OUT} + (R_4 V_1 / R_1) + (R_4 V_3 / R_3)}$$

$$= \frac{-20 \times 1}{-12 + (20 \times 2/10) + (20 \times 3/15)}$$

$$= \frac{-20}{-4} = 5 \text{ k}\Omega$$

```
10 PRINT "PROG 246"
20 PRINT "SUMMING AMPLIFIER"
30 INPUT "ENTER R1" ; R1
40 INPUT "ENTER R3" ; R3
50 INPUT "ENTER R4" ; R4
60 INPUT "ENTER VOUT" ; VOUT
70 INPUT "ENTER V1" ; V1
80 INPUT "ENTER V2" ; V2
90 INPUT "ENTER V3" ; V3
100 LET R2=-(R4*V2)/(VOUT+(R4*V1/R1)+(R4*V3/R3))
110 PRINT "R2=" ; R2
```

16.4 Subtracting amplifiers

Figure 16.4 shows a differential amplifier which behaves as a subtractor if $R_1 = R_1$ and $R_2 = R_2$. The gain of the amplifier is equal to R_2/R_1 and the output voltage is proportional to the difference between the two input voltages, i.e.

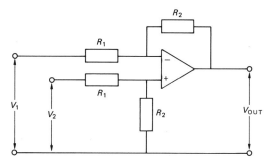

Figure 16.4

$$V_{OUT} = \frac{R_2}{R_1}(V_2 - V_1) = A_v(V_2 - V_1)$$

Example 13 A differential amplifier identical to Figure 16.4 has a gain of 12. The input signals are 3.8 and 3.4 V. Find the output voltage.

$$V_{OUT} = 12(3.8 - 3.4) = 4.8 \text{ V}$$

```
10 PRINT "PROG 247"
20 PRINT "SUBTRACTING AMPLIFIER"
30 INPUT "ENTER AV" ; AV
40 INPUT "ENTER V2" ; V2
50 INPUT "ENTER V1" ; V1
60 LET VOUT=AV*(V2-V1)
70 PRINT "VOUT=" ; VOUT
```

Example 14 A differential amplifier identical to Figure 16.4 has a gain of 10. If the output voltage is 3.5 V and V_1 is 6 V, find the value of V_2.

$$V_{OUT} = A_v(V_2 - V_1)$$

$$\frac{V_{OUT}}{A_v} = V_2 - V_1$$

$$V_2 = \frac{V_{OUT}}{A_v} + V_1 = \frac{3.5}{10} + 6 = 6.35 \text{ V}$$

```
10 PRINT "PROG 248"
20 PRINT "SUBTRACTING AMPLIFIER"
30 INPUT "ENTER AV" ; AV
40 INPUT "ENTER VOUT" ; VOUT
50 INPUT "ENTER V1" ; V1
60 LET V2=(VOUT/AV)+V1
70 PRINT "V2=" ; V2
```

Problems

1 In an inverting amplifier identical to Figure 16.2(a) the feedback resistance R_2 is 47 kΩ and R_1 is 2.7 kΩ. Calculate the gain A_v.

2   A non-inverting amplifier identical to Figure 16.2(b) consists of $R_1 = 12$ kΩ and $R_2 = 33$ kΩ. Calculate the gain.
3   A non-inverting amplifier identical to Figure 16.2(b) has a gain of 11. If R_1 is 4700 Ω, find the value of R_2.
4   A non-inverting amplifier identical to Figure 16.2(b) has a feedback resistor R_2 of 270 Ω. If R_1 is 1350 Ω, calculate the gain.
5   A summing amplifier identical to Figure 16.3(a) has the following components: $R_1 = 6$ kΩ, $R_2 = 8$ kΩ, $R_3 = 24$ kΩ. If $V_1 = 1.5$ V and $V_2 = 3$ V, calculate the value of the output voltage.
6   A two-input summing amplifier identical to Figure 16.3(a) has the following parameters: $R_1 = 25$ kΩ, $R_2 = 30$ kΩ, $V_1 = 5$ V, $V_2 = 15$ V, $V_{OUT} = -14$ V. Calculate the value of R_3.
7   A three-input summing amplifier identical to Figure 16.3(b) has the following parameters: $R_1 = 100$ kΩ, $R_2 = 50$ kΩ, $R_3 = 20$ kΩ and $R_4 = 30$ kΩ. If all the input voltages are 2 V, calculate the output voltage.
8   If the circuit in Figure 16.3(b) has the parameters $R_1 = 10$kΩ, $R_3 = 2$ kΩ, $R_4 = 6$ kΩ, $V_1 = 3$ V, $V_2 = 5$ V, $V_3 = 1.4$ V, $V_{OUT} = -20$ V, find the value of V_2.
9   A differential amplifier, identical to Figure 16.4, has a gain of 7.8. The input signals are 6.75 and 6.25 V. Find the output voltage.
10 A differential amplifier, identical to Figure 16.4, has a gain of 8.48. If the output voltage is 2.12 V and V_1 is 4.6 V, find the value of V_2.

17 Oscillators

17.1 Introduction

Oscillator circuits generate alternating voltages. Circuits such as the Hartley (Figure 17.1) and the Colpitts (Figure 17.2) generate sinusoidal waveforms. Other circuits, such as the R–C phase shift oscillator, the Wien bridge oscillator and the twin-T oscillator, also generate sine waves.

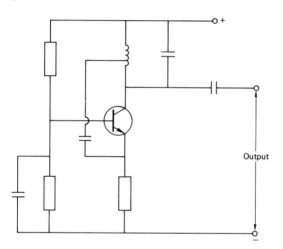

Figure 17.1

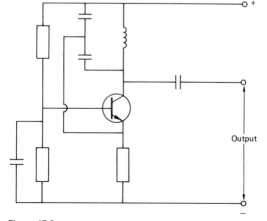

Figure 17.2

Relaxation oscillators generate waveforms that are square, sawtooth and triangular in shape.

A sinusoidal oscillator comprising inductance, capacity and resistance in series has a resonant frequency given by

$$f_R = \frac{1}{2\pi\sqrt{(LC)}}$$

This is the frequency at which the applied voltage produces the maximum output amplitude. Resonance was introduced in Chapter 11.

A parallel circuit comprising inductance, capacity and resistance has a natural frequency given by

$$f_N = \frac{1}{2\pi}\sqrt{\frac{1}{LC} - \frac{R^2}{4L^2}} \quad (R \text{ in series with } L)$$

The natural frequency is the frequency of free oscillations, i.e. the frequency at which the circuit will oscillate if left to do so. If the resistance value is small the simple resonant formula can be used for the natural frequency.

In all formulas capacitance is expressed in farads, and inductance in henrys.

Example 1 Find the natural frequency f_N of a circuit where $C = 0.05\ \mu F$, $L = 2\ \mu H$, and $R = 1\ \Omega$.

$$f_N = \frac{1}{2\pi}\sqrt{\left(\frac{1}{LC} - \frac{R_2}{4L^2}\right)}$$

Since

$$\frac{1}{L} = \frac{1}{2 \times 10^{-6}} = \frac{10^6}{2} \text{ and } \frac{1}{L^2} = \frac{10^{12}}{4}$$

and

$$\frac{1}{C} = \frac{1}{0.05 \times 10^{-6}} = \frac{10^6}{0.05}$$

then

$$f_N = \frac{1}{2\pi}\sqrt{\left(\frac{10^6 \times 10^6}{2 \times 0.05} - \frac{10^{12}}{4 \times 4}\right)}$$

$$= \frac{3\ 152\ 380}{2\pi}$$

$$= 501\ 717\ Hz = 501.717\ kHz$$

```
10 PRINT "PROG 249"
20 PRINT "NATURAL FREQUENCY"
30 INPUT "ENTER CAPACITANCE IN MICROFARADS" ; C
40 INPUT "ENTER INDUCTANCE IN MICROHENRYS" ; L
50 INPUT "ENTER RESISTANCE IN OHMS" ; R
60 LET A=1/(2*3.141593)
70 LET B=10^12/(L*C)
80 LET D=R^2*10^12/(4*L^2)
90 LET F=A*SQR(B-D)
100 PRINT "NATURAL FREQUENCY = " F "HERTZ"
```

Oscillators 297

Example 2 Calculate the frequency f_R for the circuit in Example 1.

$$f_R = \frac{1}{2\pi\sqrt{(LC)}}$$

$$f_R^2 = \frac{1}{4\pi^2 LC} = \frac{10^6 \times 10^6}{4 \times \pi^2 \times 2 \times 0.05} = 2.53 \times 10^{11}$$

$$f_R = \sqrt{(2.53 \times 10^{11})}$$

$$= 503\,292 \text{ Hz} = 503.292 \text{ kHz}$$

These first two examples show that the difference between the two frequencies is quite small since

$$\frac{R^2}{4L^2} \ll \frac{1}{LC}$$

```
10 PRINT "PROG 2 50"
20 PRINT "RESONANT FREQUENCY"
30 INPUT "ENTER CAPACITANCE IN MICROFARADS" ; C
40 INPUT "ENTER INDUCTANCE IN MICROHENRYS" ; L
50 LET A=10^12/(4*3.141593^2*L*C)
60 LET F=SQR(A)
70 PRINT "RESONANT FREQUENCY = " F "HERTZ"
```

Example 3 Find the periodic time of an oscillator having $C = 10$ μF and $L = 2$ H.

Since periodic time t (s) $= 1/f$ (see Chapter 6) then

$$t = \frac{1}{\dfrac{1}{2\pi\sqrt{(LC)}}} = 2\pi\sqrt{(LC)}$$

$$= 2\pi\sqrt{\left(\frac{2 \times 10}{10^6}\right)} = 0.028 \text{ s}$$

$$= 0.028 \times 10^3 = 28 \text{ ms}$$

This means that the resonant frequency is

$$f_R = \frac{1}{t} = \frac{1}{0.028} = 35.6 \text{ Hz}$$

```
10 PRINT "PROG 251"
20 PRINT "PERIODIC TIME"
30 INPUT "ENTER CAPACITANCE IN MICROFARADS" ; C
40 INPUT "ENTER INDUCTANCE IN HENRYS" : L
50 LET T=2*3.141593*SQR(L*C/10^6)
60 PRINT "PERIODIC TIME = " T*10^3 "MILLISECS"
```

Example 4 An oscillator is required to produce a 1 kHz sine wave. If the value of C is 0.02 μF, calculate the value of L.

$$f_R^2 = \frac{1}{4\pi^2 LC}$$

$$L = \frac{1}{f_R^2 \times 4 \times \pi^2 \times C}$$

$$= \frac{10^6}{(1000)^2 \times 4 \times \pi^2 \times 0.02}$$

$$= 1.27 \text{ H}$$

```
10 PRINT "PROG 252"
20 PRINT "INDUCTANCE"
30 INPUT "ENTER CAPACITANCE IN MICROFARADS" ; C
40 INPUT "ENTER FREQUENCY" ; F
50 LET L=10^6/(F^2*4*3.141593^2*C)
60 PRINT "INDUCTANCE = " L "HENRYS"
```

Example 5 Given that an oscillator has a resonant frequency of 1 MHz, and that $L = 20$ μH, calculate the value of C.

Since

$$f_R^2 = \frac{1}{4\pi^2 LC}$$

then

$$\begin{aligned}
C &= \frac{1}{f_R^2 \times 4 \times \pi^2 \times L} \text{ F} \\
&= \frac{10^6 \times 10^{12}}{(10^6)^2 \times 4 \times \pi^2 \times 20} \text{ pF} \\
&= 1266.5 \text{ pF} \\
&= 1266.5 \times 10^{-3} \text{ nF} \\
&= 1.27 \text{ nF}
\end{aligned}$$

```
10 PRINT "PROG 253"
20 PRINT "CAPACITANCE"
30 INPUT "ENTER INDUCTANCE IN MICROHENRYS" ; L
40 INPUT "ENTER FREQUENCY" ; F
50 LET C=10^15/(F^2*4*3.141593^2*L)
60 PRINT "CAPACITANCE = " C "NANOFARADS"
```

17.2 R–C phase shift oscillator

A typical circuit is shown in Figure 17.3. The frequency of oscillation f_o is given by the formula

$$f_o = \frac{1}{2\pi CR\sqrt{6}}$$

when

$$C = C_1 = C_2 = C_3 \text{ F}$$

and

$$R = R_1 = R_2 = R_{IN} = RL \text{ } \Omega$$

R_{IN} is the input resistance of the transistor TR1. The required conditions for oscillation are met by:

1 A 180° phase change in the R–C network
2 A 180° phase change in the transistor
3 A gain > 29 in the transistor

Example 6 In Figure 17.3 if $C = 0.1$ μF and $R = 2000$ Ω, calculate the frequency of oscillation, f_o.

$$f_o = \frac{1}{2 \times \pi \times 0.1 \times 10^{-6} \times 2000 \times \sqrt{6}}$$

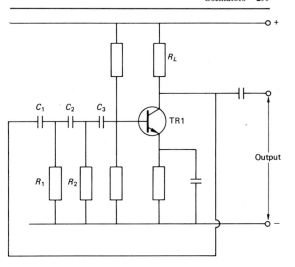

Figure 17.3

$$= \frac{10^6}{2 \times \pi \times 0.1 \times 2000 \times \sqrt{6}}$$

$$= 325 \text{ Hz}$$

```
10 PRINT "PROG 254"
20 PRINT "PHASE SHIFT OSCILLATOR"
30 INPUT "ENTER CAPACITANCE IN MICROFARADS" ; C
40 INPUT "ENTER RESISTANCE" ; R
50 LET F=10^6/(2*3.141593*C*R*SQR(6))
60 PRINT "OSCILLATION FREQUENCY = " F "HERTZ"
```

Example 7 If an R–C oscillator resonates at 65 Hz, with all the capacitors equal to 1 μF, calculate the value of R.

$$R = \frac{1}{2\pi C f \sqrt{6}} \ \Omega$$

$$= \frac{10^6}{2 \times \pi \times 1 \times 65 \times \sqrt{6}}$$

$$= 1000 \ \Omega$$

```
10 PRINT "PROG 255"
20 PRINT "PHASE SHIFT OSCILLATOR"
30 INPUT "ENTER CAPACITANCE IN MICROFARADS" ; C
40 INPUT "ENTER FREQUENCY" ; F
50 LET R=10^6/(2*3.141593*C*F*SQR(6))
60 PRINT "RESISTANCE = " R "OHMS"
```

Example 8 If the frequency of oscillation is 6912 Hz and the resistances equal 4.7 kΩ, calculate the value of C in nanofarads.

$$C = \frac{1}{2\pi f R \sqrt{6}} \ \text{F}$$

$$= \frac{10^9}{2 \times \pi \times 6912 \times 4700 \times \sqrt{6}} \text{ nF}$$

$$= 2 \text{ nF}$$

```
10 PRINT "PROG 256"
20 PRINT "PHASE SHIFT OSCILLATOR"
30 INPUT "ENTER RESISTANCE" ; R
40 INPUT "ENTER FREQUENCY" ; F
50 LET C=10^9/(2*3.141593*F*R*SQR(6))
60 PRINT "CAPACITANCE = " C "NANOFARADS"
```

17.3 Wien bridge oscillator

A typical circuit is shown in Figure 17.4 using an operational amplifier. The frequency of oscillation f_o is given by the formula

$$f_o = \frac{1}{2\pi RC}$$

where R is in ohms
C is in farads

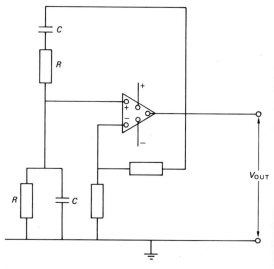

Figure 17.4

Example 9 Calculate the frequency of oscillation in the circuit of Figure 17.4 if the resistance R is 120 Ω and the capacitor C is 10 μF.

$$f_o = \frac{10^6}{2 \times \pi \times 120 \times 10}$$

$$= 133 \text{ Hz}$$

```
10 PRINT "PROG 257"
20 PRINT "WIEN BRIDGE OSCILLATOR"
30 INPUT "ENTER CAPACITANCE IN MICROFARADS" ; C
40 INPUT "ENTER RESISTANCE" ; R
50 LET F=10^6/(2*3.141593*R*C)
60 PRINT "OSCILLATION FREQUENCY = " F "HERTZ"
```

Example 10 A Wien bridge oscillator oscillates at a frequency of 3928 Hz. If the resistance is 270 Ω, calculate the value of the capacitance in microfarads.

$$C = \frac{10^6}{2 \times \pi \times 270 \times 3928}$$

$$= 0.15 \ \mu F$$

```
10 PRINT "PROG 258"
20 PRINT "WIEN BRIDGE OSCILLATOR"
30 INPUT "ENTER FREQUENCY" ; F
40 INPUT "ENTER RESISTANCE" ; R
50 LET C=10^6/(2*3.141593*R*F)
60 PRINT "CAPACITANCE = " C "MICROFARADS"
```

Example 11 In a Wien bridge oscillator, if the frequency is 500 Hz and the capacitance is 3.183 μF, calculate the resistance.

$$R = \frac{10^6}{2 \times \pi \times 3.183 \times 500}$$

$$= 100 \ \Omega$$

```
10 PRINT "PROG 259"
20 PRINT "WIEN BRIDGE OSCILLATOR"
30 INPUT "ENTER FREQUENCY" ; F
40 INPUT "ENTER CAPACITANCE IN MICROFARADS" ; C
50 LET R=10^6/(2*3.141593*C*F)
60 PRINT "RESISTANCE = " R "OHMS"
```

17.4 Twin-T oscillator

This oscillator is shown in Figure 17.5. The circuit is similar to the Wien bridge oscillator, and the frequency of oscillation f_o is given by

$$f_o = \frac{1}{2\pi RC}$$

where C is in farads
R is in ohms

Example 12 In a twin-T oscillator, if the capacitance is 2 nF, and the resistance 2 MΩ, calculate the frequency.

$$f_o = \frac{10^9}{2 \times \pi \times 2 \times 10^6 \times 2}$$

$$= 40 \ Hz$$

```
10 PRINT "PROG 260"
20 PRINT "TWIN-T OSCILLATOR"
30 INPUT "ENTER CAPACITANCE IN NANOFARADS" ; C
40 INPUT "ENTER RESISTANCE" ; R
50 LET F=10^9/(2*3.141593*R*C)
60 PRINT "OSCILLATION FREQUENCY = " F "HERTZ"
```

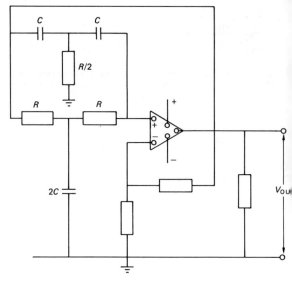

Figure 17.5

17.5 Non-sinusoidal oscillators

Multivibrators

Figure 17.6 shows a typical circuit which generates square waves. The period for one cycle is represented by

$$t = 0.7C_1R_2 + 0.7C_2R_1 \text{ s } (C \text{ in farads})$$
$$= 0.7(C_1R_2 + C_2R_1)$$

If

$$C = C_1 = C_2 \text{ and } R = R_1 = R_2$$

then

$$t = 0.7 \times 2 \times CR$$

since

$$f_o = \frac{1}{t}$$

then

$$f_o = \frac{1}{0.7(C_1R_2 + C_2R_1)} \text{ Hz}$$
$$= \frac{1}{0.7 \times 2 \times CR} \text{ Hz}$$

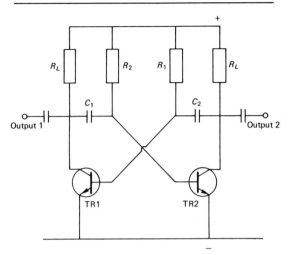

Figure 17.6

Example 13 Calculate the frequency of a multivibrator when $C = C_1 = C_2 = 0.01$ μF and $R = R_1 = R_2 = 15$ kΩ.

$$f_o = \frac{10^6}{0.7 \times 2 \times 0.01 \times 15000}$$

$$= 4762 \text{ Hz}$$

```
10 PRINT "PROG 261"
20 PRINT "MULTIVIBRATOR"
30 INPUT "ENTER CAPACITANCE IN MICROFARADS" ; C
40 INPUT "ENTER RESISTANCE" ; R
50 LET F=10^6/(.7*2*C*R)
60 PRINT "OSCILLATION FREQUENCY = " F "HERTZ"
```

Example 14 If a multivibrator is running at 50 Hz and $C = C_1 = C_2 = 14.286$ μF, calculate the value of R where $R = R_1 = R_2$.

$$R = \frac{10^6}{0.7 \times 2 \times 14.286 \times 50}$$

$$= 1000 \text{ Ω}$$

```
10 PRINT "PROG 262"
20 PRINT "MULTIVIBRATOR"
30 INPUT "ENTER CAPACITANCE IN MICROFARADS" ; C
40 INPUT "ENTER FREQUENCY" ;F
50 LET R=10^6/(.7*2*C*F)
60 PRINT "RESISTANCE = " R "OHMS"
```

Example 15 If the value of R in Example 14 is 500 Ω and the frequency is 952 Hz, calculate the value of C in nanofarads.

$$C = \frac{10^9}{0.7 \times 2 \times 500 \times 952}$$

$$= 1500 \text{ nF}$$

```
10 PRINT "PROG 263"
20 PRINT "MULTIVIBRATOR"
30 INPUT "ENTER RESISTANCE" ; R
40 INPUT "ENTER FREQUENCY" ; F
50 LET C=10^9/(.7*2*R*F)
60 PRINT "CAPACITANCE = " C "NANOFARADS
```

Example 16 A multivibrator has the following circuit components: $R_1 = 10$ kΩ, $R_2 = 20$ kΩ, $C_1 = 0.5$ μF, $C_2 = 0.1$ μF. Calculate the frequency of oscillation.

$$f = \frac{1}{0.7(C_1 R_2 + C_2 R_1)} \text{ Hz}$$

$$C_1 R_2 = 0.5 \times 10^{-6} \times 20 \times 10^3 = 0.01$$

$$C_2 R_1 = 0.1 \times 10^{-6} \times 10 \times 10^3 = 0.001$$

$$f_o = \frac{1}{0.7 \times (0.01 + 0.001)}$$

$$= 130 \text{ Hz}$$

```
10 PRINT "PROG 264"
20 PRINT "MULTIVIBRATOR"
30 INPUT "ENTER C1 IN MICROFARADS" ; C1
40 INPUT "ENTER R1" ; R1
50 INPUT "ENTER C2 IN MICROFARADS" ; C2
60 INPUT "ENTER R2" ; R2
70 LET X=(C1*R2/10^6)+(C2*R1/10^6)
80 LET F=1/(.7*X)
90 PRINT "FREQUENCY = " F "HERTZ"
```

Example 17 A multivibrator runs at a frequency of 159 Hz. The component values are $C_1 = 2$ μF, $R_1 = 1$ kΩ, $C_2 = 5$ μF. Calculate the value of R_2.

$$f_o = \frac{1}{0.7 C_1 R_2 + 0.7 C_2 R_1}$$

$$0.7 f_o C_1 R_2 + 0.7 f_o C_2 R_1 = 1$$

$$0.7 f_o C_1 R_2 = 1 - 0.7 f_o C_2 R_1$$

$$R_2 = \frac{1 - 0.7 f_o C_2 R_1}{0.7 f_o C_1}$$

$$0.7 \times 159 \times 5 \times 10^{-6} \times 10^3 = 0.5565$$

$$0.7 \times 159 \times 2 \times 10^{-6} = 2.226 \times 10^{-4}$$

$$R_2 = \frac{(1 - 0.5565) \times 10^4}{2.226}$$

$$= 1992 \text{ Ω}$$

```
10 PRINT "PROG 265"
20 PRINT "MULTIVIBRATOR"
30 INPUT "ENTER C1 IN MICROFARADS" ; C1
40 INPUT "ENTER R1" ; R1
50 INPUT "ENTER C2 IN MICROFARADS" ; C2
60 INPUT "ENTER FREQUENCY" ; F
70 LET X=.7*F*R1*C2/10^6
80 LET Y=.7*F*C1/10^6
90 LET R2=(1-X)/Y
100 PRINT "RESISTANCE R2 = " R2 "OHMS"
```

Example 18 Calculate the value of C_1 for a multivibrator having $R_1 = 50\ \Omega$, $R_2 = 200\ \Omega$, $C_2 = 0.2\ \mu F$, $f_o = 12\ 987$ Hz.

$$0.7 f_o C_1 R_2 = 1 - 0.7 f_o C_2 R_1$$

$$C_1 = \frac{1 - 0.7 f_o C_2 R_1}{0.7 f_o R_2}$$

$$0.7 \times 12\ 987 \times 0.2 \times 10^{-6} \times 50 = 0.09$$

$$0.7 \times 12\ 987 \times 200 = 1\ 818\ 180$$

$$C_1 = \frac{(1 - 0.09) \times 10^6}{1\ 818\ 180}\ \mu F$$

$$= 0.5\ \mu F$$

```
10 PRINT "PROG 266"
20 PRINT "MULTIVIBRATOR"
30 INPUT "ENTER R1" ; R1
40 INPUT "ENTER R2" ; R2
50 INPUT "ENTER C2 IN MICROFARADS" ; C2
60 INPUT "ENTER FREQUENCY" ; F
70 LET X=.7*F*R1*C2/10^6
80 LET Y=.7*F*R2
90 LET C1=(1-X)/Y
100 PRINT "CAPACITANCE C1 = " C1*10^6 "MICROFARADS"
```

17.6 Sawtooth waveform

A sawtooth waveform is used extensively in time base circuits in oscilloscopes, in television sets, in radar sets and in many other applications. A sawtooth can be generated by charging and discharging a capacitor at constant current.

Figure 17.7 shows an example of a circuit that could be added to the output of the multivibrator of Figure 17.6. C_1 will charge through R_1, and discharge through D1 and TR2 to produce a sawtooth wave form across C_1.

17.7 Triangular waveform

When a square wave drives an operational amplifier such as the one shown in Figure 17.8 the output is a triangular wave. A suitable square wave input source could be a multivibrator, or an R–C oscillator followed by a Schmitt trigger. The relationship between the input and output voltages is given by

$$V_{out}\ (p/p) = \frac{V_{in}\ (p/p)}{4 f_o RC}\ \text{where } C \text{ is in farads}$$

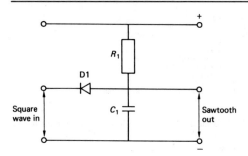

Figure 17.7

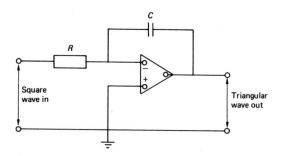

Figure 17.8

Rearranging the formula

$$f_o = \frac{V_{in}}{4V_{out}RC} \quad (f_o = \text{frequency of input square wave})$$

Example 19 Calculate the frequency of a triangular waveform generator given that $V_{in} = 10$ V, $V_{out} = 2$ V, $R = 10$ kΩ, $C = 0.5$ μF.

$$f = \frac{10 \times 10^6}{4 \times 2 \times 10 \times 10^3 \times 0.5}$$

$$= 250 \text{ Hz (the frequency of the input square wave)}$$

```
10 PRINT "PROG 267"
20 PRINT "TRIANGULAR WAVEFORM"
30 INPUT "ENTER INPUT VOLTAGE" ; VIN
40 INPUT "ENTER OUTPUT VOLTAGE" ; VOUT
50 INPUT "ENTER RESISTANCE R" ; R
60 INPUT "ENTER CAPACITANCE IN MICROFARADS" ; C
70 LET F=VIN*10^6/(4*VOUT*R*C)
80 PRINT "FREQUENCY = " F "HERTZ"
```

Oscillators 307

Example 20 If the input voltage in Example 19 is 6 V, calculate the peak to peak output voltage.

$$V_{out} = \frac{6 \times 10^6}{4 \times 250 \times 10 \times 10^3 \times 0.5}$$

$$= 1.2 \text{ V}$$

This output voltage is obtained therefore with a 6-V input at 250 Hz.

```
10 PRINT "PROG 268"
20 PRINT "TRIANGULAR WAVEFORM"
30 INPUT "ENTER INPUT VOLTAGE" ; VIN
40 INPUT "ENTER FREQUENCY" ; F
50 INPUT "ENTER RESISTANCE R" ; R
60 INPUT "ENTER CAPACITANCE IN MICROFARADS" ; C
70 LET VOUT=VIN*10^6/(4*F*R*C)
80 PRINT "OUTPUT VOLTAGE = " VOUT "VOLTS"
```

Problems

1. A circuit consists of a 0.08 µF capacitor in parallel with a 15-µH inductor having a resistance of 2 Ω. Calculate the natural frequency.
2. A series circuit consists of a 10-µF capacitor and a 10-µH inductor. Calculate the resonant frequency.
3. Find the periodic time of an oscillator having a capacitance of 24 µF and an inductance of 1.0 H. Express the answer in milliseconds.
4. A sinusoidal oscillator resonates at a frequency of 2500 Hz. If the capacitor value is 0.05 µF, find the value of the inductor in henrys.
5. Given that an oscillator has a resonant frequency of 100 kHz, and that the inductance is 150 µH, calculate the value of the capacitor in picofarads.
6. A phase shift oscillator has a value of 1.6 µF for C, and 1500 Ω for R. Calculate the resonant frequency.
7. If, in Problem 6, C is 0.1 µF and the resonant frequency is 100 Hz, find R.
8. If, in Problem 6, the resistance is 5000 Ω, and the frequency is 5000 Hz, calculate the value of C in nanofarads.
9. A Wien bridge oscillator uses a 4.2 µF capacitor and a 47 Ω resistance. Calculate the frequency of oscillation.
10. If a Wien bridge oscillator resonates at 2700 Hz and the resistance value is 330 Ω, calculate the value of the capacitor in microfarads.
11. Calculate the resistance value in a Wien bridge oscillator that has a frequency of 50 Hz and a 1.45-µF capacitor.
12. A twin-T oscillator consists of a 5-nF capacitor and a 68-kΩ resistor. Calculate the resonant frequency.
13. A multivibrator has $C = C_1 = C_2 = 0.0012$ µF and $R = R_1 = R_2 = 27$ kΩ. Calculate the resonant frequency.
14. If a multivibrator is running at 100 Hz and $C = C_1 = C_2 = 3.5$ µF, calculate the value of R given that $R = R_1 = R_2$.
15. If the value of R in Problem 14 is 560 Ω and the frequency is 1800 Hz, find the value of the capacitance.
16. A multivibrator has the following circuit components:

$R_1 = 15$ kΩ, $R_2 = 20$ kΩ, $C_1 = 1.0$ μF, $C_2 = 2$ μF. Calculate the resonant frequency.

17 A multivibrator runs at a frequency of 180 Hz. If $C_1 = 1.8$ μF, $C_2 = 4.5$ μF and $R_1 = 910$ Ω, calculate the value of R_2.

18 Calculate the value of C_1 for a multivibrator having $R_1 = 100$ Ω, $R_2 = 100$ Ω, $C_2 = 1$ μF, $f_o = 2000$ Hz.

19 Calculate the input frequency for the circuit in Figure 17.8 given $V_{IN} = 12$, $V_{OUT} = 4$, $R = 12$ kΩ, $C = 0.2$ μF.

20 If the input voltage to the circuit in Figure 17.8 is 10 V, the frequency 300 Hz, the resistance 10 000 Ω, and the capacitance 0.3 μF, calculate the output voltage.

18 Filters and attenuators

18.1 Introduction to filters

A network which attenuates certain frequencies but passes others without loss is called a filter. A filter therefore possesses at least one 'pass band', i.e. a band of frequencies where the attenuation is zero, and at least one 'attenuation band', where the frequencies have finite attenuation.

The frequencies that separate the various 'pass' and 'attenuation' bands are called 'cut-off' frequencies, usually denoted by f_1, f_2, etc. or by f_c if there is only one 'cut-off' frequency.

All filters are constructed from reactive elements; otherwise the attenuation would never become zero.

18.2 The decibel

The 'pass' and 'attenuated' bands in filters and amplifiers are often illustrated graphically. These frequency responses are plotted with the frequency parameter along the x axis and the output parameter along the y axis, using the decibel system. A graph will show therefore the gain or loss in an electronic network over a given range of frequencies.

It has always been a requirement to measure the gain or loss of electronic networks, and a unit called the decibel was introduced for this purpose.

Consider the network shown in Figure 18.1, where P_i is the input power and P_o the output power. The power gain or loss in decibels (dB)

$$= 10 \log_{10} \frac{P_o}{P_i}$$

The decibel is a logarithmic unit. If P_o is less than P_i, the network has introduced a power loss and the answer will be negative. If P_o is

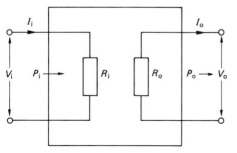

Figure 18.1

greater than P_i, the network has introduced a gain, as in an amplifier for example, and the answers will be positive.

The decibel is fundamentally a unit of power ratio, and the expressions for the voltage and current ratios can be derived from the expression for the power ratio.

Consider again Figure 18.1 where V_i is the input voltage and V_o the output voltage, and therefore

$$10 \log \frac{P_o}{P_i} = 10 \log \frac{V_o^2 R_o}{V_i^2 R_i} = 10 \log \frac{V_o^2}{V_i^2} \text{ (assuming } R_i = R_o)$$

$$= 10 \log \left(\frac{V_o}{V_i}\right)^2 = 2\left(10 \log \frac{V_o}{V_i}\right)$$

$$= 20 \log \frac{V_o}{V_i} = \text{voltage ratio in dB}$$

Note that the input resistance R_i is taken to be equal to the output resistance R_o.

I_i is the input current and I_o is the output current and therefore

$$10 \log \frac{P_o}{P_i} = 10 \log \frac{I_o^2 R_o}{I_i^2 R_i} = 10 \log \frac{I_o^2}{I_i^2}$$

$$= 10 \log \left(\frac{I_o}{I_i}\right)^2 = 2\left(10 \log \frac{I_o}{I_i}\right)$$

$$= 20 \log \frac{I_o}{I_i} = \text{current ratio in dB}$$

Example 1 If the output and input powers of an electronic network are 16 and 1.0 W respectively, calculate the gain in decibels.

$$\text{Power gain} = 10 \log \frac{16}{1} = 10 \times 1.2 = 12 \text{ dB}$$

```
10 PRINT "PROG 269"
20 PRINT "NETWORK GAIN IN DECIBELS"
30 INPUT "ENTER OUTPUT POWER IN WATTS" ; PO
40 INPUT "ENTER INPUT POWER IN WATTS" ; PI
50 LET N=PO/PI
60 LET G=10*LOG(N)/LOG(10)
70 PRINT "GAIN = " G "DBS"
```

Example 2 A network has a power gain of 18 dB. If the output power is 128 mW, calculate the input power.

```
10 PRINT "PROG 270"
20 PRINT "INPUT POWER"
30 INPUT "ENTER POWER GAIN IN DBS" ; G
40 INPUT "ENTER OUTPUT POWER IN WATTS" ; PO
50 LET N=G/10
60 LET PI=PO/(10^N)
70 PRINT "INPUT POWER = " PI "WATTS"
```

$$10 \log \frac{128}{P_i} = 18$$

$$\log \frac{128}{P_i} = \frac{18}{10} = 1.8$$

$$\frac{128}{P_i} = \text{antilog } 1.8 = 63.1$$

$$P_i = \frac{128}{63.1} = 2.03 \text{ mW}$$

Example 3 A network has an input of 8 W and an output of 2 W. Calculate the loss in decibels.

$$\text{Power loss} = 10 \log \frac{2}{8} = 10 \log 0.25$$

$$= -6 \text{ dB}$$

(See program 269, page 310.)

Example 4 A filter has a power gain of -40 dB. If the input power is 100 W, calculate the output power.

$$10 \log \frac{P_o}{100} = -40$$

$$\log \frac{P_o}{100} = \frac{-40}{10} = -4$$

$$\frac{P_o}{100} = \text{antilog } -4 = 0.0001$$

$$P_o = 100 \times 0.0001 = 0.01 \text{ W}$$

Note that the accuracy of the answer can be checked by verifying that

$$10 \log \frac{0.01}{100} = -40$$

```
10 PRINT "PROG 271"
20 PRINT "OUTPUT POWER"
30 INPUT "ENTER POWER GAIN IN DBS" ; G
40 INPUT "ENTER INPUT POWER IN WATTS" ; PI
50 LET N=G/10
60 LET PO=PI*10^N
70 PRINT "OUTPUT POWER = " PO "WATTS"
```

Example 5 A network has a voltage loss of 6 dB. If the input voltage is 10 V calculate the output voltage.

$$20 \log \frac{V_o}{10} = -6 \text{ dB}$$

$$\log \frac{V_o}{10} = -\frac{6}{20} = -0.3$$

$$\frac{V_o}{10} = \text{antilog } -0.3 = 0.5$$

$$V_o = 10 \times 0.5 = 5 \text{ V}$$

```
10 PRINT "PROG 272"
20 PRINT "OUTPUT VOLTAGE"
30 INPUT "ENTER VOLTAGE GAIN IN DBS" ; G
40 INPUT "ENTER INPUT VOLTAGE IN VOLTS" ; VI
50 LET N=G/20
60 LET VO=VI*10^N
70 PRINT "OUTPUT VOLTAGE = " VO "VOLTS"
```

Example 6 Measurements on a filter network showed that the input current was 256 µA and the output current was 2 µA. Calculate the loss in decibels.

$$20 \log \frac{2}{256} = 42 \text{ dB}$$

```
10 PRINT "PROG 273"
20 PRINT "NETWORK LOSS IN DBS"
30 INPUT "ENTER INPUT CURRENT IN MICROAMPS" ; II
40 INPUT "ENTER OUTPUT CURRENT IN MICROAMPS" ; IO
50 LET N=IO/II
60 LET LO=20*LOG(N)/LOG(10)
70 PRINT "LOSS = " LO "DBS"
```

Tables of decibel values for a range of output/input ratios exist. A few values are shown in Table 18.1.

Table 18.1   Decibel values

Output/input ratio	Power (dB)	Voltage and Current (dB)
16 : 1	+12	+24
8 : 1	+ 9	+18
4 : 1	+ 6	+12
2 : 1	+ 3	+ 6
1 : 1	0	0
1 : 2	− 3	− 6
1 : 4	− 6	−12
1 : 8	− 9	−18
1 : 16	−12	−24

Note that when the input has the same amplitude as the output, a network has no gain and no loss, i.e.

$$20 \log \frac{1}{1} = 0 \text{ (since the log of 1 is zero)}$$

Some of the advantages of using decibels in complex networks include:

1. Convenient numbers are obtained when the output/input ratios are large, e.g. a power gain of 1 000 000 = 10 log 10^6 = 60 dB.
2. If a number of networks are connected in series, and the gain of each individual network is expressed in decibels, the overall gain is the sum of the individual stages.

Example 7 If four networks connected in series have gains of 18 dB, −9 dB, +4 dB and −7 dB, what is the overall gain?

Overall gain = 18 − 9 + 4 − 7 = 6 dB

Note from Table 18.1 that a gain of 6 dB equals an overall gain of 2 for voltages and currents, and a gain of 4 for power.

```
10 PRINT "PROG 274"
20 PRINT "OVERALL GAIN IN DBS"
30 INPUT "ENTER N1" ; N1
40 INPUT "ENTER N2" ; N2
50 INPUT "ENTER N3" ; N3
60 INPUT "ENTER N4" ; N4
70 LET G=N1+N2+N3+N4
80 PRINT "OVERALL GAIN = " G "DBS"
```

Example 8 Figure 18.2 shows a two-stage network. The output of stage 1 forms the input of stage 2. Calculate the individual gains and the overall gain.

$$\text{Gain of stage 1} = \frac{V_{\text{OUT}}}{V_{\text{IN}}} = \frac{20}{10} = 2$$

$$\text{Gain of stage 2} = \frac{V_{\text{OUT}}}{V_{\text{IN}}} = \frac{80}{20} = 4$$

$$\text{Overall gain} = \frac{V_{\text{OUT}} \text{ (stage 2)}}{V_{\text{IN}} \text{ (stage 1)}} = \frac{80}{10} = 8$$

Note gain of stage 1 × gain of stage 2 = overall gain

Gain of stage 1 = 20 log 2 = 6 dB

Gain of stage 2 = 20 log 4 = 12 dB

Overall gain = 20 log 8 = 18 dB

Note gain of stage 1 + gain of stage 2 = overall gain.

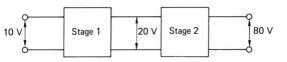

Figure 18.2

```
10 PRINT "PROG 275"
20 PRINT "OVERALL GAIN"
30 INPUT "ENTER V1" ; V1
40 INPUT "ENTER V2" ; V2
50 INPUT "ENTER V3" ; V3
60 LET N1=V2/V1
70 LET N2=V3/V2
80 LET N=N1*N2
90 LET G1=20*LOG(N1)/LOG(10)
100 LET G2=20*LOG(N2)/LOG(10)
110 LET G=G1+G2
120 PRINT "STAGE 1 GAIN = " N1
130 PRINT "STAGE 2 GAIN = " N2
140 PRINT "OVERALL GAIN = " N
150 PRINT "STAGE 1 GAIN = " G1 "DBS"
160 PRINT "STAGE 2 GAIN = " G2 "DBS"
170 PRINT "OVERALL GAIN = " G "DBS"
```

Example 9 Repeat Example 8 for the three-stage network shown in Figure 18.3.

$$\text{Gain of stage 1} = \frac{8}{1} = 8$$

$$\text{Gain of stage 2} = \frac{4}{8} = 0.5 \text{ (i.e. a loss)}$$

$$\text{Gain of stage 3} = \frac{32}{4} = 8$$

$$\text{Overall gain} = \frac{32}{1} = 32$$

Note $32 = 8 \times 0.5 \times 8$

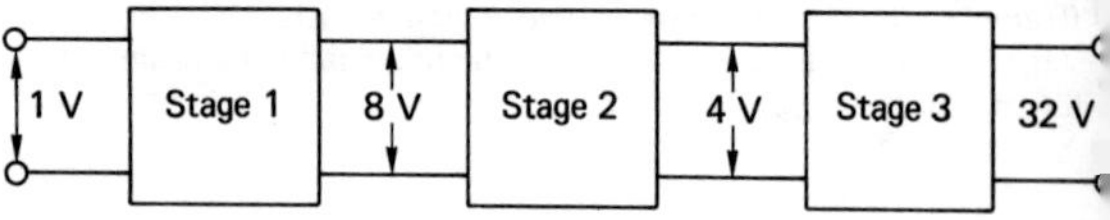

314 Circuit calculations pocket book

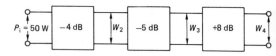

Figure 18.3

$$\text{Gain of stage 1} = 20 \log 8 = 18 \text{ dB}$$
$$\text{Gain of stage 2} = 20 \log 0.5 = -6 \text{ dB}$$
$$\text{Gain of stage 3} = 20 \log 8 = 18 \text{ dB}$$
$$\text{Overall gain} = 20 \log 32 = 30 \text{ dB}$$
$$\text{Note} \quad 30 = 18 - 6 + 18$$

```
10 PRINT "PROG 276"
20 PRINT "OVERALL GAIN"
30 INPUT "ENTER V1" ; V1
40 INPUT "ENTER V2" ; V2
50 INPUT "ENTER V3" ; V3
55 INPUT "ENTER V4" ; V4
60 LET N1=V2/V1
70 LET N2=V3/V2
75 LET N3=V4/V3
80 LET N=N1*N2*N3
90 LET G1=20*LOG(N1)/LOG(10)
100 LET G2=20*LOG(N2)/LOG(10)
105 LET G3=20*LOG(N3)/LOG(10)
110 LET G=G1+G2+G3
120 PRINT "STAGE 1 GAIN = " N1
130 PRINT "STAGE 2 GAIN = " N2
135 PRINT "STAGE 3 GAIN = " N3
140 PRINT "OVERALL GAIN = " N
150 PRINT "STAGE 1 GAIN = " G1 "DBS"
160 PRINT "STAGE 2 GAIN = " G2 "DBS"
165 PRINT "STAGE 3 GAIN = " G3 "DBS"
170 PRINT "OVERALL GAIN = " G "DBS"
```

Example 10 Find the value of W_2, W_3 and W_4 in Figure 18.4.

For stage 1

$$10 \log \frac{W_2}{50} = -4$$

$$\log \frac{W_2}{50} = \frac{-4}{10} = -0.4$$

$$\frac{W_2}{50} = \text{antilog} - 0.4 = 0.398$$

$$W_2 = 50 \times 0.398 = 19.9 \text{ W}$$

Figure 18.4

Filters and attenuators

For stage 2

$$10 \log \frac{W_3}{19.9} = -5$$

$$\log \frac{W_3}{19.9} = -\frac{5}{10} = -0.5$$

$$\frac{W_3}{19.9} = \text{antilog} - 0.5 = 0.316$$

$$W_3 = 19.9 \times 0.316 = 6.29 \text{ W}$$

For stage 3

$$10 \log \frac{W_4}{6.29} = 8$$

$$\log \frac{W_4}{6.29} = \frac{8}{10} = 0.8$$

$$\frac{W_4}{6.29} = \text{antilog } 0.8 = 6.3$$

$$W_4 = 6.29 \times 6.3 = 39.69 \text{ W}$$

Note that

$$\text{Overall gain} = 10 \log \frac{39.69}{50} = -1.0 \text{ dB}$$

which agrees with

$$-4 - 5 + 8 = -1 \text{ dB}$$

(See program 271, page 311.)

18.3 Low pass filters

A low pass filter transmits low frequencies and attenuates high frequencies. A typical frequency response is shown in Figure 18.5. At low frequencies the output voltage is constant, but as the frequency is increased a point will be reached when the output will begin to fall. The cut-off frequency, f_c, is where the output voltage

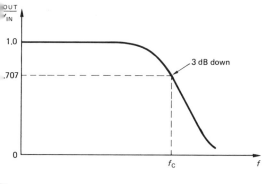

Figure 18.5

has fallen to 0.707 of the input voltage. This is the −3 dB point. If the input voltage is set at 1.0 V, then the gain at frequency

$$f_c = 20 \log \frac{0.707}{1} = -3.0 \text{ dB}$$

Figure 18.6 shows a simple low pass filter employing a resistor and a capacitor. The output voltage is derived across the capacitor.

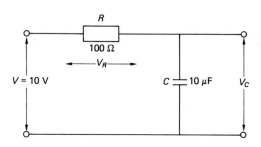

Figure 18.6

Example 11 Using the circuit shown in Figure 18.6, calculate: (i) the output voltage at the −3 dB point, (ii) the voltage across R at the −3 dB point, (iii) the cut-off frequency at the −3 dB point.

(i) Output voltage = $10 \times 0.707 = 7.07$ V

(ii) $V^2 = V_c^2 + V_R^2$ (see Section 11.6, Chapter 11)

$V_R^2 = V^2 - V_c^2 = 10^2 - 7.07^2$

$V_R = \sqrt{(10^2 - 7.07^2)} = 7.07$ V

(iii) since $V_R = V_C$

$IR = IX_C$

$R = X_C = \dfrac{1}{2\pi fC}$

$f = \dfrac{1}{2\pi RC} = \dfrac{1}{2 \times \pi \times 100 \times 10 \times 10^{-6}}$

$= 159 \text{ Hz} = f_c$

```
10 PRINT "PROG 277"
20 PRINT "LOW PASS FILTER"
30 INPUT "ENTER INPUT VOLTAGE" ; V
40 INPUT "ENTER RESISTOR VALUE" ; R
50 INPUT "ENTER CAPACITOR IN MICROFARADS" ; C
60 LET VO=.707*V
70 LET VR=SQR((V^2)-(VO^2))
80 LET F=10^6/(2*3.141593*R*C)
90 PRINT "OUTPUT VOLTAGE = " VO "VOLTS"
100 PRINT "VOLTAGE ACROSS R = " VR "VOLTS"
110 PRINT "CUT OFF FREQUENCY = " F "HERTZ"
```

Figures 18.7 and 18.8 shows T and π low pass filters respectively. These filters are more efficient and provide better attenuation than the R–C filter considered so far. These are constructed using

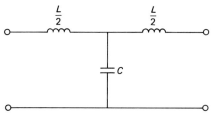

Figure 18.7

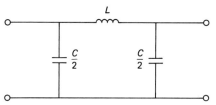

Figure 18.8

inductors and capacitors and the π section is used as a filter in power supply units. Both types are designed using the formulas

$$\text{Cut-off frequency } f_c = \frac{1}{\pi \sqrt{(LC)}}$$

$$\text{Design resistance } R_o = \sqrt{\frac{L}{C}}$$

$$R_o^2 = Z_1 Z_2$$

where Z_1 = reactance of the inductor

Z_2 = reactance of the capacitor

Example 12 A low pass filter is required with a resistance of 600 Ω and a cut-off frequency of 1000 Hz. Calculate the values of L and C.

Since

$$R_o = \sqrt{\frac{L}{C}}$$

$$R_o^2 = \frac{L}{C}$$

and

$$L = CR_o^2 \tag{18.1}$$

$$f_c = \frac{1}{\pi \sqrt{(LC)}} = \frac{1}{\pi \sqrt{(CCR_o^2)}} = \frac{1}{\pi \sqrt{(C^2 R_o^2)}}$$

$$f_c = \frac{1}{\pi C R_o}$$

$$C = \frac{1}{\pi R_o f_c} = \frac{10^6}{\pi \times 600 \times 1000} \, \mu F = 0.5305 \, \mu F$$

From equation 18.1

$$C = \frac{L}{R_o^2}$$

$$f_c = \frac{1}{\pi \sqrt{[L(L/R_o^2)]}} = \frac{1}{\pi \sqrt{(L^2/R_o^2)}} = \frac{1}{\pi (L/R_o)}$$

Transposing

$$f_c = \frac{R_o}{\pi L}$$

and

$$L = \frac{R_o}{\pi f_c} = \frac{600 \times 10^3}{\pi \times 1000} \, mH = 191 \, mH$$

Note that

$$Z_1 = 2 \times \pi \times 1000 \times 191 \times 10^{-3} = 1200 \, \Omega$$

and

$$Z_2 = \frac{10^6}{2 \times \pi \times 1000 \times 0.5304} = 300 \, \Omega$$

This confirms that

$$R_o^2 \quad (600)^2 = Z_1(1200) \times Z_2(300)$$

Note also that

$$f_c = \frac{1}{\pi \sqrt{(LC)}}$$

$$f_c^2 = \frac{1}{\pi^2 L C}$$

$$f_c = \sqrt{\left(\frac{10^6}{\pi^2 \times 0.191 \times 0.5304}\right)} = 1000 \, Hz$$

The results mean that

1 The T filter will have inductances of

$$\frac{191}{2} \, mH = 90.5 \, mH, \, C \text{ will be } 0.5305 \, \mu F$$

2 The π filter will have capacitors of

$$\frac{0.5305}{2} \, \mu F = 0.2653 \, \mu F, \, L \text{ will be } 191 \, mH$$

```
10 PRINT "PROG 278"
20 PRINT "LOW PASS T AND PI FILTERS"
30 INPUT "ENTER DESIGN IMPEDANCE" ; RO
40 INPUT "ENTER CUT OFF FREQUENCY" ; FC
50 LET C=10^6/(3.141593*RO*FC)
60 LET L=RO/(3.141593*FC)
70 PRINT "CAPACITANCE C = " C "MICROFARADS"
80 PRINT "INDUCTANCE L = " L "HENRYS"
```

18.4 High pass filters

High pass filters attenuate low frequencies and transmit high frequencies. The simple R–C filter is identical to the circuit in Figure 18.6, the only difference being that the output is developed across the resistance as shown in Figure 18.9.

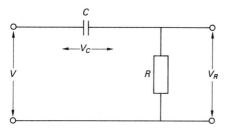

Figure 18.9

A typical frequency response is shown in Figure 18.10 and the formula for calculating the cut-off frequency is identical to that in Section 18.3.

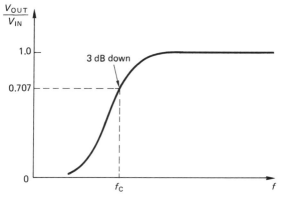

Figure 18.10

Figures 18.11 and 18.12 show T and π high pass filters respectively. Both types are designed using the formulas

$$\text{Cut-off frequency } f_c = \frac{1}{4\pi\sqrt{(LC)}}$$

$$\text{Design impedance } R_o = \sqrt{\frac{L}{C}}$$

$$Z_1 Z_2 = \frac{L}{C}$$

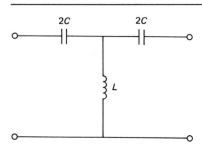

Figure 18.11

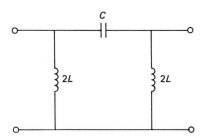

Figure 18.12

where Z_1 = reactance of the inductor
Z_2 = reactance of the capacitor

Example 13 A high pass filter is required with an impedance of 600 Ω and a cut-off frequency of 10 000 Hz. Calculate the values of L and C.

Since

$$R_o = \sqrt{\frac{L}{C}}$$

$$R_o^2 = \frac{L}{C}$$

and

$$L = CR_o^2$$

$$f_c = \frac{1}{4\pi\sqrt{(LC)}} = \frac{1}{4\pi\sqrt{(CCR_o^2)}} = \frac{1}{4\pi\sqrt{(C^2R_o^2)}}$$

$$= \frac{1}{4\pi CR_o}$$

$$C = \frac{1}{4\pi R_o f_c} = \frac{10^6}{4 \times \pi \times 600 \times 10^4} = 0.01326 \ \mu F$$

$$C = \frac{L}{R_o^2}$$

$$f_c = \frac{1}{4\pi\sqrt{[L(L/R_o^2)]}} = \frac{1}{4\pi\sqrt{(L^2/R_o^2)}} = \frac{1}{4\pi(L/R_o)}$$

Transposing

$$f_c = \frac{R_o}{4\pi L}$$

and

$$L = \frac{R_o}{4\pi f_c} = \frac{600 \times 10^3}{4 \times \pi \times 10^4} \text{ mH} = 4.774 \text{ mH}$$

Note that

$$Z_1 = 2 \times \pi \times 10^4 \times 4.774 \times 10^{-3} = 300 \text{ }\Omega$$

$$Z_2 = \frac{10^6}{2 \times \pi \times 10^4 \times 0.01326} = 1200 \text{ }\Omega$$

$$\frac{L}{C} = \frac{4.774 \times 10^6}{0.01326 \times 10^3} = 360\,000$$

Note

$$300 \times 1200 = 360\,000$$

Note also that since

$$f_c = \frac{1}{4\pi\sqrt{(LC)}} \quad f_c^2 = \frac{1}{16\pi^2 LC}$$

$$f_c = \sqrt{\left(\frac{10^3 \times 10^6}{16 \times \pi^2 \times 4.774 \times 0.01326}\right)} = 10^4 \text{ Hz}$$

```
10 PRINT "PROG 279"
20 PRINT "HIGH PASS T AND PI FILTERS"
30 INPUT "ENTER DESIGN IMPEDANCE" ; RO
40 INPUT "ENTER CUT OFF FREQUENCY" ; FC
50 LET C=10^6/(4*3.141593*RO*FC)
60 LET L=RO/(4*3.141593*FC)
70 PRINT "CAPACITANCE C = " C "MICROFARADS"
80 PRINT "INDUCTANCE L = " L "HENRYS"
```

18.5 Introduction to attenuators

It is often a requirement in electronic circuitry to attenuate voltages and currents. Attenuator pads are used for this purpose. They are constructed from purely resistive components, in order to attenuate all frequencies by the same degree and hence avoid distortion.

These networks can be designed to have any required attenuation but they must have:

1. The correct input impedance
2. The correct output impedance
3. The correct specified attenuation

The attenuation of an attenuator is usually quoted in decibels.

18.6 Symmetrical T attenuator

Figure 18.13 shows a symmetrical T network. In this type of network the input and output impedances R_o are equal, and the two resistances in the arm AB are also equal.

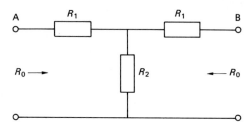

Figure 18.13

The network is designed using the formulas

$$R_o = \sqrt{(R_1^2 + 2R_1R_2)}$$

where R_o = characteristic impedance

$$R_o = \sqrt{(R_{oc}R_{sc})}$$

where R_{oc} = open circuit impedance looking into the network

R_{sc} = short circuit impedance, i.e. the impedance looking into the network with a short circuit across the output terminals

$$R_1 = R_o \frac{(N-1)}{(N+1)}$$

$$R_2 = R_o \left(\frac{2N}{N^2 - 1}\right)$$

where N = network attenuation as a voltage or current ratio

Example 14 If resistances R_1 and R_2 in Figure 18.13 are 20 Ω and 30 Ω respectively, calculate the characteristic impedance.

$$R_o = \sqrt{[20^2 + (2 \times 20 \times 30)]} = 40 \text{ Ω}$$

Note $R_{oc} = R_1 + R_2 = 20 + 30 = 50$ Ω

$$R_{sc} = R_1 + \frac{R_1 R_2}{R_1 + R_2} = 20 + \frac{20 \times 30}{20 + 30} = 32 \text{ Ω}$$

$$R_o = \sqrt{(50 \times 32)} = 40 \text{ Ω}$$

This attenuator matches a 40-Ω source to a 40-Ω load as shown in Figure 18.14. Figure 18.15 shows the equivalent circuit where

$$R_{IN} = R_o = 20 + \frac{30 \times 60}{30 + 60} = 40 \text{ Ω}$$

Similarly the resistance seen by the load with a 40-Ω source will also be 40 Ω.

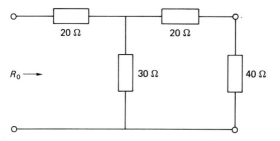

Figure 18.14

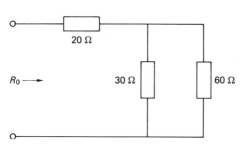

Figure 18.15

```
10 PRINT "PROG 280"
20 PRINT "SYMMETRICAL T ATTENUATOR"
30 INPUT "ENTER R1" ; R1
40 INPUT "ENTER R2" ; R2
50 LET RO=SQR((R1^2)+(2*R1*R2))
60 LET N=(R1+RO)/(RO-R1)
70 LET NDB=20*LOG(N)/LOG(10)
80 PRINT "CHARACTERISTIC IMPEDANCE = " RO "OHMS"
90 PRINT "ATTENUATION N = " N
100 PRINT "ATTENUATION IN DECIBELS = " NDB
```

Example 15 Find the attenuation in Example 14.

$$R_1 = R_o \frac{(N-1)}{(N+1)}$$

$$R_1(N+1) = R_o(N-1)$$

$$20(N+1) = 40(N-1)$$

$$20N + 20 = 40N - 40$$

$$40N - 20N = 20 + 40$$

$$20N = 60$$

$$N = 3$$

Attenuation in decibels = 20 log 3 = 9.54 dB

(See program 280, above.)

Example 16 Design a T-type attenuator to give 25 dB attenuation and have a characteristic impedance of 600 Ω.

$$20 \log N = 25$$

$$\log N = \frac{25}{20} = 1.25$$

$$N = \text{antilog } 1.25 = 17.8$$

$$R_1 = R_o \frac{(N-1)}{(N+1)} = \frac{600 \times 16.8}{18.8} = 536 \text{ Ω}$$

$$R_2 = R_o \frac{(2N)}{(N^2-1)} = \frac{600 \times 35.6}{(17.8)^2 - 1} = 67.6 \text{ Ω}$$

```
10 PRINT "PROG 281"
20 PRINT "SYMMETRICAL T ATTENUATOR"
30 INPUT "ENTER ATTENUATION IN DECIBELS" ; NDB
40 INPUT "CHARACTERISTIC IMPEDANCE" ; RO
50 LET X=NDB/20
60 LET N=10^X
70 LET R1=RO*(N-1)/(N+1)
80 LET R2=RO*2*N/((N^2)-1)
90 PRINT "R1 = " R1 "OHMS"
100 PRINT "R2 = " R2 "OHMS"
```

18.7 Asymmetrical T attenuator

This type of attenuator is designed to match a source impedance to a load impedance which is different in value. This means that the input and output impedances of the network have different values, and all three resistances will have different values.

Figure 18.16 shows an attenuator where R_1 and R_2 are the source and load impedance respectively. The design formulas are

$$R_A = R_1 \frac{(N^2+1)}{(N^2-1)} - 2\sqrt{(R_1 R_2)}\left(\frac{N}{N^2-1}\right)$$

$$R_B = 2\sqrt{(R_1 R_2)}\left(\frac{N}{N^2-1}\right)$$

$$R_C = R_2 \frac{(N^2+1)}{(N^2-1)} - 2\sqrt{(R_1 R_2)}\left(\frac{N}{N^2-1}\right)$$

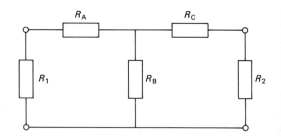

Figure 18.16

Example 17 Given that R_1 and R_2 in Figure 18.16 are 450 and 600 Ω respectively, and that the attenuation N is 10, calculate the values of R_A, R_B and R_C.

$$R_A = \left(450 \times \frac{101}{99}\right) - 2\sqrt{(450 \times 600)} \times \frac{10}{99}$$

$$= 459.09 - 104.97 = 354.12 \text{ Ω}$$

$$R_B = 104.97 \text{ Ω}$$

$$R_C = \left(600 \times \frac{101}{99}\right) - 104.97 = 507.15 \text{ Ω}$$

```
10 PRINT "PROG 282"
20 PRINT "ASYMMETRICAL T ATTENUATOR"
30 INPUT "ENTER R1" ; R1
40 INPUT "ENTER R2" ; R2
50 INPUT "ENTER ATTENUATION" ; N
60 LET X=R1*((N^2+1)/(N^2-1))
70 LET Y=2*SQR(R1*R2)*(N/(N^2-1))
80 LET Z=R2*((N^2+1)/(N^2-1))
90 LET RA=X-Y
100 LET RB=Y
110 LET RC=Z-Y
120 PRINT "RA = " RA "OHMS"
130 PRINT "RB = " RB "OHMS"
140 PRINT "RC = " RC "OHMS"
```

18.8 Symmetrical π attenuator

Figure 18.17 shows a symmetrical π network. The input and output impedances are equal and the resistances between AC and BC are equal. The network is designed using the formulas

$$R_o = \sqrt{\left(\frac{R_1 R_2^2}{R_1 + 2R_2}\right)}$$

$$R_o = \sqrt{(R_{oc} R_{sc})}$$

$$R_1 = R_o \frac{(N^2 - 1)}{2N}$$

$$R_2 = R_o \left(\frac{N + 1}{N - 1}\right)$$

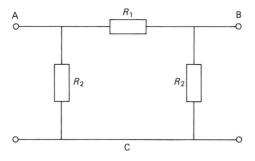

Figure 18.17

Example 18 If the resistances R_1 and R_2 in Figure 18.17 are 20 and 30 Ω respectively, calculate the characteristic impedance.

$$R_o = \sqrt{\left(\frac{20 \times 30^2}{20 + (2 \times 30)}\right)} = 15 \, \Omega$$

```
10 PRINT "PROG 283"
20 PRINT "SYMMETRICAL PI ATTENUATOR"
30 INPUT "ENTER R1" ; R1
40 INPUT "ENTER R2" ; R2
50 LET RO=SQR((R1*(R2^2))/(R1+(2*R2)))
60 LET N=(RO+R2)/(R2-RO)
70 PRINT "CHARACTERISTIC IMPEDANCE = " RO "OHMS"
80 PRINT "ATTENUATION = " N
```

Example 19 Find the attenuation in the network in Example 18.

$$R_2 = R_o\left(\frac{N+1}{N-1}\right)$$

$$30 = 15\frac{(N+1)}{(N-1)}$$

$$30(N-1) = 15(N+1)$$

$$30N - 30 = 15N + 15$$

$$30N - 15N = 15 + 30$$

$$15N = 45$$

$$N = 3 \, (9.5 \, \text{dB})$$

(See program 283, above.)

18.9 Asymmetrical π attenuator

The purpose and the characteristics of this type of attenuator are the same as those of the asymmetrical T. Figure 18.18 shows an attenuator where R_1 and R_2 are the source and load impedances respectively. The design formulas are

$$R_A = R_1\left(\frac{N^2 - 1}{N^2 - 2NS + 1}\right)$$

$$R_B = \frac{\sqrt{(R_1R_2)}}{2}\left(\frac{N^2 - 1}{N}\right)$$

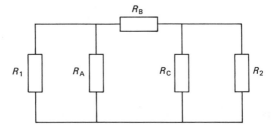

Figure 18.18

Filters and attenuators 327

$$R_C = R_2\left(\frac{N^2 - 1}{N^2 - (2N/S) + 1}\right)$$

where $S^2 = \dfrac{R_1}{R_2}$

Example 20 Given that R_1 and R_2 in Figure 18.18 are 40 and 10 Ω respectively, and that the attenuation N is 10, calculate the values of R_A, R_B and R_C.

$$S = \sqrt{\frac{40}{10}} = 2$$

$$R_A = 40\left(\frac{100 - 1}{100 - 40 + 1}\right) = 64.91 \text{ Ω}$$

$$R_B \frac{\sqrt{(40 \times 10)}}{2}\left(\frac{100 - 1}{10}\right) = 99 \text{ Ω}$$

$$R_C = 10\left(\frac{100 - 1}{100 - 10 + 1}\right) = 10.879 .\text{Ω}$$

```
10 PRINT "PROG 284"
20 PRINT "ASYMMETRICAL PI ATTENUATOR"
30 INPUT "ENTER R1" ; R1
40 INPUT "ENTER R2" ; R2
50 INPUT "ENTER ATTENUATION" ; N
60 LET S=SQR(R1/R2)
70 LET X=((N^2)-1)/((N^2)-(2*N*S)+1)
80 LET Y=SQR(R1*R2)*((N^2)-1)/(2*N)
90 LET Z=((N^2)-1)/((N^2)-(2*N/S)+1)
100 LET RA=R1*X
105 LET RC=R2*Z
110 LET RB=Y
120 PRINT "RA = " RA "OHMS"
130 PRINT "RB = " RB "OHMS"
140 PRINT "RC = " RC "OHMS"
```

Problems

1 If the input and output powers of an electronic network are 0.5 and 25 W respectively, calculate the gain in decibels.
2 A network has a power gain of 9 dB. If the output power is 64 mW, calculate the input power.
3 If a network has an input of 1.6 W and an output of 0.4 W calculate the loss in decibels.
4 A filter unit has a power loss of 24 dB. If the input power is 1.28 W, calculate the output power.
5 A network has a voltage loss of −30 dB. If the input voltage is 8 V, calculate the output voltage.
6 A filter unit has an input current of 1.414 μA and an output current of 0.5 μA. Calculate the loss in decibels.
7 Four networks connected in series have gains of 100 dB, −30 dB, +6 dB and −48 dB. What is the overall gain?
8 A network identical to Figure 18.2 has an input voltage of 3.5 V. If the output of stage 1 is 10.5 V, and the output of stage 2 is 21 V, calculate the gain of each stage and the overall gain of the network.
9 A three-stage network identical to Figure 18.3 is fed with a voltage of 1.5 V. The inputs to stages 2 and 3 are 6 V and 3 V

and the output from stage 3 is 24 V. Calculate the individual stage gains and the overall network gain.

10 The individual gains of a three-stage network identical to Figure 18.4 are +12 dB, −15 dB and +9 dB. Calculate the power output of each stage if the input to stage 1 is 4 W.

11 A low pass R–C network identical to Figure 18.6 has a 120 Ω resistor and a 12-μF capacitor. If the input voltage is 20 V calculate: (a) the output voltage at the −3 dB point, (b) the voltage across R at the −3 dB point, (c) the cut-off frequency at the −3 dB point.

12 A low pass T filter identical to Figure 18.7 is required to have an impedance of 400 Ω and a cut-off frequency of 1500 Hz. Calculate the values of L and C.

13 A high pass T filter identical to Figure 18.11 has a design impedance of 200 Ω and a cut-off frequency of 3000 Hz. Calculate the values of L and C.

14 A symmetrical T attenuator identical to Figure 18.13 has the following parameters: $R_1 = 25\ \Omega$, $R_2 = 45\ \Omega$. Calculate the characteristic impedance.

15 Find the attenuation in Problem 14.

16 Design a symmetrical T-type attenuator to give 40 dB voltage attenuation, and a characteristic impedance of 300 Ω.

17 Given that the values of R_1 and R_2 in an asymmetrical T attenuator identical to Figure 18.16 are 100 and 200 Ω, and given that the attenuation N is 5, calculate the values of R_A, R_B and R_C.

18 If R_1 and R_2 in a symmetrical π attenuator identical to Figure 18.17 are 40 and 60 Ω respectively, calculate the characteristic impedance.

19 Find the attenuation in Problem 18.

20 Given that R_1 and R_2 in the asymmetrical π attenuator of Figure 18.18 are 90 and 10 Ω respectively and that the attenuation N is 12, calculate the values of R_A, R_B and R_C.

19 Denary, binary and logic

19.1 The denary system

(a) Introduction
In order to understand the binary system it is necessary to consider first some aspects of the denary (decimal) number system which is in common use.

Counting in the denary system is achieved using ten integers, 0 through 9. The system is said to have a base of 10, and all numbers are formed by multiplying the powers of 10 by the integers 0 to 9. Example 1 shows three numbers being formed in this way.

Example 1

$$10^3 \quad 10^2 \quad 10^1 \quad 10^0 \quad 10^{-1} \quad 10^{-2} \quad 10^{-3}$$

$$\uparrow \text{Denary point}$$

$$\begin{aligned}
402 &= 4 \times 10^2 + 0 \times 10^1 + 2 \times 10^0 \\
&= 4 \times 100 + 0 \times 10 + 2 \times 1 = 402 \\
61.4 &= 6 \times 10^1 + 1 \times 10^0 + 4 \times 10^{-1} \\
&= 6 \times 10 + 1 \times 1 + 0.4 = 61.4 \\
0.022 &= 0 \times 10^{-1} + 2 \times 10^{-2} + 2 \times 10^{-3} \\
&= 0 \times 0.1 + 2 \times 0.01 + 2 \times 0.001 \\
&= 0 + 0.2 + 0.002 = 0.022
\end{aligned}$$

(b) Other bases
It is possible, of course, to count using other bases. Base 9, for example, would use nine integers, 0 to 8, base 8 would use eight integers, 0 to 7, etc. Table 19.1 shows how counting up to 20 would be done in nine different bases using the rules that are familiar in the denary system.

(c) Base conversion of integers
Numbers in base 10 can be converted into other bases by dividing successively by the new base number until no remainder exists. The various remainders taken in reverse order will give the number in the new scale.

Example 2 Convert 20_{10} to bases 9, 7, 5 and 3 and check the answers against Table 19.1.

$$\begin{array}{l} 9 \underline{|20} \\ 9 \underline{|2} \quad 2 \uparrow \\ \underline{} \quad 2 \end{array} \quad 20_{10} = 22_9 \qquad \begin{array}{l} 7 \underline{|20} \\ 7 \underline{|2} \quad 6 \uparrow \\ \underline{} \quad 2 \end{array} \quad 20_{10} = 26_7$$

$$\begin{array}{l} 5 \underline{|20} \\ 5 \underline{|4} \quad 0 \uparrow \\ \underline{} \quad 4 \end{array} \quad 20_{10} = 40_5 \qquad \begin{array}{l} 3 \underline{|20} \\ 3 \underline{|6} \quad 2 \uparrow \\ 3 \underline{|2} \quad 0 \\ \underline{} \quad 2 \end{array} \quad 20_{10} = 202_3$$

Table 19.1 Counting in bases 2 to 10

10	9	8	7	6	5	4	3	2
0	0	0	0	0	0	0	0	0
1	1	1	1	1	1	1	1	1
2	2	2	2	2	2	2	2	10
3	3	3	3	3	3	3	10	11
4	4	4	4	4	4	10	11	100
5	5	5	5	5	10	11	12	101
6	6	6	6	10	11	12	20	110
7	7	7	10	11	12	13	21	111
8	8	10	11	12	13	20	22	1000
9	10	11	12	13	14	21	100	1001
10	11	12	13	14	20	22	101	1010
11	12	13	14	15	21	23	102	1011
12	13	14	15	20	22	30	110	1100
13	14	15	16	21	23	31	111	1101
14	15	16	20	22	24	32	112	1110
15	16	17	21	23	30	33	120	1111
16	17	20	22	24	31	100	121	10000
17	18	21	23	25	32	101	122	10001
18	20	22	24	30	33	102	200	10010
19	21	23	25	31	34	103	201	10011
20	22	24	26	32	40	110	202	10100

Numbers can be converted back into base 10 by using the method shown in Example 1.

```
10 PRINT "PROG 285"
20 PRINT "BASE 10 TO BASE 4"
30 INPUT "ENTER THE DENARY VALUE OF Y" ; Y
40 LET X=Y/4
50 LET Z=INT(X)
60 IF X=(Z+.75) THEN PRINT "3"
70 IF X=(Z+.5) THEN PRINT "2"
80 IF X=(Z+.25) THEN PRINT "1"
90 IF X=Z THEN PRINT "0"
100 LET X=Z/4
110 IF X>.75 THEN GOTO 50
120 IF X=.75 THEN PRINT "3"
130 IF X=.5 THEN PRINT "2"
140 IF X=.25 THEN PRINT "1"
150 PRINT "READ ANSWER FROM BOTTOM TO TOP"
160 END
```

Example 3 Convert the following to base 10, and check the answers against Table 19.1: (i) 122_3, (ii) 31_5, (iii) 11_7.

(i) 3^2 3^1 3^0

      1   2   2

$$122_3 = 1 \times 3^2 + 2 \times 3 + 2 \times 1 = 17_{10}$$

(ii) 5^2 5^1 5^0

         3   1

$$31_5 = 3 \times 5 + 1 \times 1 = 16_{10}$$

(iii) 7^2 7^1 7^0

         1   1

$$11_7 = 1 \times 7 + 1 \times 1 = 8_{10}$$

(d) Base conversion of values less than unity

To convert fractions from base 10 to another base, the procedure is as follows. Multiply by the new base and record a carry in the integer position. The integer carries are taken in the forward order to give the new base number.

Example 4 Convert 0.125_{10} to base 6.

$$0.125 \times 6 = 0.75 = 0.75 \text{ carry } 0$$
$$0.75 \times 6 = 4.5 = 0.5 \text{ carry } 4$$
$$0.5 \times 6 = 3.0 = 0 \text{ carry } 3$$

$$0.125_{10} = 0.043_6$$

Example 5 Convert 0.0625_{10} to base 4.

$$0.0625 \times 4 = 0.25 = 0.25 \text{ carry } 0$$
$$0.25 \times 4 = 1.0 = 0 \text{ carry } 1$$

$$0.0625_{10} = 0.01_4$$

Example 6 Convert the answers to Examples 4 and 5 to base 10 using the method shown in Example 1.

(i) $\cdot\; 6^{-1} \quad 6^{-2} \quad 6^{-3}$
     $\cdot\; 0 \quad\quad 4 \quad\quad 3$

$.043_6 = 0 \times 6^{-1} + 4 \times 6^{-2} + 3 \times 6^{-3}$
$ = 0 \quad\quad +0.111 \quad +0.014 = 0.125_{10}$

(ii) $\cdot\; 4^{-1} \quad 4^{-2}$
     $\cdot\; 0 \quad\quad 1$

$.01_4 = 0 \times 4^{-1} + 1 \times 4^{-2} = 0.0625_{10}$

19.2 The binary system

(a) Introduction

A number system that only uses two digits, 0 and 1, is called a binary system. Table 19.1 showed the denary numbers 0 to 20 expressed in base 2.

Computers calculate using the binary notation. The two digits correspond to the 'on' and 'off' states of the electronic circuitry.

To convert denary integers and fractions into binary, and binary integers and fractions into denary, follow the rules outlined in Section 19.1.

Example 7 Convert 20.625_{10} into binary.

```
2 | 20
2 | 10   0
2 |  5   0
2 |  2   1
2 |  1   0
  | ·    1
```

$0.625 \times 2 = 1.25 = 0.25 \text{ carry } 1$
$0.25 \times 2 = 0.5 = 0.5 \text{ carry } 0$
$0.5 \times 2 = 1.0 = 0 \text{ carry } 1$

$$20.625_{10} = 10100.101_2$$

```
10 PRINT "PROG 286"
20 PRINT "DENARY NUMBERS TO BINARY"
30 INPUT "ENTER DENARY NUMBER Y" ; Y
40 LET X=Y/2
50 LET Z=INT(X)
60 IF Z=X THEN PRINT "0"
70 IF Z<X THEN PRINT "1"
80 LET X=Z/2
90 IF X>.5 THEN GOTO 50
100 IF X=.5 THEN PRINT "1"
110 PRINT "READ ANSWER FROM BOTTOM TO TOP"


10 PRINT "PROG 287"
20 PRINT "DENARY FRACTION TO BINARY"
30 INPUT "ENTER FRACTION Y" ; Y
40 LET X=2*Y
50 IF X<1 THEN PRINT "0"
60 IF X=1 THEN GOTO 100
70 IF X>1 THEN GOTO 120
80 LET X=2*X
90 GOTO 50
100 IF X=1 THEN PRINT "1"
110 STOP
120 X=X-1
130 PRINT "1"
140 GOTO 80
150 STOP
```

Example 8 Convert 110.0011_2 to denary.

$2^2 \quad 2^1 \quad 2^0 \;\cdot\; 2^{-1} \quad 2^{-2} \quad 2^{-3} \quad 2^{-4}$

$1 \quad\; 1 \quad\; 0 \;\cdot\; 0 \quad\;\; 0 \quad\;\; 1 \quad\;\; 1$

$1 \times 2^2 + 1 \times 2^1 + 0 \times 2^0 = 6$

$0 \times 2^{-1} + 0 \times 2^{-2} + 1 \times 2^{-3} + 1 \times 2^{-4}$

$\quad = 0 + 0 + 0.125 + 0.625 = 0.1875$

$\quad\quad 110.0011_2 = 6.1875_{10}$

```
10 PRINT "PROG 288"
20 PRINT "BINARY NUMBERS TO DENARY"
30 INPUT "ENTER A" ; A
40 INPUT "ENTER B" ; B
50 INPUT "ENTER C" ; C
60 INPUT "ENTER D" ; D
70 INPUT "ENTER E" ; E
80 INPUT "ENTER F" ; F
90 INPUT "ENTER G" ; G
100 LET H=A*64
110 LET I=B*32
120 LET J=C*16
130 LET K=D*8
140 LET L=E*4
150 LET M=F*2
160 LET N=G
170 LET P=H+I+J+K+L+M+N
180 PRINT "DENARY NUMBER = " P
```

(b) Addition

The rules for binary addition are

$\quad 0 + 0 = 0$
$\quad 0 + 1 = 1$
$\quad 1 + 0 = 1$
$\quad 1 + 1 = 10$ (or zero carry one)

Example 9 Add 101 and 100.

```
    101     first column    0 + 1 = 1
  + 100     second column   0 + 0 = 0
  ─────
   1001     third column    1 + 1 = 10
```

```
10 PRINT "PROG 289"
20 PRINT "BINARY ADDITION"
30 INPUT "ENTER A" ; A
40 INPUT "ENTER B" ; B
50 INPUT "ENTER C" ; C
60 INPUT "ENTER D" ; D
70 INPUT "ENTER E" ; E
80 INPUT "ENTER F" ; F
90 INPUT "ENTER G" ; G
100 LET H=A*64
110 LET I=B*32
120 LET J=C*16
130 LET K=D*8
140 LET L=E*4
150 LET M=F*2
160 LET N=G
170 LET P=H+I+J+K+L+M+N
180 PRINT "DENARY NUMBER = " P
190 INPUT "ENTER A2" ; A2
200 INPUT "ENTER B2" ; B2
210 INPUT "ENTER C2" ; C2
220 INPUT "ENTER D2" ; D2
230 INPUT "ENTER E2" ; E2
240 INPUT "ENTER F2" ; F2
250 INPUT "ENTER G2" ; G2
260 LET H2=A2*64
270 LET I2=B2*32
280 LET J2=C2*16
290 LET K2=D2*8
300 LET L2=E2*4
310 LET M2=F2*2
320 LET N2=G2
330 LET P2=H2+I2+J2+K2+L2+M2+N2
340 PRINT "DENARY NUMBER = " P2
350 LET Y=P+P2
360 LET X=Y/2
370 LET Z=INT(X)
380 IF Z=X THEN PRINT "0"
390 IF Z<X THEN PRINT "1"
400 LET X=Z/2
410 IF X>.5 THEN GOTO 370
420 IF X=.5 THEN PRINT "1"
430 PRINT "READ ANSWER FROM TOP TO BOTTOM"
```

Example 10 Add 111 and 111.

```
    111     first column    1 + 1 = 10 (zero carry one)
  + 111     second column   1 + 1 + 1 = 10 + 1 = 11 (one
                                                     carry one)
  ─────
   1110     third column    1 + 1 + 1 = 10 + 1 = 11 (one
                                                     carry one)
```

(c) Subtraction

The rules for binary subtraction are

$0 - 0 = 0$
$0 - 1 = 1$ (after borrow 1)
$1 - 0 = 1$
$1 - 1 = 0$
$10 - 1 = 1$

Example 11 Subtract 100 from 111.

$$111 = 7 \text{ first column} \quad 1 - 0 = 1$$
$$-100 = -4 \text{ second column} \quad 1 - 0 = 1$$
$$\overline{011} = \overline{3} \text{ third column} \quad 1 - 1 = 0$$

```
10 PRINT "PROG 290"
20 PRINT "BINARY SUBTRACTION"
30 INPUT "ENTER A" ; A
40 INPUT "ENTER B" ; B
50 INPUT "ENTER C" ; C
60 INPUT "ENTER D" ; D
70 INPUT "ENTER E" ; E
80 INPUT "ENTER F" ; F
90 INPUT "ENTER G" ; G
100 LET H=A*64
110 LET I=B*32
120 LET J=C*16
130 LET K=D*8
140 LET L=E*4
150 LET M=F*2
160 LET N=G
170 LET P=H+I+J+K+L+M+N
180 PRINT "DENARY NUMBER = " P
190 INPUT "ENTER A2" ; A2
200 INPUT "ENTER B2" ; B2
210 INPUT "ENTER C2" ; C2
220 INPUT "ENTER D2" ; D2
230 INPUT "ENTER E2" ; E2
240 INPUT "ENTER F2" ; F2
250 INPUT "ENTER G2" ; G2
260 LET H2=A2*64
270 LET I2=B2*32
280 LET J2=C2*16
290 LET K2=D2*8
300 LET L2=E2*4
310 LET M2=F2*2
320 LET N2=G2
330 LET P2=H2+I2+J2+K2+L2+M2+N2
340 PRINT "DENARY NUMBER = " P2
350 LET Y=P-P2
360 LET X=Y/2
370 LET Z=INT(X)
380 IF Z=X THEN PRINT "0"
390 IF Z<X THEN PRINT "1"
400 LET X=Z/2
410 IF X>.5 THEN GOTO 370
420 IF X=.5 THEN PRINT "1"
430 PRINT "READ ANSWER FROM TOP TO BOTTOM"
```

Example 12 Subtract 1010 from 1101.

$$1101 = 13 \text{ first column} \quad 1 - 0 = 1$$
$$-1010 = -10 \text{ second column} \quad 10 \text{ (after borrow)} - 1 = 1$$
$$\overline{0011} = \overline{3} \text{ third column} \quad 1 - 1 \text{ (after return)} = 0$$
$$ \text{ fourth column} \quad 1 - 1 = 0$$

Note that during the subtraction process, the columns look like this.

col. 4	col. 3	col. 2	col. 1
1	1	10	1
−1	−1	−1	−0
0	0	1	1

Example 13 Subtract 111 from 1000.

$$\begin{aligned} 1000 &= 8 \\ -\ 111 &= -7 \\ \hline 0001 &= 1 \end{aligned}$$

During subtraction the columns look like this.

col. 4	col. 3	col. 2	col. 1
1	10	10	10
−1	−10	−10	−1
0	0	0	1

Example 14 Subtract 1011 from 101. The answer to this problem will be negative. The rule is to subtract the smaller number from the larger number.

$$\begin{aligned} 101 &= 5 \\ -1011 &= -11 \\ \hline -0110 &= -6 \end{aligned}$$

first column $1 - 1 = 0$
second column $1 - 0 = 0$
third column $10 - 1 = 1$
fourth column $1 - 1 = 0$

(d) The 1's complement

This is an alternative method of subtraction which required less circuitry in digital computers. The 1's complement of a binary number is found when each 0 is changed to 1, and each 1 is changed to 0, thus

Binary number	1's complement
1010	0101
1110	0001
10110	01001

In a subtraction problem, instead of subtracting a number, the 1's complement of the number is added and the last carry is added to the final answer.

Example 15 Subtract 10111 from 11001 using the 1's complement method.

1's complement of 10111 = 01000

$$\begin{array}{r} 11001 \\ +\ 01000 \\ \hline 100001 \\ 00001 \\ +1 \\ \hline 00010 \end{array}$$

Example 16 Subtract 1111 from 11111, i.e. subtract 01111 from 11111.

```
   11111
 + 10000   1's complement
  ──────
┌─101111
│
│  01111
└→  +1
   ─────
   10000
```

If there is no end around carry, the answer is negative, and the answer is in the 1's complement form. Taking the 1's complement and adding the negative prefix will give the answer, as the next example shows.

Example 17 Subtract 1101 from 1001.

```
1001              9
0010  1's complement  −13
────             ───
1011             −4
```

The answer is the 1's complement of 1011 = −0100 = −4.

(e) The 2's complement

This is another method of subtracting binary numbers. The 2's complement of a binary number is found by first finding the 1's complement, and then adding 1, i.e.

2's complement = 1's complement + 1

Binary number	2's complement
0001	1110 + 1 = 1111
1010	0101 + 1 = 0110
11111	00000 + 1 = 00001

A number can be subtracted by adding the 2's complement and then disregarding the last carry.

Example 18 Subtract 100 from 111 using the 2's complement method.

2's complement of 100 = 011 + 1 = 100

```
   111                  7
 + 100                 −4
  ────                 ──
  1̸011   Answer = 011   3
```

Example 19 Subtract 11101 from 101111, i.e. subtract 011101 from 101111.

```
  101111                 47
  100011  2's complement  −29
  ──────                 ───
 1̸010010                  18
```

Example 20 Subtract 1110 from 1001.

```
    1110                           9
 + 0111   2's complement        −14
 ──────                         ────
 −⟨1⟩0101                        − 5
```

Note that the answer is negative. The smaller number is subtracted from the larger number.

(f) Multiplication
The rules for binary multiplication are

$0 \times 0 = 0$
$0 \times 1 = 0$
$1 \times 0 = 0$
$1 \times 1 = 1$

The method of multiplying two numbers follows the 'long multiplication' method used with denary number.

Example 21 Multiply 1101 by 110.

```
    1101              13
  ×  110             ×  6
  ──────             ────
    0000              78
   1101               ──
  1101
  ───────
  1001110
```

```
10 PRINT "PROG 291"
20 PRINT "BINARY MULTIPLICATION"
30 INPUT "ENTER A"  ; A
40 INPUT "ENTER B"  ; B
50 INPUT "ENTER C"  ; C
60 INPUT "ENTER D"  ; D
70 INPUT "ENTER E"  ; E
80 INPUT "ENTER F"  ; F
90 INPUT "ENTER G"  ; G
100 LET H=A*64
110 LET I=B*32
120 LET J=C*16
130 LET K=D*8
140 LET L=E*4
150 LET M=F*2
160 LET N=G
170 LET P=H+I+J+K+L+M+N
180 PRINT "DENARY NUMBER = " P
190 INPUT "ENTER A2" ; A2
200 INPUT "ENTER B2" ; B2
210 INPUT "ENTER C2" ; C2
220 INPUT "ENTER D2" ; D2
230 INPUT "ENTER E2" ; E2
240 INPUT "ENTER F2" ; F2
250 INPUT "ENTER G2" ; G2
260 LET H2=A2*64
270 LET I2=B2*32
280 LET J2=C2*16
290 LET K2=D2*8
300 LET L2=E2*4
310 LET M2=F2*2
320 LET N2=G2
330 LET P2=H2+I2+J2+K2+L2+M2+N2
340 PRINT "DENARY NUMBER = " P2
350 LET Y=P*P2
360 LET X=Y/2
370 LET Z=INT(X)
380 IF Z=X THEN PRINT "0"
390 IF Z<X THEN PRINT "1"
400 LET X=Z/2
410 IF X>.5 THEN GOTO 370
420 IF X=.5 THEN PRINT "1"
430 PRINT "READ ANSWER FROM TOP TO BOTTOM"
```

Note that the multiplication by 0 in row 3 can be omitted provided that the required shift of numbers is correctly made as shown in the next example.

Example 22 Multiply 10001 by 101

```
    10001              17
  ×   101            ×  5
   ───────           ─────
    10001              85
   10001               ──
  ────────
  1010101
```

(g) Division

Division of binary numbers follows the same pattern as in denary numbers.

Example 23 Divide 1100 by 100.

```
          11               4 |12
   100 ┌──────               3
       │ 1100               ──
         100
         ───
         100
         100
         ───
         000   Answer = 11
```

```
1 LIST
10 PRINT "PR 292"
20 PRINT "BINARY DIVISION"
30 INPUT "ENTER A" ; A
40 INPUT "ENTER B" ; B
50 INPUT "ENTER C" ; C
60 INPUT "ENTER D" ; D
70 INPUT "ENTER E" ; E
80 INPUT "ENTER F" ; F
90 INPUT "ENTER G" ; G
100 LET H=A*64
110 LET I=B*32
120 LET J=C*16
130 LET K=D*8
140 LET L=E*4
150 LET M=F*2
160 LET N=G
170 LET P=H+I+J+K+L+M+N
180 PRINT "DENARY NUMBER = " P
190 INPUT "ENTER A2" ; A2
200 INPUT "ENTER B2" ; B2
210 INPUT "ENTER C2" ; C2
220 INPUT "ENTER D2" ; D2
230 INPUT "ENTER E2" ; E2
240 INPUT "ENTER F2" ; F2
250 INPUT "ENTER G2" ; G2
260 LET H2=A2*64
270 LET I2=B2*32
280 LET J2=C2*16
290 LET K2=D2*8
300 LET L2=E2*4
310 LET M2=F2*2
320 LET N2=G2
330 LET P2=H2+I2+J2+K2+L2+M2+N2
340 PRINT "DENARY NUMBER = " P2
350 LET Y=P/P2
360 LET X=Y/2
370 LET Z=INT(X)
380 IF Z=X THEN PRINT "0"
390 IF Z<X THEN PRINT "1"
400 LET X=Z/2
410 IF X>.5 THEN GOTO 370
420 IF X=.5 THEN PRINT "1"
430 PRINT "READ ANSWER FROM TOP TO BOTTOM"
```

Example 24 Divide 11011 by 1001.

```
         11              9 | 27
1001 ) 11011                 3
       1001
       ----
        1001
        1001
        ----
        0000
```

19.3 Octal numbers

(a) Introduction
The octal system has a base of eight, and uses the digits 0 to 7. It is important in digital work. The conversion of denary to octal and octal to denary for both integers and fractions follows the rules given in Section 19.1. It will be shown later in this section that large denary numbers are more easily converted to binary if the numbers are first converted to octal and then to binary.

```
10 PRINT "PROG 293"
20 PRINT "BASE 10 TO BASE 8"
30 INPUT "ENTER THE DENARY VALUE OF Y" ; Y
40 LET X=Y/8
50 LET Z=INT(X)
60 IF X=(Z+.875) THEN PRINT "7"
70 IF X=(Z+.75) THEN PRINT "6"
80 IF X=(Z+.625) THEN PRINT "5"
90 IF X=(Z+.5) THEN PRINT "4"
91 IF X=(Z+.375) THEN PRINT "3"
92 IF X=(Z+.25) THEN PRINT "2"
93 IF X=(Z+.125) THEN PRINT "1"
94 IF X=Z THEN PRINT "0"
100 LET X=Z/8
110 IF X>.875 THEN GOTO 50
120 IF X=.875 THEN PRINT "7"
130 IF X=.75 THEN PRINT "6"
140 IF X=.625 THEN PRINT "5"
141 IF X=.5 THEN PRINT "4"
142 IF X=.375 THEN PRINT "3"
143 IF X=.25 THEN PRINT "2"
144 IF X=.125 THEN PRINT "1"
150 PRINT "READ ANSWER FROM BOTTOM TO TOP"
160 END
```

(b) Octal to binary conversion
The most important use of octal numbers lies in the conversion of octal numbers into binary. Direct conversion is obtained by first producing a table showing the count to 7 in both octal and binary thus

000	001	010	011	100	101	110	111
0	1	2	3	4	5	6	7

Equivalent binary numbers can now be read directly from the table, e.g.

$3_8 = 011_2$, $4_8 = 100_2$ etc.

For numbers larger than 7 one octal digit is converted at a time.

Example 25 Convert 26_8 to binary.

```
  2      6
 010    110
```

i.e. octal 26 = 010110 in binary.

Example 26 Convert 42.631_8 to binary.

$$42.631_8 = 100010 \cdot 110011001_2$$

(c) Binary to octal conversion
The method is to group the binary numbers into groups of three starting at the binary point, and then convert each group into octal. It may be necessary to add 0's at each end to complete the groups. This is shown in the next example.

Example 27 Convert $1101 \cdot 10111_2$ to octal.

First form three groups by adding two zeros at the beginning, and one zero at the end.

```
001   101 · 101   110
 1     5  ·  5     6   = 15.56₈
```

$= 15.56_8$

Example 28 Convert 352_{10} into binary by: (i) direct conversion, (ii) converting to octal and then to binary.

```
(i)  2 |352                    (ii)  8 |352
     2 |176  0                       8 | 44   0
     2 | 88  0                       8 |  5   4
     2 | 44  0                           ·    5
     2 | 22  0
     2 | 11  0                             5    4    0
     2 |  5  1                             ↓    ↓    ↓
     2 |  2  1                            101  100  000
     2 |  1  0
         ·   1
```

Note how the direct conversion needs many more divisions than the denary–octal–binary method.

19.4 Hexadecimal numbers

(a) Introduction
This counting system uses a base of 16. The symbols are 0 to 9 and A to F (see Table 19.2). The next number after F is 10, followed by 11, 12, 13, 14, 15, 16, 17, 18, 19, 1A, 1B, 1C, 1D, 1E, 1F, 20, 21, 22, 23, 24, 25, 26, 27, 28, 29, 2A, etc.

The hexadecimal system has an advantage over the octal system. In that it requires less digits to represent a binary number. The rules for base conversions follow those already given in Section 19.1.

(b) Denary to hex conversion
Example 29 Convert 335_{10} to hex.

```
16 |335
16 | 20   remainder 15 = F
16 |  1   remainder  4
    ·     remainder  1
```

$335_{10} = 14F$

Denary, binary and logic 341

Table 19.2 Hexadecimal numbers

Denary	Binary	Hexadecimal
0	0000	0
1	0001	1
2	0010	2
3	0011	3
4	0100	4
5	0101	5
6	0110	6
7	0111	7
8	1000	8
9	1001	9
10	1010	A
11	1011	B
12	1100	C
13	1101	D
14	1110	E
15	1111	F

Example 30 Convert 29.75_{10} to hex.

$$\begin{array}{l} 16\underline{|29}\\ 16\;\underline{|1}\quad 13 = D\;\uparrow\\ \underline{\cdot}\quad 1\\ 29.75_{10} = 1D\cdot C\end{array} \qquad 0.75 \times 16 = 12.0 = 0 \text{ carry } 12 = C$$

(c) Hex to denary conversion

Example 31 Convert $4AB_{16}$ to denary.

$$16^2 \quad 16^1 \quad 16^0$$
$$4 \quad\; 10 \quad\; 11$$

$$4AB = 4 \times 16^2 + (10 \times 16) + (11 \times 1) = 1195_{10}$$

Example 32 Convert A·BC to denary.

$$16^0 \cdot 16^{-1} \; 16^{-2}$$
$$\;A \;\cdot\; B \quad\; C$$

$$A \cdot BC = 10 \times 1 + (11 \times 0.0625) + (12 \times 0.00390625)$$
$$= 10 + 0.6875 + 0.046875$$
$$= 10 \cdot 734375_{10}$$

(d) Hex to binary conversion

Since 16 is the fourth power of 2, this conversion is simple and can be carried out directly from Table 19.2.

Example 33 Convert A2C to binary.

From Table 19.2

$$\begin{array}{ccc} A & 2 & C \\ \downarrow & \downarrow & \downarrow \\ 1010 & 0010 & 1100 \end{array}$$

$A2C = 101000101100_2$

Example 34 Convert 8.FE to binary.

From Table 19.2

$$\begin{array}{ccc} 8 & F & E \\ \downarrow & \downarrow & \downarrow \\ 1000 & 1111 & 1110 \end{array}$$

$8 \cdot FE = 1000 \cdot 11111110_2$

(e) Binary to hex conversion
This conversion can be carried out directly from Table 19.2, by first dividing the binary number into groups of four digits, starting at the binary point, and working either left or right. It may be necessary to add zeros at the beginning and the end of the binary number, in a similar manner to the binary–octal conversion.

Example 35 Convert 10001110_2 to hex.

$$\begin{array}{cc} 1000 & 1110 \\ \downarrow & \downarrow \\ 8 & E \end{array}$$

Example 36 Convert $101 \cdot 110111_2$ to hex.

$$\begin{array}{ccc} 0101 \cdot & 1101 & 1100 \\ \downarrow & \downarrow & \downarrow \\ 5 \cdot & D & C \end{array}$$

19.5 Logic gates and truth tables

(a) Introduction
The transmission of information in the form of binary signals through a computer is controlled by logic gates. The gate symbols are shown in Figure 19.1. The first column shows the system introduced in 1969, column 2 shows the 1977 revised system, and column 3 the US military standard.

(b) Gate inputs
Figure 19.1 shows gates with two inputs, but in practice all gates apart from the NOT may have a number of inputs. With two-input gates there are four possible combinations of binary inputs. These inputs are

Input 1	Input 2
0	0
0	1
1	0
1	1

A three-input gate would have eight possible combinations:

Input 1	Input 2	Input 3
0	0	0
0	0	1
0	1	0
0	1	1
1	0	0
1	0	1
1	1	0
1	1	1

Denary, binary and logic   343

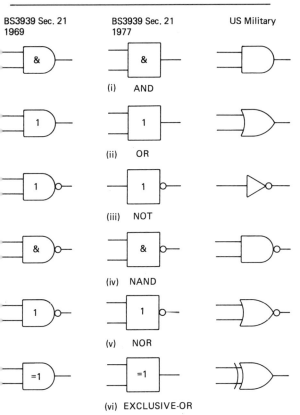

Figure 19.1

In general the number of possible input combinations to a gate is given by 2^N where N is the number of gate inputs. A four-input gate would have 2^4 input combinations, i.e. 16, which corresponds to column 2, Table 19.2.

(c) Gate outputs
A gate output is always presented in the form of a truth table. Such a table shows the state of the output for all possible input combinations. The truth tables are shown in figure 19.2. Gates behave like switches. In the AND gate, for example, a logic 1 output is only obtained when all the inputs, irrespective of how many there are, are at logic 1. A logic 0 output is obtained for all the other conditions. Inspection of the other truth tables will lead to simple conclusions governing whether the output is logic 1 or logic 0. The EXCLUSIVE-OR gate produces an output at logic 0 if all the inputs are at the same logic, and a logic 1 if the inputs are different.

The rest of the chapter will be devoted to examples of gates that are interconnected to give a variety of outputs.

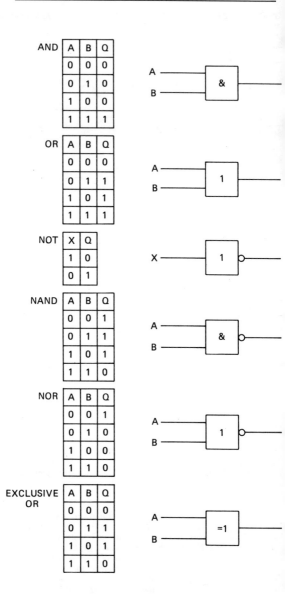

Figure 19.2

Example 37 Write the truth table for Figure 19.3. Which single gate could be used to replace the network?

A	B	C	Q
0	0	0	1
0	1	0	1
1	0	0	1
1	1	1	0

Answer. The NAND gate.

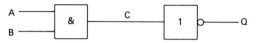

Figure 19.3

Example 38 Write the truth table for Figure 19.4. Which single gate could be used to replace the network?

A	B	C	Q
0	0	0	1
0	1	1	0
1	0	1	0
1	1	1	0

Answer. The NOR gate.

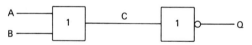

Figure 19.4

Example 39 Write the truth table for Figure 19.5. Which single gate would replace the combination?

A	B	C	D	Q
0	0	1	1	0
0	1	1	0	1
1	0	0	1	1
1	1	0	0	1

Answer. The OR gate.

Example 40 Write the truth table for Figure 19.6. Which single gate would replace the combination?

A	B	C	D	Q
0	0	1	1	0
0	1	1	0	0
1	0	0	1	0
1	1	0	0	1

Answer. The AND gate.

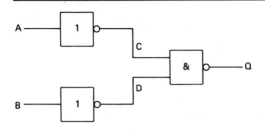

Figure 19.5

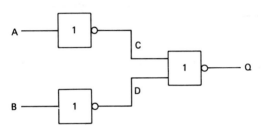

Figure 19.6

Example 41 Write the truth table for Figure 19.7. Which single gate would replace the combination?

A	B	C	D	E	F	Q
0	0	1	1	0	0	0
0	1	0	1	0	1	1
1	0	1	0	1	0	1
1	1	0	0	0	0	0

Answer. The EXCLUSIVE-OR gate.

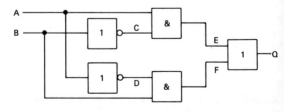

Figure 19.7

Example 42 Figure 19.8 shows NAND gate providing one of the inputs to an AND gate. Write the truth table. How many combinations give a logic 1 output?

A	B	C	D	Q
0	0	0	1	0
0	0	1	1	1
0	1	0	1	0
0	1	1	1	1
1	0	0	1	0
1	0	1	1	1
1	1	0	0	0
1	1	1	0	0

Answer. Three combinations.

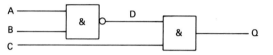

Figure 19.8

Example 43 Write the truth table for Figure 19.9. Which single gate would effect the combination?

A	B	C	D	E	Q
0	0	0	1	1	0
0	0	1	1	1	0
0	1	0	1	1	0
0	1	1	1	1	0
1	0	0	1	1	0
1	0	1	1	0	0
1	1	0	0	1	0
1	1	1	0	0	1

Answer. The AND gate.

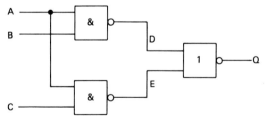

Figure 19.9

Example 44 Write the truth table for Figure 10. How many of the four input combinations give a logic 1 output?

A	B	C	D	E	F	G	Q
0	0	0	0	0	0	1	0
0	0	0	1	0	1	0	0
0	0	1	0	0	1	0	0
0	0	1	1	0	1	0	0
0	1	0	0	0	0	1	0
0	1	0	1	0	1	0	0
0	1	1	0	0	1	0	0
0	1	1	1	0	1	0	0
1	0	0	0	0	0	1	0
1	0	0	1	0	1	0	0
1	0	1	0	0	1	0	0
1	0	1	1	0	1	0	0
1	1	0	0	1	0	1	1
1	1	0	1	1	1	0	0
1	1	1	0	1	1	0	0
1	1	1	1	1	1	0	0

Answer. One combination.

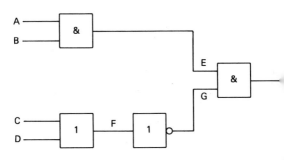

Figure 19.10

Problems

1. Convert 41_{10} to bases, 5, 7 and 9.
2. Convert 103_4, 24_5 and 210_9 to base 10.
3. Convert 0.08_{10} to base 5.
4. Convert 0.015625_{10} to base 4.
5. Convert 29_{10}, 78_{10} and 121_{10} to binary.
6. Convert 0.3125_{10} to binary.
7. Convert 111.111_2 to denary.
8. Add 10110_2 and 10111_2.
9. Subtract 11101_2 from 111101_2.
10. Subtract 10011_2 from 1011_2.
11. Use the 1's complement to subtract: (a) 10001_2 from 11011_2, (b) 1111_2 from 1001_2.
12. Use the 2's complement to subtract: (a) 11000_2 from 10110_2, (b) 1101_2 from 111_2.
13. Multiply 10010_2 by 101_2.

14 Multiply 10001_2 by 111_2
15 Divide 1110_2 by 111_2.
16 Divide 111100_2 by 1010_2.
17 Convert 16_8 and 36.231_8 to binary.
18 Convert 1100.1100_2 to octal.
19 Convert 478_{10} to hex.
20 Convert 77.875_{10} to hex.
21 Convert hex 3BC to denary.
22 Convert hex FE.D4 to denary.
23 Convert hex C4E and 5.DA to binary.
24 Convert 11001111_2 and 11.101010_2 to hex.
25 Write the truth table for Figure 19.11. Which single gate could be used to replace the combination

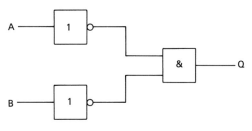

Figure 19.11

26 Write the truth table for Figure 19.12. Which single gate could be used to replace the combination?

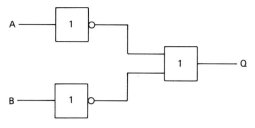

Figure 19.12

27 Write the truth table for Figure 19.13. Which binary input gives a logic 1 output?

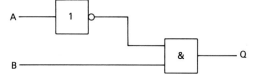

Figure 19.13

28 Write the truth table for Figure 19.14. Which binary input gives a logic 1 output?

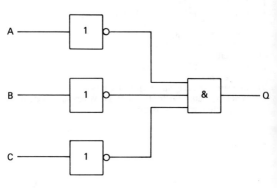

Figure 19.14

29 Write the truth table for Figure 19.15. Which single gate could be used to replace the combination?

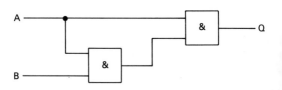

Figure 19.15

30 Write the truth table for Figure 19.16. Express in denary form the input combinations that produce a logic 1 output. A is the most significant bit.

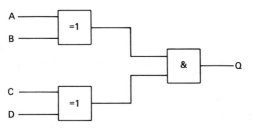

Figure 19.16

20 Two- and three-phase systems

20.1 Introduction

Chapter 6 considered the generation of an a.c. voltage by means of a single coil rotating in a magnetic field. Such an arrangement produces a single sinusoidal voltage and is referred to as a single-phase a.c. supply. Single-phase supplies have certain disadvantages and these disadvantages can be overcome by increasing the number of alternator windings.

20.2 Two-phase supply

In a two-phase alternator there are two equal voltage sources. Two coils displaced by 90° (see Figure 20.1) are rotated in a magnetic field. This arrangement generates two voltages with a phase difference of 90° ($\pi/2$ radians); see Figure 20.2.

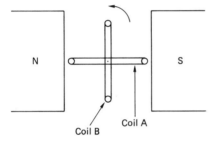

Figure 20.1

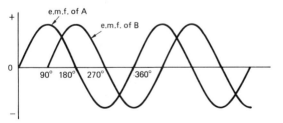

Figure 20.2

The two voltages can be supplied separately in a four-wire system as shown in Figure 20.3, or by joining the phases and using a common return wire as shown in Figure 20.4. The latter method is referred to as the two-phase, three-wire system.

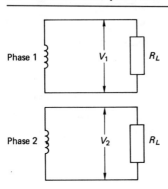

Figure 20.3

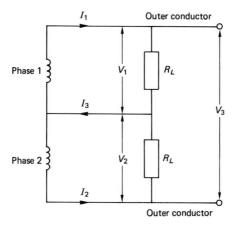

Figure 20.4

Given that V_1 is the phase 1 voltage and V_2 the phase 2 voltage, the voltage between the outer conductors is the phasor sum of both voltages. See Figure 20.5 where

$$V_3^2 = V_1^2 + V_2^2$$

and

$$V_3 = \sqrt{(V_1^2 + V_2^2)}$$

since

$$V_1 = V_2$$
$$V_3^2 = V_1^2 + V_1^2 = 2V_1^2$$

and

$$V_3 = \sqrt{(2V_1^2)} = \sqrt{(2)}V_1$$

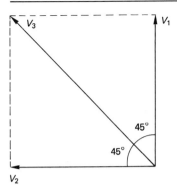

Figure 20.5

Similarly

$$V_3 = \sqrt{(2)}V_2$$

If the current in phase 1 is I_1, and the current in phase 2 is I_2, and the currents differ in phase by 90°, then the current I_3 in the neutral wire is given by

$$I_3 = \sqrt{(I_1^2 + I_2^2)}$$

See phasor diagram in Figure 20.6.

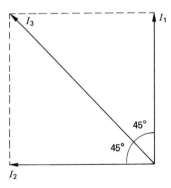

Figure 20.6

If, however, the loads are balanced where $I_1 = I_2$ then

$$I_3 = \sqrt{(2)}I_1 = \sqrt{(2)}I_2$$

It should be remembered that the theory and calculations apply equally to a system of stationary coils and a rotating magnetic field.

Example 1 A two-phase, three-wire system has a phase voltage of 2500 V. Calculate the voltage V_3 between the outer conductors.

$$V_3 = \sqrt{2} \times 2500 = 3535 \text{ V}$$

If the two currents are 250 A and displaced by 90°, calculate the current I_3 in the neutral wire.

$$I_3 = \sqrt{2} \times 250 = 353.6 \text{ A}$$

```
10 PRINT "PROG 294"
20 PRINT "TWO PHASE SUPPLY"
30 INPUT "PHASE VOLTAGE V2" ; V2
40 LET V3=1.414*V2
50 PRINT "VOLTAGE BETWEEN OUTER CONDUCTORS = " V3
```

Example 2 A two-phase, three-wire system with balanced loads has a voltage of 2828 V between the outer conductors. Calculate the phase voltage.

$$\text{phase voltage} = \frac{2828}{\sqrt{2}} = 2000 \text{ V}$$

```
10 PRINT "PROG 295"
20 PRINT "TWO PHASE SUPPLY"
30 INPUT "ENTER OUTER CONDUCTOR VOLTAGE" ; V3
40 LET V2=V3/1.414
50 PRINT "PHASE VOLTAGE = " V2
```

Example 3 The phase currents in a two-phase, three-wire system are 30 A and 40 A. Given that the currents are in phase with their respective voltages, calculate the current I_3 in the third wire.

$$I_3 = \sqrt{30^2 + 40^2} = 50 \text{ A}$$

```
10 PRINT "PROG 296"
20 PRINT "TWO PHASE SUPPLY"
30 INPUT "ENTER PHASE CURRENT I1" ; I1
40 INPUT "ENTER PHASE CURRENT I2" ; I2
50 LET I3=SQR((I1^2)+(I2^2))
60 PRINT "CURRENT IN NEUTRAL = " I3
```

20.3 Three-phase supply

Three similar coils mounted 120° apart rotating in a magnetic field will generate a three-phase supply, the voltages differing in phase by 120°, see Figure 20.7. If the three phases are designated R (red), & (yellow), and B (blue), and assuming that the instantaneous voltage in phase R is

$$e_R = E_m \sin \phi \text{ (see Chapter 6)}$$

then the instantaneous voltage in phases Y and B are

$$e_Y = e_m \sin (\phi - 120°)$$
$$e_B = E_m \sin (\phi - 240°)$$

The supplies from the three phases can be used separately as shown in Figure 20.8. Each phase is isolated and each one loaded separately. This arrangement is rather unwieldly and expensive, and it is more convenient to wire the phases in a mesh connection (which is also called delta) shown in Figure 20.9, or in a star connection which is shown in Figure 20.10. In both arrangements the supplies are carried by three conductors, called lines (Figure 20.10 shows four wires and three phases plus neutral).

Alternators are usually connected in star, and this arrangement together with a neutral line provides line voltages between

Two- and three-phase systems 355

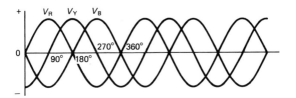

Figure 20.7

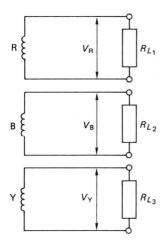

Figure 20.8

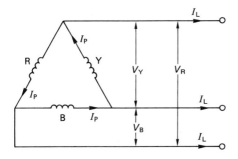

Figure 20.9

356 Circuit calculations pocket book

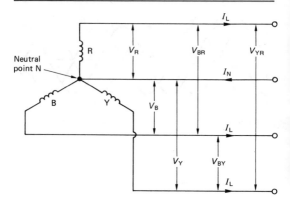

Figure 20.10

conductors R, Y and B, and phase voltages between R, Y and B and the neutral line as shown in Figure 20.10.

Both delta and star connected loads can be supplied by the line voltages. In addition individual loads such as lighting appliances, fans and heaters can be supplied by the phase voltages of the four-wire star system of Figure 20.10.

20.4 Three-phase star connected load

Figure 20.11 shows a star connected load. One end of each load is connected to a line conductor, and the other ends are connected to form a neutral point N (or star point).

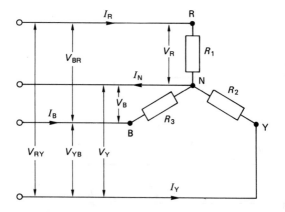

Figure 20.11

The line voltages, which by definition are the voltages between any two lines, are indicated by V_{RY}, V_{YB} and V_{BR}.

The voltages between any one line and the neutral conductor are called phase voltages and are indicated by V_R, V_Y and V_B.

Examination of Figure 20.11 shows that the line currents I_R, I_Y and I_B will be equal to the respective phase currents. IN this system, if

V_L = line voltage and V_p = phase voltage

then

$V_L = \sqrt{(3)}V_P$ ($\sqrt{3}$ is taken as 1.73 throughout this chapter)

and

$I_L = I_P$

where I_L = line current
I_P = phase current

If the load is balanced, i.e.

$R_1 = R_2 = R_3$

then

$V_R = V_Y = V_B$
$I_R = I_Y = I_B$
$V_{RY} = V_{YB} = V_{BR}$ (each shifted by $2\pi/3$ radians)

and the current in the neutral conductor I_N will be zero.

The total powers dissipated in a three-phase system with balanced loads are

Apparent power = $\sqrt{(3)}V_L I_L = 3V_P I_P$ VA

Active power = $\sqrt{(3)}V_L I_L \times$ power factor
= $3V_P I_P \cos \phi$ W

Reactive power = $\sqrt{(3)}V_L I_L \sin \phi$
= $3V_P I_P \sin \phi$ W

ϕ is the angle between the phase voltage and the phase current. If the loads are purely resistive the power factor = $\cos \phi = 1$.

20.5 Three-phase delta connected load

Figure 20.12 shows a delta (or mesh) connected load. The line voltages are indicated by V_{RY}, V_{YB} and V_{BR} and these are equal to the respective phase voltages, i.e.

$V_L = V_P$

The line currents are indicated by I_R, I_Y and I_B and in this system, if

I_L = line current and I_P = phase current

then

$I_L = \sqrt{(3)}I_P$

The formulas for calculating the apparent, active and reactive powers are identical to the formulas for the star system in Section 20.4.

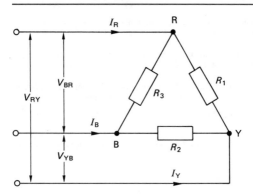

Figure 20.12

Example 4 A balanced delta connected load carries phase currents of 120 A, and the voltage across each phase is 240 V. What is the value of the line current and the value of the line voltage?

$$I_L = \sqrt{(3)}I_P = \sqrt{(3)} \times 120 = 207.6 \text{ A}$$
$$V_L = V_P = 240 \text{ V}$$

```
10 PRINT "PROG 297"
20 PRINT "BALANCED DELTA LOAD"
30 INPUT "ENTER PHASE CURRENT" ; IP
40 INPUT "ENTER PHASE VOLTAGE" ; VP
50 LET IL=1.73*IP
60 LET VL=VP
70 PRINT "LINE CURRENT = " IL "AMPS"
80 PRINT "LINE VOLTAGE = " VL "VOLTS"
```

Example 5 The line voltage in a balanced delta connected load is 600 V. If the line current is 20 A, and the power factor 0.98, calculate: (i) the apparent and active powers, (ii) the value of each phase current.

(i) Apparent power $= \sqrt{(3)}V_L I_L$
$= 1.73 \times 600 \times 20 = 20.76$ kVA

Active power $= \sqrt{(3)}V_L I_L \times$ p.f.
$= 1.73 \times 600 \times 20 \times 0.98 = 20.35$ kW

(ii) Phase current $I_P = \dfrac{I_L}{\sqrt{3}} = \dfrac{20}{1.73} = 11.56$ A

```
10 PRINT "PROG 298"
20 PRINT "BALANCED DELTA LOAD"
30 INPUT "ENTER LINE VOLTAGE" ; VL
40 INPUT "ENTER LINE CURRENT" ; IL
45 INPUT "ENTER POWER FACTOR" ; PF
50 LET AP=1.73*VL*IL
60 LET ACP=1.73*VL*IL*PF
70 LET IP=IL/1.73
80 PRINT "APPARENT POWER = " AP "WATTS"
90 PRINT "ACTIVE POWER = " ACP "WATTS"
100 PRINT "PHASE CURRENT = " IP "AMPS"
```

Two- and three-phase systems 359

Example 6 A three-phase star connected alternator delivers a line current of 100 A to a balanced resistive load at a line voltage of 400 V. Calculate: (i) the phase to neutral voltage, (ii) the output apparent power.

(i) $V_p = \dfrac{V_L}{\sqrt{3}} = \dfrac{400}{1.73} = 231$ V

(ii) Apparent power = $\sqrt{(3)} V_L I_L = 1.73 \times 400 \times 100$
  = 69.2 kW

```
10 PRINT "PROG 299"
20 PRINT "STAR CONNECTED ALTERNATOR"
30 INPUT "ENTER LINE CURRENT" ; IL
40 INPUT "ENTER LINE VOLTAGE" ; VL
50 LET VP=VL/1.73
60 LET AP=1.73*VL*IL
70 PRINT "PHASE TO NEUTRAL VOLTAGE = " VP
80 PRINT "APPARENT POWER = " AP "WATTS"
```

Example 7 A three-phase, star connected alternator delivers a power of 2.7 MW at a power factor of 0.9. The line voltage is 11.0 kV. Calculate: (i) the apparent output power, (ii) the line current.

(i) The 2.7 MW is the active power and the formulas in Section 20.4 show that

Active power = apparent power × power factor.

$\dfrac{\text{Active power}}{\text{p.f.}}$ = Apparent power = $\dfrac{2.7 \times 10^6}{0.9}$

  = 3×10^6 = 3000 kVA = 3 MVA

(ii) Apparent power = $\sqrt{(3)} V_L I_L$

$I_L = \dfrac{\text{Apparent power}}{\sqrt{(3)} V_L}$

  = $\dfrac{3000 \times 10^3}{1.73 \times 11 \times 10^3}$ = 157.6 A

```
10 PRINT "PROG 300"
20 PRINT "STAR CONNECTED ALTERNATOR"
30 INPUT "ENTER POWER IN WATTS" : P
40 INPUT "ENTER POWER FACTOR" ; PF
50 INPUT "ENTER TERMINAL VOLTAGE IN VOLTS" ; V
60 LET AP=P/PF
70 LET IL=AP/(1.73*V)
80 PRINT "APPARENT POWER = " AP "VOLT-AMPS"
90 PRINT "LINE CURRENT = " IL "AMPS"
```

Example 8 A three-phase star connected alternator has a line voltage of 11.0 kV. The output of the alternator is 20 MVA at a power factor of 0.92. Calculate: (i) the line to neutral voltage, (ii) the output power, (iii) the line current.

(i) Line to neutral voltage = phase voltage V_p

$V_p = \dfrac{V_L}{\sqrt{3}} = \dfrac{11 \times 10^3}{\sqrt{3}} = 6.36$ kV

(ii) Output power = apparent power × power factor
  = $20 \times 0.92 = 18.4$ MW

(iii) $I_L = \dfrac{\text{Apparent power}}{\sqrt{(3)}V_L} = \dfrac{20 \times 10^6}{1.73 \times 11 \times 10^3} = 1051$ A

```
10 PRINT "PROG 301"
20 PRINT "STAR CONNECTED ALTERNATOR"
30 INPUT "ENTER LINE VOLTAGE IN VOLTS" ; VL
40 INPUT "ENTER APPARENT POWER IN VOLT-AMPS" ; AP
50 INPUT "ENTER POWER FACTOR" : PF
60 LET VP=VL/1.73
70 LET P=AP*PF
75 LET IL=AP/(1.73*VL)
80 PRINT "PHASE VOLTAGE = " VP "VOLTS"
90 PRINT "OUTPUT POWER = " P "WATTS"
100 PRINT "LINE CURRENT = " IL "AMPS"
```

Example 9 A three-phase, 240-V, delta connected induction motor takes a line current of 50 A at a power factor of 0.9. Calculate: (i) the input power, (ii) the input apparent power, (iii) the current in each phase.

(i) Input power = active power = $\sqrt{(3)}V_L I_L \times$ p.f.

$\qquad = 1.73 \times 240 \times 50 \times 0.9 = 18.7$ kW

(ii) Input apparent power = $\sqrt{(3)}V_L I_L$

$\qquad = 1.73 \times 240 \times 50 = 20.76$ kVA

(iii) $\qquad I_P = \dfrac{I_L}{\sqrt{3}} = \dfrac{50}{1.73} = 28.9$ A

```
10 PRINT "PROG 302"
20 PRINT "DELTA CONNECTED INDUCTION MOTOR"
30 INPUT "ENTER LINE VOLTAGE IN VOLTS" ; VL
40 INPUT "ENTER LINE CURRENT IN AMPS" ; IL
50 INPUT "ENTER POWER FACTOR" ; PF
60 LET ACP=1.73*VL*IL*PF
70 LET APP=1.73*VL*IL
80 LET IP=IL/1.73
90 PRINT "INPUT POWER = " ACP "WATTS"
100 PRINT "INPUT APPARENT POWER = " APP "WATTS"
110 PRINT "PHASE CURRENT = " IP "AMPS"
```

Example 10 A three-phase alternator when fully loaded has a phase current of 30 A. The phase voltage is 240 V. Calculate: (i) the line voltage, the line current and the apparent power output, for a star connection, (ii) the line voltage, the line current and the apparent power output for a delta connection.

(i) $V_L = \sqrt{(3)}V_P = 1.73 \times 240 = 415.2$ V

$I_L = I_P \quad = 30$ A

Apparent power = $\sqrt{(3)} \times 415.2 \times 30 = 21.6$ kVA

(ii) $V_L = V_P = 240$ V

$I_L = \sqrt{(3)}I_P = 1.73 \times 30 = 52$ A

Apparent power = $\sqrt{(3)} \times 240 \times 51.9 = 21.55$ kVA

```
10 PRINT "PROG 303"
20 PRINT "THREE PHASE ALTERNATOR"
30 INPUT "ENTER PHASE CURRENT IN AMPS" ; IP
40 INPUT "ENTER PHASE VOLTAGE IN VOLTS" ; VP
50 LET VLS=1.73*VP
60 LET ILS=IP
70 LET APPS=1.73*VLS*ILS
80 LET VLD=VP
90 LET ILD=1.73*IP
```

```
100 LET APPD=1.73*VLD*ILD
110 PRINT "STAR LINE VOLTAGE = " VLS "VOLTS"
120 PRINT "STAR LINE CURRENT = " ILS "AMPS"
130 PRINT "STAR APPARENT POWER = " APPS "VOLT-AMPS"
140 PRINT "DELTA LINE VOLTAGE = " VLD "VOLTS"
150 PRINT "DELTA LINE CURRENT = " ILD "AMPS"
160 PRINT "DELTA APPARENT POWER = " APPD "VOLT-AMPS"
```

Example 11 A three-phase system supplies 30 kW at a power factor of 0.8. If the line voltage is 415 V, calculate: (i) the line and phase currents when the load is star connected, (ii) the line and load phase currents when the load is mesh connected.

(ii) Active power = $\sqrt{(3)} V_L I_L \times$ p.f.

$$I_L = \frac{30 \times 10^3}{1.73 \times 415 \times 0.8} = 52.2 \text{ A}$$

$$I_P = I_L = 52.2 \text{ A}$$

(ii) Line current 52.2 A

$$I_P = \frac{I_L}{\sqrt{3}} = \frac{52.2}{1.73} = 30.2 \text{ A}$$

```
10 PRINT "PROG 304"
20 PRINT "THREE PHASE GENERATOR"
30 INPUT "ENTER ACTIVE POWER IN WATTS" ; ACP
40 INPUT "ENTER POWER FACTOR" ; PF
50 INPUT "ENTER LINE VOLTAGE IN VOLTS" ; VL
60 LET ILS=ACP/(1.73*VL*PF)
70 LET IPS=ILS
80 LET ILM=ILS
90 LET IPM=ILM/1.73
100 PRINT "STAR LINE CURRENT= " ILS "AMPS"
110 PRINT "STAR PHASE CURRENT = " IPS "AMPS"
120 PRINT "MESH LINE CURRENT = " ILM "AMPS"
130 PRINT "MESH PHASE CURRENT = " IPM "AMPS"
```

Problems

1. A two-phase three-wire system has a phase voltage of 1500 V. Calculate the voltage between the outer conductors. If the two-phase currents are 55 A and displaced by 90°, calculate the current in the neutral wire.

2. A two-phase three-wire system with balanced loads has a voltage of 1600 V between the outer conductors. Calculate the phase voltage.

3. The phase currents in a two-phase three-wire system are 17 A and 19 A. Given that the currents are in phase with their respective voltages, calculate the current in the third wire.

4. A balanced delta connected load carries phase currents of 10 A, and the voltage across each phase is 200 V. Find the value of the line current and the value of the line voltage.

5. The line voltage in a balanced delta connected load is 450 V. If the line current is 22 A and the power factor 0.97, calculate: (a) the apparent and active power, (b) the value of each phase current.

6. A three-phase star connected generator delivers a line current of 60 A to a balanced resistive load at a line voltage of 415 V. Calculate: (a) the phase to neutral voltage, (b) the output apparent power.

7. A three-phase star connected alternator delivers a power of 1.2 MW at a power factor of 0.96. The terminal line voltage is

2.5 kV. Calculate: (a) the apparent output power, (b) the line current.

8   A three-phase star connected alternator has a line voltage of 12 kV. The output of the alternator is 4.2 MVA at a power factor of 0.93. Calculate: (a) the line to neutral voltage, (b) the output power, (c) the line current.

9   A three-phase 240-V delta connected induction motor takes a line current of 12.5 A at a power factor of 0.92. Calculate: (a) the active power, (b) the apparent power, (c) the current in each phase.

10 A three-phase alternator when fully loaded has a phase current of 16 A. The phase voltage is 440 V. Calculate: (a) the line voltage, the line current and the apparent output power for a star connection, (b) the line voltage, the line current and the apparent output power for a delta connection.

11 A three-phase system supplies 20 kW at a power factor of 0.9. If the line voltage is 415 V, calculate: (a) the line and phase currents when the load is star connected, (b) the line and phase currents when the load is mesh connected.

Appendix

Symbols, abbreviations and definitions

Multiples and submultiples		
T	tera	10^{12}
G	giga	10^{9}
M	mega	10^{6}
k	kilo	10^{3}
d	deci	10^{-1}
c	centi	10^{-2}
m	milli	10^{-3}
μ	micro	10^{-6}
n	nano	10^{-9}
p	pico	10^{-12}

Greek letters used as symbols

Letter	Capital	Small
Alpha		α Angle, temperature coefficient of resistance
Delta	Δ increment	δ small increment
Epsilon		ε permittivity
Theta		θ angle
Lambda		λ wavelength
Mu		μ micro, permeability
Pi		π circumference/diameter
Rho		ρ resistivity
Phi	Φ magnetic flux	φ angle
Omega	Ω ohm	ω angular velocity

Table of quantities and units

Quantity	Quantity symbol	Unit	Unit symbol
Capacitance	C	farad	F
Charge or quantity of electricity	Q	coulomb	C
Current	I	ampere	A
		milliampere	mA
		microampere	μa
Electric field strength	E	volts/metre	V/m
Electromotive force	E	volt	V
Energy	W	joule	J
Flux density, electric	D	coulombs/square metre	C/m²
Force	F	newton	N
Frequency	f	hertz	Hz
Impedance	Z	ohm	Ω
Inductance, self	L	henry	H
Inductance, mutual	M	henry	H
Magnetic field strength	H	ampere/metre	A/m
Magnetic flux	Φ	weber	Wb
Magnetic flux density	B	tesla	T
Magnetomotive force	F	ampere	A
Permeability of free space	μ_o	henry/metre	H/m
Permeability, relative	μ_r		
Permeability, absolute	μ	henry/metre	H/m
Permittivity of free space	ε_o	farad/metre	F/m

Table of quantities and units *continued*

Quantity	Quantity symbol	Unit	Unit symbol
Permittivity, relative	ε_r		
Permittivity, absolute	ε	farad/metre	F/m
Power	P	watt	W
Reactance	X	ohm	Ω
Reluctance	R_m	ampere/weber	A/Wb
Resistance	R	ohm	Ω
Resistivity	ρ	ohm metre	Ω m
Time	t	second	s
Wavelength	λ	metre	m

Transistor amplifier symbols

Parameter	Common emitter	Common base	Common collector
d.c. input resistance	h_{IE}	h_{IB}	h_{IC}
d.c. output resistance	h_{OE}	h_{OB}	h_{OC}
d.c. current gain	h_{FE}	h_{FB}	h_{FC}
a.c. input resistance	h_{ie}	h_{ib}	h_{ic}
a.c. output resistance	h_{oe}	h_{ob}	h_{oc}
a.c. current gain	h_{fe}	h_{fb}	h_{fc}

A_v = a.c. voltage gain
A_i = a.c. current gain
A_p = a.c. power gain

Miscellaneous

Quantity	Unit
Length, l	metre, m
Area, A	square metres, m^2
Velocity, V	metres/second, m/s
Attenuation	decibel (dB)
Operator j	j

Answers to problems

Chapter 1

1	0.75 kV	7.5 kV	8.25 kV	18.25 kV
2	6100 V	200 V	25 250 V	
3	600 mV	3 100 mV	2.5 mV	
	600 000 µV	3 100 000 µV	2 500 µV	
4	6×10^{-5} V	2×10^{-4} V	1.6×10^{-3} V	
	0.06 mV	0.2 mV	1.6 mV	

Chapter 2

1. 7.79×10^{-3} Ω
2. 5.84×10^{-3} Ω
3. 9.8×10^{-5} Ω m, 9.8×10^{-3} Ω cm, 9.8×10^{-2} Ω mm
4. 36 530 m
5. 7.4 mm
6. 2.08 Ω
7. 46.4 Ω
8. 5.26×10^{-3}/°C
9. 32.5°C
10. −57.72°C
11. 24 300 to 29 700 Ω
12. 10 000 Ω
13. 680 Ω
14. 1 000 000 Ω
15. 748 Ω
16. 1 202 248 Ω
17. 19 387.5 Ω
18. 10 305.65 Ω
19. 120 Ω
20. 2460 Ω, 1 986.667 Ω, 473.68 Ω

Chapter 3

1. 0.024 V
2. 900 C, 0.25 A h
3. 8.33 min
4. 1 A
5. 6.67×10^{-2} A, 0.667 W
6. 1 V, 50 Ω
7. (a) 5.56×10^{-2} A (b) 18.33 V, 26.11 V, 55.56 V (c) 5.55 W (d) 1.02 W, 1.46 W, 3.09 W
8. 35 Ω, 0.4 W, 0.6 W, 1.4 W
9. 1.75 A, 1.96 Ω, 3.92 Ω, 7.84 Ω
10. 0.2 A, 1.2 W, 0.15 A, 0.05 A, 0.9 W, 0.3 W
11. (a) 1.5 A (b) 1 A, 0.33 A, 0.167 A (c) 150 W (d) 100 W, 33.33 W, 16.67 W
12. 2 Ω
13. (a) 1 A (b) 120 V, 80 V, 40 V (c) 0.8 A, 0.2 A (d) 240 W (e) 120 W, 64 W, 16 W, 40 W
14. (a) 0.125 A (b) 0.025 A, 0.1 A (c) 8 V, 16 V (d) 3 W (e) 0.6 W, 0.8 W, 1.6 W
15. 70 Ω
16. 500 Ω

17 (a) 2500 Ω (b) 5000 Ω
18 3.33 V, 4 V
19 2.02×10^{-2} Ω, 1.0×10^{-2} Ω, 2.00×10^{-3} Ω
20 720 Ω

Chapter 4

1 2.36 A, 1.82 A, 0.55 A
2 18.2 V

Chapter 5

2 3.6 s, 3.6 s
3 0.033 ms, 33 000 ns

Chapter 6

1 (a) 229.18° (b) 85.93° (c) 45.83°
2 (a) 1.309 radians (b) 3.142 radians (c) 8.03×10^{-2} radians
  (d) 5.24 radians
4 200 V, 70.71 V
5 0.877 mV
6 46.67 kV
7 22.22 μs, 6666 m
8 667 kHz, 450 m
9 3.41 m, 2.77 m

Chapter 7

1 1×10^{-9} F, 1 nF, 1000 pF
2 8.5×10^{-3} μF
3 4.25×10^{-4} mm
4 35.39 cm²
5 18 nC
6 1.67 kV
7 400 μF
8 44 mJ
9 8.94 V
10 555.56 μF
11 750 000 V/m
12 0.75 mm
13 130 V
14 400 μC/m²
15 5000 mm²
16 9.05×10^{-4} μC
17 (a) 1.875 μF (b) 15.625 V, 9.375 V (c) 46.875 μC, 46.875 μC
   (d) 366.211 μJ, 219.727 μJ
18 (a) 40 μF, 800 μC (b) 8 V, 2 V, 10 V (c) 3200 μJ, 800 μJ,
   4000 μJ
19 (a) 32 μF (b) 144 μC, 48 μC (c) 432 μJ, 144 μJ
20 (a) 10 μF (b) 30 μC, 42 μC, 48 μC (c) 180 μJ, 252 μJ, 288 μJ
21 (a) 5 V, 5 V (b) 50 μC, 25 μC, 25 μC (c) 125 μJ, 62.5 μJ,
   62.5 μJ
22 (a) 15 V, 10 V (b) 45 μC, 105 μC, 80 μC, 70 μC (c) 337.5 μJ,
   787.5 μJ, 400 μJ, 350 μJ
23 (a) 120 V, 80 V (b) 2800 μC, 1200 μC, 1200 μC
   (c) 280 000 μJ, 72 000 μJ, 48 000 μJ
24 198.94 Ω
25 10.61 μF
26 4980 Hz

Appendix **367**

Chapter 8

1. 6 mH, 6000 μH
2. 0.96 mH, 0.00096 H
3. 3375 V
4. 250
5. 122.5 mW
6. 2500 V
7. 2750 V
8. 1.19 A
9. 16 H
10. 0.0396 H, 9.9 V
11. 42.11 μH
12. 87 mH
13. 2.7 J
14. 0.308 H
15. 20 A
16. 24 V
17. 100 ms
18. (a) 8 H (b) 20 A/s (c) 36 V, 44 V, 80 V (d) 100 A (e) 9000 J, 11 000 J, 20 000 J (f) 40 kJ
19. (a) 1.38 H (b) 40 A/s, 10 A/s, 8 A/s (c) 58 A/s (d) 480 A, 120 A, 96 A (e) 696 A (f) 1600 J, 400 J, 320 J (g) 2320 J
20. 102.1 Ω
21. 22.7 kHz
22. 0.159 H

Chapter 9

1. (a) 176.74 V, 223.32 V (b) 63.26 V, 16.68 V (c) 123.22 μA, 32.48 μ, 8.56 μA
2. (a) 63.26 V, 8.56 V (b) −32.48 V, −16.68 V (c) −123.22 μA, −32.48 μA, −8.56 μA
3. (a) 31.6 V, 49.08 V (b) 18.39 V, 0.92 V (c) 18.39 mA, 6.77 mA, 0.92 mA
4. (a) 18.39 V, 0.92 V (b) −18.39 V, −0.92 V (c) −18.39 mA, −6.77 mA, −0.92 mA
5. (a) 34.59 V, 39.73 V (b) 14.72 V, 1.99 V (c) 1.35 mA
6. (a) 5.41 V, 0.27 V (b) −14.72 V, −1.99 V (c) −27.07 mA, −1.35 mA
7. (a) 1.39 s (b) 100 μA
8. (a) 2.77 s (b) 120 μA
9. (a) 6.77 V, 0.34 V (b) 43.23 V, 49.66 V (c) 0.32 A, 0.43 A, 0.48 A
10. (a) −18.39 V, −2.49 V (b) 18.39 V, 2.49 V (c) 0.18 A, 0.068 A, 0.025 A
11. (a) 0.2 s (b) 0.11 s (c) 2.38 A
12. (a) 0.12 s (b) −12 V
13. (a) 250 A (b) 3000 V (c) 0.38 A (d) 4.5 V

Chapter 10

1. 18.75 N
2. 57.69 T
3. 83.33 mm
4. 22.86 A
5. 1.8 V
6. 2.08 T
7. 0.045 m

368 Circuit calculations pocket book

8 1.14 m/s
9 15 V
10 3900 A/m
11 62.5
12 13.33 A
13 525 mm
14 (a) 8000 A/m (b) 10.05 mT (c) 9.65 μWb
15 11 936.62 mm^2
16 7957.75 A/m
17 3.77 mT, 10mT

Chapter 11

1 1.167 A
2 17.59 kV
3 0.0088 H
4 50.9 kHz
5 1.63 A
6 49.5 V
7 37.9 μF
8 21.2 kHz
9 (a) 56.55 Ω (b) 57.43 Ω (c) 1.92 A (d) 19.16 V, 108.32 V (e) 36.69 W, 207.49 VA, 210.71 VA (f) 0.174 (g) 79.97°
10 (a) 138.23 Ω (b) 600 Ω (c) 583.9 Ω (d) 175.16 V, 41.47 V (e) 52.55 W, 12.44 VA, 54 VA (f) 0.973 (g) 13.32°
11 (a) 80 Ω (b) 100 Ω (c) 0.255 H (d) 180 V, 240 V (e) 540 W, 720 VA, 900 VA (f) 0.6 (g) 53.13°
12 (a) 10.7 kΩ (b) 11.7 kΩ (c) 467 V (d) 188 V, 427 V (e) 7.52 W, 17.1 VA, 18.7 VA (f) 0.403 (g) 66.25°
13 (a) 700 Ω (b) 750 Ω (c) 2784 Hz (d) 162 V, 420 V (e) 97.2 W, 252 VA, 270 VA (f) 0.36 (g) 68.9°
14 28.3 Ω, 40.9 mH
15 75 V, 11.1 A, 4.05 Ω, 0.017 H
16 (a) 176.8 Ω (b) 252.3 Ω (c) 0.238 A (d) 42.8 V, 42 V (e) 10.2 W, 10 VA, 14.3 VA (f) 0.713, 44.5°
17 (a) 106 Ω (b) 600 Ω (c) 591 Ω (d) 207 V, 37.1 V (e) 72.3 W, 13 VA, 73.5 VA (f) 0.984, 10.19°
18 30.7 μF
19 1148 V
20 82.4 Hz
21 (a) 5.3 μf (b) 68.9 mA (c) 420 Ω
22 (a) 128 V (b) 324 Hz (c) 0.915 A
23 (a) 25.1 Ω (b) 11.4 Ω (c) 15 Ω (d) 10 A (e) 60 V, 251 V, 114 V (f) 599 W, 1374 VA, 1498 VA (g) 0.4, 66.4°
24 (a) 9.42 Ω (b) 16.57 Ω (c) 16.61 Ω (d) 7.22 A (e) 108 V, 68.1 V, 120 V (f) 782 W, 373 VA, 867 VA (g) 0.903, 25.5°
25 159 Hz
26 (a) 5 A (b) 628 Ω (c) 1.59 A (d) 5.25 A (e) 17.7° (f) 191 Ω (g) 5000 W, 1592 VA, 5247 VA
27 (a) 0.96 A (b) 15.9 Ω (c) 15.1 Ω (d) 15.1 A (e) 86.4° (f) 15.88 Ω (g) 230 W, 3619 VA, 3626 VA
28 (a) 82.9 Ω (b) 133 Ω (c) 1.33 A (d) 0.829 A (e) 0.497 A (f) 221 Ω
29 (a) 628 Ω (b) 75.8 Ω (c) 632 Ω (d) 0.396 A (e) 84.1° (f) 3.3 A (g) 2.9 A (h) 86.1 Ω (i) 89.2° (j) 10.2 W, 726 VA
30 1779 Hz

Chapter 12

1. (a) 26.83 $\angle 63.43°$ (b) 10 $\angle -36.87°$ (c) 28.28 $\angle 45°$ (d) 4.47 $\angle 116.57°$ (e) 35.36 $\angle 135°$ (f) 1.58 $\angle 108.44°$ (g) 6.7 $\angle -153.44°$ (h) 71 $\angle -135°$ (i) 2.92 $\angle -120.96°$
2. (a) 7.07 + j7.07 (b) 86.6 + j50 (c) 8.49 − j8.49 (d) 103.92 − j60 (e) 1.74 + j9.85
3. (a) 6.08 $\angle -9.46°$ (b) 9.06 $\angle 96.34°$ (c) 15.62 $\angle 129.81°$ (d) 21.21 $\angle -135°$
4. (a) 9 $\angle 30°$ (b) 53.71 $\angle 138.65°$ (c) 40 $\angle -110°$ (d) 25.78 $\angle -38.08°$
5. (a) 2 $\angle 180°$ (b) 11.05 $\angle 174.81°$ (c) 1.41 $\angle -135°$ (d) 33.54 $\angle -153.44°$
6. (a) 7.57 $\angle -22.48°$ (b) 5 $\angle -180°$ (c) 5.29 $\angle -79.11°$ (d) 4.93 $\angle 132.02°$
7. (a) −19 + j26 (b) 46 + j20 (c) −66 − j34 (d) 640 + j150 (e) −12 + j66
8. (a) 1 (b) 1.08 − j0.38 (c) −0.38 + j0.45 (d) 0.5

Chapter 13

1. Primary 160, secondary 80
2. Primary 4.5, secondary 45
3. 37.54 Hz
4. 62.56 Hz
5. 5.16 mWb
6. 1.18 mWb
7. 960 V
8. 24 V
9. 20 mV
10. 2.7 V
11. 260
12. 20
13. 15
14. 190.5
15. 30 : 1
16. (a) 500 (b) 400 Ω (c) primary 0.6 A, secondary 1.2 A
17. (a) 2000 (b) 11.25 Ω (c) primary 4.44 A, secondary 1.11 A
18. 9
19. 0.2
20. 400
21. 120
22. 150 Ω
23. 150
24. 30
25. (a) 0.87 A (b) 2.62 A (c) 3.81 W, 205.65 W
26. (a) 0.192 A (b) 0.048 A (c) 36.7 W, 2.304 W
27. (a) 40 A, 1000 A (b) 25 000
28. (a) 50 A, 30 A (b) 72

Chapter 14

1. 23.9 V
2. 0.286 Ω
3. 1.105 V
4. 1.0 Ω, 55 V
5. 23.6 V
6. 4.43 V

7   (a) 0.34 A (b) 5.66 V (c) 11.32 V
8   (a) 2.14 A (b) 23.31 V (c) 8.79 V
9   644.4 W
10 (a) 3.87 A (b) 0.97 A
11 1.45 Ω, 0.5 A
12 11.52 W, 311.04 W
13 (a) 100 V (b) 5.56 A (c) 0.89 A
14 (a) 56.3 V (b) 62.3 V
15 61.54%
16 77%
17 54.29%
18 (a) 0.33 A (b) 5.73 V (c) 9 V (d) 0.167 A (e) 0.106 A
19 (a) 3 A (b) 76.39 V (c) 84.85 V (d) 2.12 A (c) 1.9 A
20 (a) 14.14 V (b) 0.004 A (c) 13.94 V
21 (a) 70.7 V (b) 0.02 A (c) 69.7 V
22 76.92 Ω, 0.325 W
23 42.50 mA
24 680 mA
25 9.76 Ω, 6250 mA
26 99.9 Ω, 3.0 mA

Chapter 15

1   (a) 7.035 mA (b) 6.6 V (c) 411 kΩ (d) 17.14 kΩ (e) 943 Ω (f) 200
2   (a) 6.56 V (b) 170 kΩ (c) 937 Ω
3   $R_1 = 76$ kΩ, $R_2 = 5$ kΩ,
4   $R_1 = 60.24$ kΩ, $R_2 = 24.7$ kΩ
5   (a) 3 V (b) 7 V (c) 29.5 kΩ (d) 1167 Ω (e) 300 (f) 6.02 mA
6   (a) 1 kΩ (b) 317 Ω (c) 300 (d) 95 (e) 28 500
7   (a) 20 kΩ (b) 2 kΩ (c) 100 (d) 3.03 mA (e) 4.44 kΩ (f) 12 kΩ (g) 90 (h) 36 (i) 3240 (j) 2 kΩ

Chapter 16

1   −17.4
2   3.75
3   47 kΩ
4   1.2
5   −15 V
6   20 kΩ
7   −4.8 V
8   2 kΩ
9   3.9 V or −3.9 V depending which input is at 6.75 V
10 4.85 V

Chapter 17

1   144 900 Hz
2   15 915.49 Hz
3   30.78 ms
4   0.081 H
5   16.8 nF
6   27 Hz
7   6497.47 Ω
8   2.6 nF
9   806.26 Hz
10 0.179 μF
11 2195.24 Ω

Answers to problems 371

12 468.1 Hz
13 2204.586 Hz
14 2040.82 Ω
15 708.62 nF
16 28.57 Hz
17 2134.17 Ω
18 6.14 μF
19 312 Hz
20 2.77 V

Chapter 18

1 17 dB
2 8.06 mW
3 6 dB
4 5.1 mW
5 0.253 V
6 9 dB
7 28 dB
8 3, 2, 6, 9.54 dB, 6 dB, 15.54 dB
9 4, 0.5, 8, 16, 12 dB, −6 dB, 18 dB, 24 dB
10 63.4 W, 2 W, 16 W
11 (a) 14.14 V (b) 14.14 V (c) 111 Hz
12 0.53 μF, 84.9 mH
13 0.133 μF, 5.31 mH
14 53.62 Ω
15 2.75
16 $R_1 = 294$ Ω, $R_2 = 6$ Ω
17 49.43 Ω, 58.9 Ω, 157.76 Ω
18 30 Ω
19 3
20 176 Ω, 179 Ω, 10.5 Ω

Chapter 19

1 131, 56, 45
2 19, 14, 171
3 0.02
4 0.001
5 11101, 1001110, 1111001
6 .0101
7 7.875
8 101101
9 100000
10 −1000
11 (a) 1010 (b) −110
12 (a) −10 (b) −110
13 1011010
14 1110111
15 10
16 110
17 001110, 011110.010011001
18 14.60
19 1DE
20 4D.E
21 956
22 254.828125
23 110001001110, 0101.11011010
24 CF, 3.A8

25 NOR
26 NAND
27 01
28 000
29 AND
30 10, 9, 6, 5

Chapter 20

1 2121 V, 77.77 A
2 1132 V
3 25.5 A
4 17.3 A, 200 V
5 (a) 17.13 kVA, 16.6 kW (b) 12.72 A
6 (a) 240 V (b) 43 kVA
7 (a) 1250 kVA (b) 289 A
8 (a) 6.94 kV (b) 3.9 MW (c) 202 A
9 (a) 4.77 kW (b) 5.19 kVA (c) 7.23 A
10 (a) 761 V, 16 A, 21.1 kVA (b) 440 V, 27.7 A, 21.1 kVA
11 (a) 31 A, 31 A (b) 31 A, 17.9 A

Index

Alternating sinusoidal voltage, 89
  instantaneous values, 91
  peak and peak to peak values, 92
  r.m.s value, 92
  frequency, 94
  wavelength, 95
  average value, 97
Ammeters, 44
Ampere hour efficiency, 252
Ampere turns, 163
Apparent power, 179
Attenuators, 321
  symmetrical T, 322
  asymmetrical T, 324
  symmetrical, π 325
  asymetrical, π 326

Batteries, 244
Battery charger, 249
Binary numbers, 331
  1's complement, 335
  2's complement, 336

Capacitive reactance, 122
Common base charcteristics, 276
Common collector characteristics, 281
Common emitter characteristics, 263
Constant current charging, 251
Constant voltage charging, 249
Coulombs, 26

DC circuits, 26
  series, 29
  parallel, 34
  series parallel, 37
Decibels, 309
Degrees, 90
Delta connection, 354
Denary system, 329

Electromagnetism, 160
Electromotive force, 161

Filters, 309
  low pass, 315
  high pass, 319
Flux density, 160
Force on a conductor, 160
Full wave rectifier, 254

Generated e.m.f, 161

Half wave rectifier, 253
Hexadecimal numbers, 340

Inductive reactance, 135
Inductors, 125
  induced voltage, 126
  self inductance, 128
  energy stored, 130
  mutual inductance, 131
  circuits, 132
  reactance, 135

J notation, 213
Joules, 29

Kilowatt–hour, 29
Kirchhoff's laws, 50

Load matching, 235
Logic gates, 342

Magnetic field strength, 163
Magnetizing current, 163
Magnetomotive force, 163
Maximum power transfer, 83
Measurements, 41
  voltage, 41
  current, 44
Multivibrator, 302

Natural frequency, 296
Non-sinusoidal waveforms, 302
Norton's theorem, 77

Octal numbers, 339
Operational amplifiers *et seq*, 286
Oscillators, 295

Periodic time, 94
Permeability, 164
Phase angle, 178
Phase shift oscillator, 298
Phasors, 213
Polar notation, 214
Power factor, 178
Power supply unit, 255

Radians, 90
Reactive power, 178
Rectangular notation, 214

Reflected resistance, 234
Reluctance, 170
Resistivity, 5
Resistors, 5
  colour code, 13
  tolerances, 14
  wattage, 16
  series circuit, 16
  parallel circuit, 17
  series parallel circuit, 21
Resonance, 200

Sawtooth waveform, 305
Sine wave, 89
Star connection, 354
Superposition theorem, 58

Temperature coefficient of
  resistance, 10
Thévenin's theorem, 62
Three phase supply, 354
Time, 86

Time constant, 140
Total flux, 164
Transformers, 229
Transients, 139
Triangular waveform, 305
Trigonometric notation, 214
True power, 179
Truth tables, 342
Twin-T oscillator, 301
Two phase supply, 351

Voltage, 1, 89
  units, 1, 5, 26, 86, 102, 125
Voltage magnification, 202
Voltage regulator, 257
Voltmeters, 41

Watt–hour efficiency, 253
Wattmeter, 45
Wheatstone bridge, 45
Wien bridge oscillator, 300